Statistical Techniques in Geographical Analysis

Second Edition

Gareth Shaw
Senior Lecturer in Geography, University of Exeter

Dennis Wheeler
Reader in Geography, University of Sunderland

David Fulton Publishers
London

David Fulton Publishers Ltd
2 Barbon Close, London WC1N 3JX

First published in Great Britain by
David Fulton Publishers 1994
Reprinted 1996, 1997

British Library Cataloguing in Publication Data

A catalogue record for this book is available from the British Library

ISBN 1–85346–229–2

Typeset by Action Typesetting Limited, Gloucester
Printed in Great Britain by the Cromwell Press, Melksham

Contents

Preface to the Second Edition

Since the publication of the first edition of this text there have been many advances in computer software and hardware. Combined with the fall in the costs of desktop PCs in particular, these changes have brought within the convenient reach of many geography teachers and students computing power far beyond what was considered possible even ten years ago when 'mainframe' systems were the principal option. This Second Edition attempts to bring geography students to a familiarity not only with the majority of those techniques introduced in the first edition but also with the two currently most popular software systems for dealing with them. We refer to the MINITAB and SPSS (Statistical Package for the Social Sciences) systems. In order to include the worked examples in applying the software some material has been excluded from the first edition. In particular the chapters on classification and trend-surface analysis have not reappeared, neither have the exercises that formed the conclusion of each chapter. We trust that the exchange of material is a profitable one for those who purchase the second edition.

Many people have contributed to the completion of the Second Edition but particular thanks go to Malcolm Farrow of Sunderland University's Maths and Computing School who was kind enough to review and offer advice on the first draft of the text. Any lingering errors remain the sole responsibility of the authors. Both Clecom, the British MINITAB agents, and SPSS UK Ltd. were generous in supplying the authors with copies of their software.

Where figures illustrate the screen displays for MINITAB and SPSS, the former are given in their entirety but the latter are presented with break lines to indicate the distinctions between the instructions display and the results display. This point is discussed more fully in Chapter 2.

Dr Gareth Shaw
Geography Department
University of Exeter

Dr Dennis Wheeler
Geography Department
University of Sunderland

Acknowledgements

The statistical tables in the appendices of this book have been prepared from published items, and the authors are indebted to the following organisations and individuals for allowing us to use their material. The table of z values is reproduced by permission of William Heinemann Ltd. from *Statistics Made Simple* by H.T. Hayslett. The Biometrika Trustees gave permission for us to use the tables of chi-square statistics initially published in *Biometrika*, volume 32, and the tables of critical F values from *Biometrika Tables for Statisticians*, volume 1, by E.S. Pearson and H.O. Hartley. The tables of Kolmogorov – Smirnov and Kruskal – Wallis statistics were adapted from original material in the *Journal of the American Statistical Association*, with whose permission they are reproduced. The tables of Mann – Whitney statistics are reproduced by permission of the Institute of Mathematical Statistics from original material in the *Annals of Mathematical Statistics.* The tables of critical Spearman and Pearson product-moment correlation coefficients are reproduced by permission of George Allen and Unwin from *Statistical Tables* by H.R. Neave. The authors would also like to thank J. Koerts and A.P.J. Abrahamse of the Erasmus University Rotterdam for permission to publish the tables of Durbin – Watson statistics which originally appeared in their text *On the Theory and Application of the General Linear Model,* published by the University of Rotterdam Press. We are grateful to the Literary Executor of the late Sir Ronald A. Fisher FRS, to Dr Frank Yates FRS and to the Longman Group Ltd., London, for their permission to reprint what appears here as Appendix II, from their book *Statistical Tables for Biological, Agricultural and Medical Research* (6th edition, 1974).

Chapter 1

Introduction

With the acceptance and incorporation of statistical methods in geographical analysis, all geography students are now exposed to some form of instruction in quantitative techniques. However, for many of them this is a painful and sometimes unrewarding experience. Problems arise initially due to the fact that some students lack any comprehensive background in numerical analysis. These problems are often compounded by the complicated mathematical notation of statistical methods. Further problems centre around the disillusionment that many undergraduates feel towards statistical methods when courses and texts venture no further than univariate or bivariate methods, yet many of the problems encountered in student project work are of a multivariate character.

These difficulties are not new and are recognised by most teachers. However, few serious attempts have been made to solve them. Some organisations, such as the Quantitative Methods Study Group of the Institute of British Geographers, have attempted to fill a need in the teaching of statistical methods through the publication of a series of booklets in the Catmog series which first appeared in 1975. Unfortunately, as Gregory (1983) points out these have suffered from two major problems: first, the unstructured order in which themes have been published; and second, their different levels of complexity. Indeed, many are far too difficult to be of use for undergraduate teaching, though they remain a valuable resource to researchers. Sadly, there are too few publications that bridge the gap between current undergraduate textbooks and the purely statistical texts dealing with multivariate methods. Even fewer texts have sought to introduce geography students to the benefits of using the now readily-available statistics computer packages to help them in their work and training in statistical methods.

This book attempts to tackle these problems by presenting a wide range of statistical procedures, each amplified through its application in commercial statistics packages. The authors trust that the topics covered are varied enough to satisfy the needs of most undergraduate and many postgraduate students. But in any such text only a selection of topics can be covered and some, of necessity, have had to be excluded. To readers who feel that their favoured, and doubtless indispensable, technique or application has been neglected we can only apologise. Our selection is based on two criteria. First we looked at the methods most widely used in published research works and balanced these, secondly, with items that we have found useful in our own research. We hope that the final choices are equally practical and helpful to other geographers.

The layout of the text is progressive, from the simpler to the more advanced topics, and the reader will also be aware at the same time of four general areas – those of univariate, bivariate, multivariate and spatial techniques. It must be stressed however that the topics and sections are not each treated in isolation. We have tried to emphasise the links between them and to lend a feeling of cohesion and continuity to the programme of study.

Many geography students enrol on degree courses without having pursued mathematics since the age of 15 or 16 and apprehension, at the very least, when embarking on a course in statistics has become almost traditional. It is not uncommon to start a lecture with a group of students all of whom will have some understanding of the concept of 'the average'. Yet once express that concept in algebraic terms of $\bar{X}=\Sigma X/n$ and confusion will prevail. Unfortunately such expressions, and many far more intimidating, are unavoidable. With this problem foreseen we have included a chapter that discusses the more common algebraic and mathematical problems that arise in the teaching of geography students. As with the statistical procedures themselves, the range of topics covered is far from exhaustive. The chapter also includes an important introduction to the use of the two packages adopted in the text, those of MINITAB and SPSS. In both respects it is a chapter that can be regularly referred to at later stages and should not be skipped or read and promptly forgotten. As far as possible, all the statistical methods that we introduce and discuss are further applied using one or both of the packages cited above. We have chosen not to go into the mathematical theory underlying each of the tests other than to discuss how they operate and, in most cases, how they might be performed 'by hand'; the exceptions being some of the more advanced multivariate methods for which computational assistance is all but indispensable. In most cases students are often surprised to find that the arithmetic tasks, though perhaps repetitive, rarely go beyond the simplest functions of addition, subtraction, multiplying, dividing and taking squares or square roots. Whenever an equation is introduced we have attempted to describe it in its simplest terms showing how it acts as a series of coded instructions for these straightforward arithmetic operations.

The problems of obtaining good and reliable data are perennial in geographical statistics. To help teachers and students to overcome this difficulty we have included in Chapter 3 reference to some of the most significant data sources in Britain and in the United States. Many of the worked examples have been drawn from such sources. Hence by including not only an introduction to statistical methods but by also embracing computer applications and by directing geographers to appropriate data sources we hope that we have struck a practical balance for our readers.

References

Catmog (1975–) *Concepts and Techniques in Modern Geography*. Geo Abstracts, Norwich.

Gregory, S. (1983) Quantitative geography: the British experience and the role of the Institute. *Trans. Inst. Br. Geogrs*. (N.S.), 8, 80–89.

Chapter 2

An Introduction to Mathematical Statistics and Computer Applications

2.1 Introduction

The aim of this chapter is not to educate the reader in all or even many of the branches of mathematics and computing, but rather to provide a brief introduction to the arithmetic and computational procedures employed in the following chapters. Those who have pursued a course in mathematics to 'A' level (in England) or to High School (in the United States) or equivalent grades elsewhere may feel confident in omitting this chapter. Though all would benefit from the introductory sections that describe the two widely-available statistics packages that are used to illustrate the application of computer technology to data analysis.

Geography students, many of whom arrive at degree level from a generally 'arts' background, are rarely at ease when handling quantitative data, the successful analysis of which is not so much a question of mathematical ability as of confidence and clear thinking. All too often students allow themselves to be intimidated by the vocabulary and algebraic shorthand of statistical analysis and some never recover from the trauma of their initial encounters with this branch of their discipline. This need not be the case. In reality many equations used by statisticians are surprisingly simple and involve only the irksome repetition of the basic arithmetic procedures of addition, subtraction, multiplication and division. They are not often found to be challenging mathematical enterprises. Perhaps it is the abundant use of Greek symbols or the wealth of subscripts that prompts uncertainty in the minds of many students. But whatever the cause this chapter demonstrates that such fears are groundless. When the syntax of statistical algebra becomes familiar many of the arithmetic difficulties disappear. And, as will become increasingly clear, a great deal of the arithmetic drudgery can be left to the computer; though this is certainly not to suggest that the mathematical principles on which they operate can be overlooked – quite the opposite, they need to be very clearly understood in order that sense can be made of the figures that appear on screen or printer.

2.2 Arithmetic procedures

A common problem when using statistical equations results from misunderstanding of the sequence in which different arithmetic procedures are carried out. While such procedures may be individually correct, when not executed in the proper order they will lead to an incorrect answer. A small example will clarify this point. Unless otherwise specified, multiplication and division must always precede addition and subtraction. Thus in the expression $2.5 + 17.2 \times 3.1$, it is the multiplication that must be executed first irrespective of the order in which the two processes are listed. Only then can 2.5 be added to the product to give the answer 55.82. Had the addition $2.5 + 17.2$ been executed first, the (incorrect) answer of 61.07 would have resulted. By the same convention, divisions also precede additions and subtractions, so that $4.7/2.1 - 2.0$ gives $2.238 - 2.0 = 0.238$, and not $4.7/0.1 = 47.0$. Within strings of successive additions and subtractions alone, or of multiplications and divisions alone the order of execution is immaterial, hence:

$$2.1 + 4.7 - 3.7 + 1.4 = 4.5$$

and

$$3.2 \times 1.1 \times 7.0/2.0 = 12.32$$

irrespective of the order in which the operations are executed. Care, however, is needed when successive subtractions are indicated. For example $17 - 2 - 1$ should give 14 and not 16 as the last two items are $-2 - 1$ and not $2 - 1$. In such cases execution should proceed from left to right. This problem does not arise with successive additions.

Should any change in the conventional order of precedence be required, it is indicated by the use of brackets. Going back to $2.5 + 17.2 \times 3.1$; had we wished to add the first two items before dividing by 3.1 then the expression would have been $(2.5 + 17.2) \times 3.1$. Instructions within brackets must always be executed before that quantity is used beyond the confines of the brackets. This is an important principle and one that should be clearly understood as it has particular importance for many of the equations that appear throughout this, and other, texts.

Powers and indices, as operational indicators, take precedence over addition, subtraction, multiplication and division and must be executed first. For example:

$$2 \times 4^3 = 2 \times 64 = 128 \quad \text{and not} \quad 8^3 = 512$$

Similarly:

$$4+7^2=4+49=53 \quad \text{and not } 11^2=121$$

Notice that, especially with the use of powers, failure to follow the correct order of execution can lead to huge inaccuracies in the final answer. Should the order need to be changed brackets will again be used to denote the sequence of execution. A common example in statistical analysis is the need to square the difference between two quantities, say 7.2 and 2.4. It is incorrect to express this as $7.2-2.4^2$ because, as we have seen, the squaring would be carried out before the subtraction. The expression should read $(7.2-2.4)^2$. Now the subtraction is completed first, the result then being squared.

$$(7.2-2.4)^2=4.8^2=23.04$$

Where several pairs of brackets are present in an expression the operations within the innermost are executed first, working then outwards. It is sometimes helpful to distinguish the order by using paired rounded and squared brackets. It should be obvious that brackets can only be used in pairs. For example:

$$[2(3.2+4.0)]^2=[2(7.2)]^2=14.4^2=207.36$$

but

$$[2(3.2)+4.0^2 \text{ is incorrect}$$

The executable instructions +, −, × and /, together with powers provide the arithmetic basis of nearly all the equations used in this text. The first four should provide no difficulties, but the use of powers requires further elaboration.

At the most elementary level indices and powers indicate repetitive multiplication so that:

$$2^2=2\times 2=4$$

or

$$4.2^4=4.2\times 4.2\times 4.2\times 4.2=311.17$$

Indices are a convenient shorthand without which lengthy descriptions would have to be used. The expression 2.1^7 is far easier to represent and less likely to be misread than its longer version. Indices are a facility widely used by computers and pocket calculators to avoid the use of long sequences of digits when referring to very large, or very small, quantities.

Most machines will not display numbers of more than eight digits in their standard form. Rather than abandon numbers greater than 99,999,999 or less than 0.0000001 a system of 'floating point' presentations are used. For example $22{,}180 \times 27{,}560$ gives 611,280,800; a quantity that is more easily represented and displayed as 6.1128 08 or as 6.1128 E08. In both cases such quantities are to be interpreted as 6.1128×10^8 where the power to which ten is raised indicates the number of places that the decimal point has to be moved. Remember that to multiply any quantity by 100 (which is 10^2) is to move the decimal point two places to the right. Thus, 2.5×10^2 would become 250.0.

The floating point convention applies equally to very small numbers, but their explanation requires a brief discussion of negative indices. In the same way as 2^2 is executable so too are expressions such as 2^{-2} or 3.6^{-3}. Negative indices are best understood by remembering that they change their sign if their reciprocals are used, i.e. if the expression is placed 'below the line'. Thus:

$$2^{-2} = 1/2^2 \quad = 1/4 \qquad = 0.25$$

and

$$3.6^{-3} = 1/3.6^3 = 1/46.66 = 0.0214$$

Notice that the use of negative indices on positive quantities can never yield a negative answer and will do so only if the quantity is itself negative in the first instance. We can now see how floating point formats might be used to describe very small quantities.

The expression 7.4 -08 or 7.4 E-08 represents the quantity 7.4×10^8, which can be rewritten as $7.4 \times 1/10^8$. When expanded this becomes 7.4×0.00000001 or 0.000000074. The negative index now shows how many places the decimal point must be moved to the left. The final quantities thereby becoming smaller as the index becomes larger.

There is, however, a price to be paid for the convenience of floating point numbers. If the quantity 611,280,800 is given as 6.1128 E08 then the expansion back to a 'standard' quantity will give 611,280,000. There has been a rounding error of 800, at least in terms of the interpretation of the printed or displayed answers, though internally the machine will retain the correct quantities. Clearly such 'errors' are relatively small, but students should be aware that they may arise when interpreting quantities from computer screens or print-outs.

We can now turn to consider fractional indices. Integer (whole number) indices should present few problems, but fractional terms are less simple. How should $4^{2.5}$ or $7^{1.5}$ be evaluated? From the practical point of view, many pocket calculators will do the task. Similarly computer programs can be given simple instructions to evaluate such items and, as the following section demonstrates, logarithms can be used to solve the problem.

Table 2.1 Numerical equivalents of exponential terms

Exponential term	Result	Exponential term	Result
9^2	81.0	9^{-2}	0.0123
9^1	9.0	9^{-1}	0.1111
$9^{0.8}$	5.7990	$9^{-0.8}$	0.1724
$9^{0.5}$	3.0	$9^{-0.5}$	0.3333
$9^{0.2}$	1.5520	$9^{-0.2}$	0.6440
9^0	1.0	$9^{-0.1}$	0.8027

Table 2.1 summarises the contrasting results of using positive and negative fractional indices. Note should be made of the fact that 1.0 raised to any power remains 1.0, while any number raised to the power 0.0 is, by convention, taken as 1.0. The following points should also be noted as general guiding principles when numbers larger than 1.0 are being raised to various powers:

1. Positive indices greater than 1.0 will increase the final quantity;
2. Positive indices less than 1.0 will decrease the final quantity, but can never yield less than 1.0;
3. Negative indices of any magnitude greater than 1.0 can only yield answers of less than 1.0.

For numbers between 1.0 and zero any positive index can only yield a result which is itself between 1.0 and zero. The result can only exceed 1.0 if a negative index is used.

In this general context the reader should be aware of the various means by which the instruction for a square root are given. Most will already know that $\sqrt{X}$ indicates square root of X. But fractional indices can also be used to the same effect. Both $X^{0.5}$ and $X^{1/2}$ indicate that the square root of X is required.

This brief review can be concluded by stating the three 'laws' of operations with indices:

1. Powers may be added for multiplication from a common base:

$$X^m \times X^n = X^{m+n} \text{ e.g. } 3^2 \times 3^3 = 3^5 = 243$$

2. Powers may be subtracted for division from a common base:

$$X^m / X^n = X^{m-n} \text{ e.g. } 3^3/3^2 = 3^{3-2} = 3^1 = 3$$

3. Powers may be multiplied when they appear in sequence:

$$(X^m)^n = X^{m \times n} \text{ e.g. } (3^3)^2 = 3^{3 \times 2} = 3^6 = 729$$

2.3 Logarithms

Any number on the 'arithmetic' scale which is greater than zero can be expressed by its counterpart on the logarithmic scale. The advantages of such re-expressions are many, but at at this stage only the basic principles require discussion.

The reader will need to be familiar with the two most frequently used logarithmic systems – the natural (denoted by *ln* or $\log_e$) and the common (denoted by $\log_{10}$). The latter are the simpler and better known. Common logs are sometimes referred to as 'logarithms to the base 10'. Table 2.2 lists some of their more obvious characteristics and indicates how arithmetic quantities and their logarithmic equivalents are related.

Table 2.2 Logarithmic equivalents of numbers greater than 1.0

Number	Log	Number	Log	Number	Log	Number	Log
1.0	0	3.0	0.4771	5.0	0.6989	7.0	0.8451
10.0	1.0	30.0	1.4771	50.0	1.6989	70.0	1.8451
100.0	2.0	300.0	2.4771	500.0	2.6989	700.0	2.8451
1000.0	3.0	3000.0	3.4771	5000.0	3.6989	7000.0	3.8451

Table 2.2 also demonstrates why common logarithms have the 'base' of 10. The common logarithm of a number is the power to which 10.0 must be raised in order to achieve that number. The previous section dealt with the use of powers, and here we can see some of those principles being applied. Thus:

$$10^2 = 100 \text{ so that } \log_{10}100 = 2.0$$

and

$$10^{1.6989} = 50 \text{ so that } \log_{10}50 = 1.6989$$

No logarithmic system permits logs to made from negative numbers, but there are numbers – those between zero and 1.0 – that yield negative logarithms. Table 2.3 shows how numbers less than 1.0 relate in general to their logarithmic equivalents.

Table 2.3 Logarithmic equivalents of numbers less than 1.0

Number	Log	Number	Log	Number	Log
0.1	−1.0	0.3	−0.5229	0.5	−0.301
0.01	−2.0	0.03	−1.5229	0.05	−1.301
0.001	−3.0	0.003	−2.5229	0.005	−2.301

From the table it can be seen that there can be no log of zero and that smaller and smaller fractions merely become ever larger negative logarithms. The relationship that links such fractions and common logs is again determined by the power to which 10 must be raised to produce that number. For example:

$$10^{-1} = 0.1 \qquad \text{so that } \log_{10}0.1 = -1.0$$

and

$$10^{-0.301} = 0.5 \qquad \text{so that } \log_{10}0.5 = -0.301$$

Students familiar with the published logarithmic tables in mathematical handbooks will see that while Table 2.2 conforms to their expectations of the logarithmic equivalents of ordinary numbers Table 2.3 does not. This difference arises because published logarithmic tables use the 'bar' system when dealing with numbers of less than 1.0. By this means the log of 0.3 would be $\bar{1}.4771$ and not -0.5229. In the same manner the log of 0.03 would be $\bar{2}.4771$ and not -1.5229. Computers and pocket calculators are unable to work with this convention which is one that avoids the need to publish two sets of log tables, one for numbers greater than 1.0 and another for numbers between 0 and 1.0. The connection between the two seemingly contradictory systems is made clear if we write the 'bar' notation in its fullest form. In this way $\bar{1}.4771$ should be expressed as $-1 + 0.4771$, which is of course -0.5229.

Antilogging is the process by which logs are converted back to their appropriate arithmetic equivalents. The common antilog of 2.0, for example, is 100.0. If the base of the logarithmic system, in this case 10.0, is raised to the power indicated by the logarithm, the result is the antilog. To take another example, and referring back to Table 2.2. If we have a logarithm 0.4771 its antilogarithm is $10^{0.4771}$ which is 3.0.

Logarithms can be used to solve problems involving the use of powers in their various forms. Lengthy divisions and multiplications can also be more easily dealt with in logarithmic form. These points are conveniently summarised in the 'laws' of logarithmic operations.

First law of logarithms

When multiplying two numbers we may instead add their respective logarithms, the antilogarithms ('antilog') of the result being the correct answer. A simple example might be 28.47×39.23, which can be found as follows:

$$\log_{10}28.47 = 1.4544 \text{ and } \log_{10}39.23 = 1.5936$$

but

$$\log_{10}(28.47 \times 39.23) = \log_{10}28.47 + \log_{10}39.23$$

from which we establish that the common log of the required product is:

$$1.4544 + 1.5936 = 3.048$$

But this quantity is itself a common logarithm and must be 'antilogged' in order to give a sensible answer. In this case the antilog of 3.048 is $10^{3.048}$, which gives 1116.86.

Divisions can be performed in the same fashion but by subtraction instead of addition. Thus:

$$\log_{10}(28.47/39.23) = 1.4544 - 1.5936 = -0.1392$$

the antilog of which is given by $10^{-0.1392}$, yielding the result 0.7258.

The general algebraic expressions of these laws can be written as:

$$a \times b = \text{antilog}(\log_{10}a + \log_{10}b)$$

and

$$a/b = \text{antilog}(\log_{10}a - \log_{10}b)$$

The second law of logarithms

The value of a number raised to some power can be found by multiplying the log of that number by the unlogged power term, and antilogging the resulting product. Thus $3.67^{1.89}$ becomes:

$$(\log_{10}3.67) \times 1.89 = 0.5647 \times 1.89 = 1.0673$$

the antilog of which is 11.676.

Negative powers can be similarly dealt with and $3.67^{-1.89}$ becomes:

$$0.5647 \times (-1.89) = -1.0673$$

the antilog of which is 0.0856.

This operation can be expressed algebraically as

$$a^b = \text{antilog}(\log_{10}a \times b)$$

The examples used to illustrate these laws are extremely simple, but the principles that they convey will be found to be useful in the more practical settings found in later chapters.

Natural logarithms

Attention can now be turned to the second of the major logarithmic systems: natural logarithms. Common logs are to the base 10.0, but natural logarithms are to the base of 2.71828.... a quantity designated by the symbol *e*. The natural logarithm of 2.7183 is, thus 1.0. Other equivalents are listed in Table 2.4. All the rules demonstrated to hold for common logs apply equally to natural logs. For example $3.67^{1.89}$ would, with natural logs, become:

$$(\log_e 3.67) \times 1.89 = 1.3002 \times 1.89 = 2.457$$

the natural antilog of which is 11.67.

Table 2.4 Some natural logarithms

Number	Number as an exponent of *e*	Natural log
1	e^0	0
2.7183	e^1	1
5.0	$e^{1.609}$	1.609
7.3891	e^2	2.0
10.0	$e^{2.303}$	2.303
20.0855	e^3	3.0
1.6487	$e^{0.5}$	0.5

As before, the process of antilogging is that of raising the log base (*e* in this case and not 10) to the power specified by the log, and

$$e^{2.457} = 11.67$$

To avoid any confusion we shall henceforth adopt the usual convention of denoting common logs simply by 'log' and natural logs by '*ln*'. It is, however, easy to convert between the two systems and:

$$\log(x) = ln(x) \times 0.4343$$

and

$$ln(x) = \log(x) \times 2.303$$

2.4 Statistical procedures

Thus far attention has focused on arithmetic and mathematical operations. There is, however, a notation specific to statistical methods that employs what appears to be a bewildering wealth of subscripts and Greek symbols. In reality such symbols are merely a shorthand for a number of mathematical tasks, few of which go further than the arithmetic outlined in the preceding pages.

The most commonly encountered of these instructional symbols is the Greek letter Σ (capital sigma). No numerical quantity is associated with this symbol; it is an operational instruction that denotes the necessity of adding together the values indicated in the variable which it precedes. That variable might, for example, be temperature which can be denoted in this case as variable X (or indeed any other letter); the instruction to add a succession of individual temperature readings would appear as ΣX. In the same way ΣY would indicate the necessity of adding together the individual items that constitute the sample of a variable denoted by Y.

A more comprehensive form of this expression may be used. If a data set contained the daily mean temperature for each day of one year the statistician might express the need to add all 365 observations together by the expression:

$$\sum_{i=1}^{365} X_i$$

in which subscript i indicates the sequence number between 1 and 365, the summation taking place from X_1 through to X_{365}. Table 2.5 indicates how i, X and Σ are connected.

Table 2.5 The index notation in statistical analysis

Sequence number (i)	Numerical value	
1	$X_i = X_1 = 10.2$	
2	$X_i = X_2 = 11.3$	
3	$X_i = X_3 = 9.7$	$\sum_{i=1}^{5} X_i = 50.8$
4	$X_i = X_4 = 7.5$	
5	$X_i = X_5 = 12.1$	

The total number of observations in a data set or sample are not usually known in advance and the term n is conventionally employed to denote that quantity. This allows a more general summation instruction to be written down:

$$\sum_{i=1}^{n} X_i$$

Such expressions are widely adaptable. The instruction to add together just the tenth to the twentieth items in a sample might be given as:

$$\sum_{i=10}^{20} X_i$$

One subscript only is required when observations are listed as a single sequence, or vector, of quantities. If, however, the sample consists of rows as well as columns, i.e. if it is in the form of a matrix of numbers, then double subscripts are needed so that individuals can be located by row and by column number. The practical difference between a vector and matrix of observations is made clearer by taking a simple example. A set of daily temperature observations over a five-day period can be put together as a string or vector of quantities. If however, data have been taken for five days at five different stations, then the results can be summarised in a matrix in which each day corresponds to a row and each station to a column. All the first day's temperatures appear in the first row, while all the observations for station number 1 appear in the first column, and so on. By this means the observation found in row 2 of column 3 might be described as $X_{2,3}$. This distinction between vectors and matrix notation and the location of items within them is illustrated in Figure 2.1. If all numbers within a matrix are to be summed, then use is made of a double summation sign of the form:

$$\sum_{i=1}^{r} \sum_{j=1}^{k} X_{ij}$$

where X is the variable, r is the total number of rows and i the row number in question, and k is the total number of columns and j the column number in question.

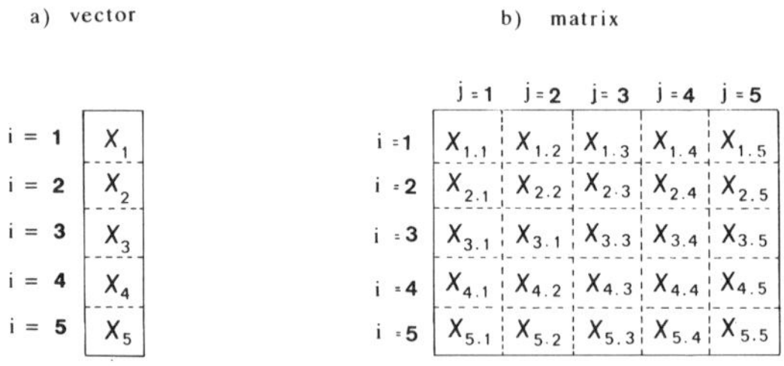

a) vector

i = 1	X_1
i = 2	X_2
i = 3	X_3
i = 4	X_4
i = 5	X_5

b) matrix

	j = 1	j = 2	j = 3	j = 4	j = 5
i = 1	$X_{1.1}$	$X_{1.2}$	$X_{1.3}$	$X_{1.4}$	$X_{1.5}$
i = 2	$X_{2.1}$	$X_{2.2}$	$X_{2.3}$	$X_{2.4}$	$X_{2.5}$
i = 3	$X_{3.1}$	$X_{3.1}$	$X_{3.3}$	$X_{3.4}$	$X_{3.5}$
i = 4	$X_{4.1}$	$X_{4.2}$	$X_{4.3}$	$X_{4.4}$	$X_{4.5}$
i = 5	$X_{5.1}$	$X_{5.2}$	$X_{5.3}$	$X_{5.4}$	$X_{5.5}$

Figure 2.1 Graphical representation of vector and matrix subscripts.

The need to add together successive items of a sample is commonplace in statistical analysis. The instruction for determining the mean or average of a number of items (see Chapter 4) is conveniently described by:

$$\bar{X} = \frac{\Sigma X}{n}$$

where n is number of observations, and $\bar{X}$ is the mean of those figures. The 'bar' over the symbol used to denote the variable (X in this example) indicates that it represents the mean of all the observations. It must not be confused with the 'bar' notation used in logarithmic expressions and already described in section 2.3. To simplify matters, particularly where all numbers in a set are to be added, the item $i=1$ to n is omitted. This practice will be generally observed henceforth unless special circumstances dictate their retention.

Because Σ is an operational instruction it is important to establish where it stands in the order of precedence for arithmetic operations. The use of brackets or the location of the sigma sign itself will usually make the position clear. In the equation for the mean the placing of the summation sign above the division line, makes it clear that summation of all Xs must take place before division by n. Indices take precedence over summation and the instruction ΣX^2 indicates the necessity of squaring all Xs before adding those squares. Table 2.6 summarises such conventions and illustrates how brackets may be necessary to modify some instructions.

Table 2.6 Conventions for the execution of mathematical statements

Instruction	Sequence of execution	
ΣX^2	1. Square all Xs	2. Sum the squares
$(\Sigma X)^2$	1. Sum all Xs	2. Square the sum of all Xs
ΣXY	1. Multiply all pairs of X and Y	2. Sum all XY products[1]
$3(\Sigma X)$	1. Sum all Xs	2. Multiply the sum by 3.0
$3[\Sigma(X-1)]$	1. Subtract 1 from all Xs	2. Sum results of all subtractions
	3. Multiply the sum by 3.0	
$\sum \frac{(X-Y)^2}{2.5}$	1. Subtract all Ys from all Xs 3. Divide all the squares by 2.5	2. Square all $X-Y$ differences 4. Sum all quotients

1. Strictly speaking, this should be presented as $\Sigma(XY)$ but is conventionally written as above and will be encountered as such in later chapters.

A final example, which occurs many times is the need to square a set of differences, sum them and then divide by the total number of observations. Most often this appears in the form of sequential differences from the mean in which the mean is subtracted, in turn, from each individual observation in the sample. Such a quantity, let us term it s, can be produced by the algebraic equation:

$$s = \frac{\Sigma(X_i - \bar{X})^2}{n}$$

The order of execution indicated here is exactly that outlined above: within the brackets the mean of X is subtracted from each of the individual observations. Those differences are each squared, then the squares are summed and, finally, that sum is divided by n (the number of observations in the data set). Had we wished, on the other hand, to have squared the sum, rather than sum the squares, then the expression would have required an additional pair of outer brackets:

$$s = \frac{[\Sigma(X_i - \bar{X})]^2}{n}$$

thereby requiring that the inner brackets are evaluated before the squaring is undertaken.

2.5 Graphs of statistical data

Graphs, maps and diagrams of statistical data can help geographers to describe and to understand the information with which they are working. In later chapters we will be dealing with a number of mathematical descriptions which attempt to summarise what might otherwise be perceived as an unwieldy mass of data. Important among such descriptions are algebraic equations, the parameters of which will change from example to example.

Such equations can often be used to describe and to summarise the behaviour of two variables and how one changes with respect to another, for example, how rainfall might increase with altitude or population density changes with distance away from city centre. The algebraic expression summarising such relationships can be plotted as a line on a sheet of graph paper. A whole variety of lines and curves can be produced depending on the form of the equation. But the principles by which they may be prepared can be introduced by taking some simple examples.

The equation of a straight line

The expression $Y = a + bX$ plots as a straight line. The equation works by having fixed values for a and b, the constants, and by allowing X to vary from case to case. The value of Y is then dependent upon the quantities attached to a, b and X. Section 9.2 will show how values can be estimated for a and b, but for the moment let us say that $a = 5.0$ and that $b = 1.5$. Remember that these are now fixed for the purposes of this one example. If we now substitute successive values of X into our equation we can estimate the corresponding values of Y to produce the results in Table 2.7 which, in turn can be plotted to produce Figure 2.2.

By convention the values for X are always plotted along the horizontal axis, and those for Y along the vertical. Notice also that if we had chosen a greater value for b, the line would have risen more steeply; a smaller value and it would have risen less steeply; a negative value and it would have fallen from left to right. It should in addition be observed that the point at which the line crosses the Y axis, i.e. when $X = 0.0$, is determined

Table 2.7 Straight line data obtained by successive substitutions in $Y = a + bX$

Value of X	$Y=5.0+1.5X$	$Y=5.0+2.5X$	$Y=2.0+1.5X$
0.0	$Y=5.0+0.0=5.0$	$Y=5.0+0.0=5.0$	$Y=2.0+0.0=2.0$
2.0	$Y=5.0+3.0=8.0$	$Y=5.0+5.0=10.0$	$Y=2.0+3.0=5.0$
4.0	$Y=5.0+6.0=11.0$	$Y=5.0+10.0=15.0$	$Y=2.0+6.0=8.0$
6.0	$Y=5.0+9.0=14.0$	$Y=5.0+15.0=20.0$	$Y=2.0+9.0=11.0$
8.0	$Y=5.0+12.0=17.0$	$Y=5.0+20.0=25.0$	$Y=2.0+12.0=14.0$
10.0	$Y=5.0+15.0=20.0$	$Y=5.0+25.0=30.0$	$Y=2.0+15.0=17.0$
12.0	$Y=5.0+18.0=23.0$	$Y=5.0+30.0=35.0$	$Y=2.0+18.0=20.0$

by the value of a as no matter what value we attach to b, when $X=0.0$ the product must also be 0.0.

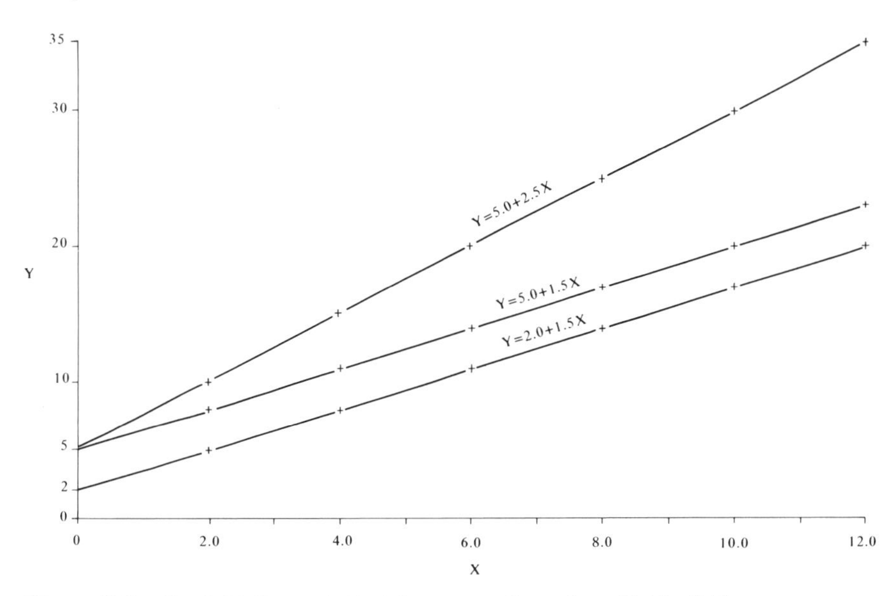

Figure 2.2 Straight lines plotted from equations (see Table 2.7).

The equation for a curve

There are an infinite variety of equations which, when plotted, will describe a curve. Consider such an expression as $Y=a+X^b$, which is similar to that used above in that it includes two constants, and an X and Y quantity. Now, however, they produce a curved line. Let us set a to 5.0 and b to 1.5, as we did before. The result of successive substitutions for X produces the figures in Table 2.8 and the curve in Figure 2.3.

Once again we should note that had we chosen a greater value for b then the curve would have risen more sharply. With a negative value for b, and in accordance with our earlier discussion of the use of indices, the curve would not have sloped downwards as negative indices do not

Table 2.8 Curved lines obtained by successive substitution on $Y = a + X^b$

Value of X	$Y = 5.0 + X^{1.5}$	$Y = 2.0 + X^{2.0}$
0.5	$Y = 5.0 + 0.353 = 5.353$	$Y = 2.0 + 0.25 = 2.25$
1.0	$Y = 5.0 + 1.0 = 6.0$	$Y = 2.0 + 1.0 = 3.0$
2.0	$Y = 5.0 + 2.83 = 7.83$	$Y = 2.0 + 4.0 = 6.0$
3.0	$Y = 5.0 + 5.20 = 10.20$	$Y = 2.0 + 9.0 = 11.0$
4.0	$Y = 5.0 + 8.0 = 13.0$	$Y = 2.0 + 16.0 = 18.0$
5.0	$Y = 5.0 + 11.18 = 16.18$	$Y = 2.0 + 25.0 = 27.0$
6.0	$Y = 5.0 + 14.70 = 19.70$	$Y = 2.0 + 36.0 = 38.0$
7.0	$Y = 5.0 + 18.52 = 23.52$	$Y = 2.0 + 49.0 = 51.0$

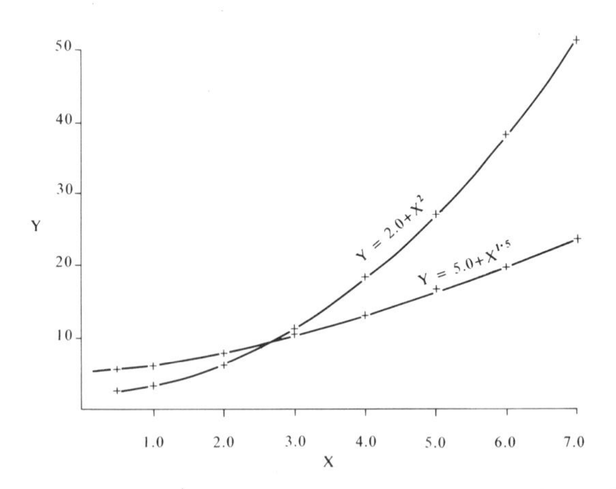

Figure 2.3 Plots derived from equations of curves (see Table 2.8).

produce negative quantities, but such a curve would be convex- rather than concave-upwards. The quantity attached to a determines the location of the curve on the coordinate system. It is 1.0 less than the point at which the curve passes through the vertical co-ordinate at $X = 1.0$. To understand this remember from section 2.2 that 1.0 raised to any power is 1.0. Hence $Y = 5.0 + 1.0^{1.5}$ becomes 6.0.

Many other forms of curve are possible, the majority of which need not be restricted to just two constants or indeed to just one predictor term. We could have used $Y = aX^3$, or $Y = aX^b$ or $Y = 1/X$. All, however could be plotted in much the same way. Quantities would have to be attached to the constants a and b and estimates of Y then produced by successive substitutions over a range of X values. Chapters 9 and 10 will be concerned with how we might determine the values of such constants, but for the moment it is sufficient to understand only how such equations might be plotted.

We can also plot a logarithmic curve using an expression such as $Y = a + b \log X$. Table 2.9 and Figure 2.4 show how such a curve might be prepared and presented. Notice in particular how the slope of the

Table 2.9 Successive substitutions in a logarithmic equation

Value of X	$Y=2.0+1.5 \log X$	$Y=5.0-1.5 \log X$
0.5	$Y=2.0+1.5\times -0.301=1.549$	$Y=5.0-(1.5\times -0.301)=5.451$
1.0	$Y=2.0+1.5\times \; 0.0 \; =2.0$	$Y=5.0-1.5\times \; 0.0 \; =5.0$
5.0	$Y=2.0+1.5\times \; 0.699=2.949$	$Y=5.0-1.5\times \; 0.699 \; =3.951$
10.0	$Y=2.0+1.5\times \; 1.0 \; =3.5$	$Y=5.0-1.5\times \; 1.0 \; =3.5$
20.0	$Y=2.0+1.5\times \; 1.301=3.952$	$Y=5.0-1.5\times \; 1.301 \; =3.049$
40.0	$Y=2.0+1.5\times \; 1.602=4.403$	$Y=5.0-1.5\times \; 1.602 \; =2.597$
60.0	$Y=2.0+1.5\times \; 1.778=4.667$	$Y=5.0-1.5\times \; 1.778 \; =2.335$
80.0	$Y=2.0+1.5\times \; 1.903=4.855$	$Y=5.0-1.5\times \; 1.903 \; =2.146$
100.0	$Y=2.0+1.5\times \; 2.0 \; =5.0$	$Y=5.0-1.5\times \; 2.0 \; =2.0$

curve changes with the sign of the product $b \log X$. The slope is upwards from left to right for positive quantities, but downwards if they are negative.

However, logarithmic curves are an unusual case insofar as specially prepared graph paper is available which allows us, in effect, to plot logged data without first calculating those log values. Semi-logarithmic graph paper has standard uniform and regular increments along one of its axes. The other axis, however, has a logarithmic scale on which numbers are plotted but separated by distances that represent their log equivalents. Figure 2.5 shows the data from Table 2.9 but now plotted on semi-log graph paper. In effect we are plotting the logs of X without having first to evaluate them. The most significant feature of such plots of logarithmic curves is that they now appear as straight lines. This linearising principle is an important one that we will return to in Chapter 10 when we look at non-linear regression.

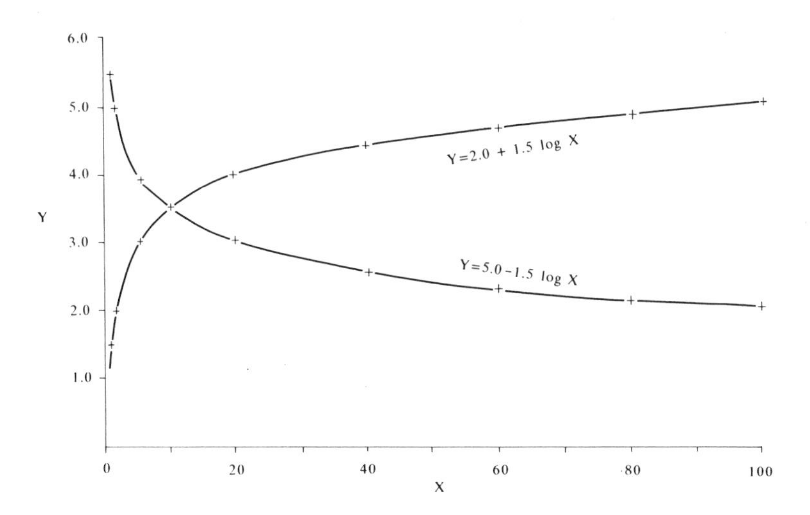

Figure 2.4 Plots derived from logarithmic equations (see Table 2.9).

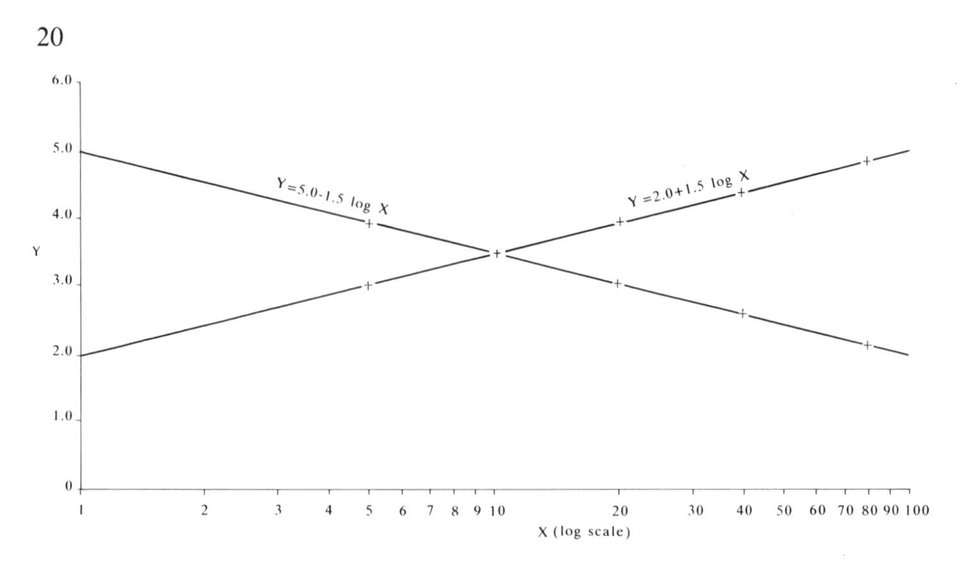

Figure 2.5 Logarithmic curves from Figure 2.4 plotted on two-cycle semi-logarithmic graph paper.

Although not displayed, double-logarithmic graph paper is also available on which both axes are plotted on the logarithmic scale. All such paper, whether semi- or double-log in form are printed with one, two or three cycles. Each cycle of 1 to 10 represents one increment on the log scale. For example one-cycle would be used if data ranged between 0.1 and 1.0, between 10 and 100 or over any such ten-fold range. Two-cycle paper would cover, for example, data ranging from 10 to 1000 or from 100 to 10000 or any other one hundred-fold (10^2) range. Three-cycle paper might be used if the data embraced a thousand fold (10^3) range such as 1.0 to 1000 or 0.1 to 100. The use of logarithmic graph paper is described in most texts on cartographic techniques, for example Monkhouse and Wilkinson (1971).

2.6 Mathematical symbols

Some mathematical symbols will already be familiar to the reader. The universal constant π (pi) has a value 3.14159 and is used in equations to describe various properties of circles and spheres. Another universal constant, e, has already been introduced in this chapter and always has the value 2.7128. These symbols have fixed values, unlike those of the previous section which vary from case to case.

Some other mathematical symbols might be less well known. The factorial sign (!) has no numerical equivalent and is an instructional sign, rather like the Σ sign. But it can appear only after integer (whole number) quantities and it instructs us to multiply together all whole numbers from 1 up to and including that indicated. Thus:

$$2! = 1 \times 2 = 2$$

$$3! = 1 \times 2 \times 3 = 6$$

$$4! = 1 \times 2 \times 3 \times 4 = 24$$

Clearly fractional quantities cannot be specified by this means. Some care is needed when the factorial instruction appears within lengthy formulae and the following examples show its order of precedence and how it should be treated:

$$5! + 2! = 120 + 2 \text{ and cannot be taken as } 7!$$

$$5! \times 3! = 120 \times 6 \text{ and does not equate to } 15!$$

$$6!/3! \quad = 720/6 \text{ and does not equate to } 2!$$

However, the expression $(6-3)!$ must be simplified to $3!$ as the brackets take precedence over the factorial.

Another instructional sign that can be encountered is that for the modulus. This is given by a pair of vertical lines. They are not brackets as such, but do indicate that the sign of any mathematical quantity produced by operations within the pair is to be ignored. For example:

$$|2-4| = 2 \text{ and not } -2$$

As with the use of brackets, the modulus sign can only be used as a matched pair.

Another sign to be familiar with is the plus or minus sign (±) which indicates that the numbers either side must be both added and subtracted give two final quantities, and not just one. For example:

$$64 \pm 2.5 = 66.5 \text{ and } 61.5$$

Some chapters also use the ‘greater than’ and ‘less than’ signs. These are a useful shorthand and appear as follows:

$$100 > 99 \text{ reads as 100 is greater than 99}$$

$$100 < 101 \text{ reads as 100 is less than 101}$$

for fairly obvious reasons this notation is more commonly used in algebraic forms:

$X > Y$ means that X is greater than Y

$Z < X$ means that Z is less than X

2.7 Computing with MINITAB and SPSS

Statistical analysis lends itself particularly well to the computer. Repetitious calculations using large quantities of data, whether those calculations are complicated or simple, is where the ‘number crunching’ computer can save time and reduce the risk of error. It is perfectly possible to write computer programs to run many of the statistical procedures described in this text. Good descriptions of how to prepare such programs in the language BASIC will be found in texts such as Sharp and Sawden (1984) and Kirkby *et al.* (1987). The use of the computer language FORTRAN is described in Davis (1973). Useful though it often is to write programs to suit specific tasks most students, and many researchers, are more likely to resort to one of the commercially-prepared programs, also known as 'packages'. A number of packages are available but we have chosen the two most widely available and comprehensive; the MINITAB and the SPSS (Statistical Package for the Social Sciences) systems. Both offer a wide range of subprograms and can be used on either mainframe or PCs. Given recent trends in computer technology and teaching techniques we will concentrate only on the PC versions of the two systems. The hardware specifications and other details of the two systems are provided in Appendix XII and should be consulted before any purchases are made. Both companies produce informative brochures outlining the work that can be undertaken on their systems.

In common with the arithmetic section of this chapter, it is not possible to go into the details of every aspect of the two systems. We will only use examples which are applicable to the techniques described in the following chapters. But, it must be emphasised, their capabilities and adaptabilities go far beyond the applications to which we will put them. Nevertheless, from this introduction students should be able to exploit with greater confidence the wider applications of MINITAB and SPSS.

Computers will differ in how they access and set up each of the two packages. We will assume that students are familiar with their own machines, the function of the different keys on the keyboard and with operating systems, of which MS-DOS in its various forms is the most widely used. The only commands and syntax to be introduced here and in the subsequent chapters are those that are specific to either MINITAB or to SPSS.

Both packages comprise a series of subprograms that can be specified to carry out tasks of data input, storage, retrieval and, of course, analysis. MINITAB and SPSS both expect data to be presented as if it were a matrix of the form specified in section 2.4. Each row of the matrix corresponds to a case or individual, and each column to a variable used to describe that set of individuals (the sample). If only one variable is used, there is only one column in the matrix, five variables gives five columns and so on. A sample of fifty individuals would give fifty rows to each column. The information may consist of simple numerical data, such as temperatures, rainfall, population data, etc. Alternatively the data may be numbers that allocate an individual observation to a class. In a questionnaire survey, for example, individuals may have ticked a box indicating their sex. This information could be coded as 0 for male and 1 for female. Such data are equally legitimate in a data file, though it must be remembered that the numbers refer to a classification scheme and have no numerical weight of their own – we could equally have coded the two groups as 99 and 100 without difficulty.

Our attention in this chapter is confined to problems of entering, storing and retrieving data sets of any form within MINITAB and SPSS. Both systems allow data to be input from the keyboard immediately before its analysis. Alternatively, and more commonly, it will have been entered at some earlier point in the research, stored in a computer file and retrieved for analysis at a later stage. In this form the data can be used again and again without need for it to be input at each stage of analysis. Data files can be stored within networks, on individual machine's hard discs or can be written to 5.25 or 3.5 inch diskettes that can be easily used on different computers. Such a use of files requires a good grasp of the local operating system so that files can be copied or read between different drives, hard disc drive to floppy drive or vice versa. The descriptions which follow show how data are input by the researcher but data files from any source can be used provided that they are stored on disc in ASCII (American Standard Code for Information Exchange) and set out as a matrix of the type described above. In this form they can be read directly by SPSS and by MINITAB. Both systems have their own detailed handbooks which should always be consulted for more detailed aspects of their operation (MINITAB, 1989 and Norusis 1990a, 1990b). In addition a number of independently produced texts also provide guidance for these two systems, for example Ryan *et al.* (1985) and Bryman and Cramer (1990).

2.8 Data management in MINITAB

When you have accessed MINITAB within your own machine (this is most commonly done by typing **MINITAB**, then by striking **RETURN**) you will get the screen prompt **MTB>**. Whenever this appears you must input a MINITAB instruction. For earlier versions of MINITAB these instructions had to be given to the computer through the keyboard. The latest, Release 8, version of MINITAB also allows instructions to be selected from pull-down menus. In the following examples, and to be as comprehensive as possible, only the keyboard instructions are used.

There are two such instructions that will indicate to the machine that you want to input data: **READ** or **SET**. Either can be used to enter data into the machine, the **READ** instruction allows the data matrix to be entered row by row, and the **SET** instruction allows entry column by column. Both must be accompanied by a specification of the column number(s) being used. An example using a simple matrix of four samples of five observations each is given in Figure 2.6 where **READ** is used to input the matrix. Notice that following the **READ** command the required number of columns are specified by indicating the range of column **(C)**

```
MTB  >  READ C1-C4
DATA >  8.5 3.5 6.7 4.5
DATA >  3.4 9.2 4.9 5.5
DATA >  1.9 7.4 4.6 8.8
DATA >  0.5 5.6 2.9 5.1
DATA >  3.5 8.9 5.6 3.4
DATA >  END
        5 ROWS READ
MTB  >  SAVE 'TESTFILE'

Worksheet saved into file: TESTFILE.MTW
MTB  >  RETRIEVE 'TESTFILE'
  WORKSHEET SAVED 8/21/1992

Worksheet retrieved from file: TESTFILE.MTW
MTB  >  SET C5
DATA >  3.4 6.7 8.4 2.2 0.7
DATA >  END
MTB  >  PRINT C1-C5

ROW    C1    C2    C3    C4    C5

  1   8.5   3.5   6.7   4.5   3.4
  2   3.4   9.2   4.9   5.5   6.7
  3   1.9   7.4   4.6   8.8   8.4
  4   0.5   5.6   2.9   5.1   2.2
  5   3.5   8.9   5.6   3.4   0.7

MTB  >  SAVE 'TESTFILE'

Worksheet saved into file: TESTFILE.MTW

MTB  >  STOP
```

Figure 2.6 MINITAB screen display illustrating use of the READ, SET, SAVE and PRINT instructions.

numbers. Following the 'return' of this instruction to the computer the screen prompt will change from **MTB>** to **DATA>**. Data must at each **DATA** prompt be entered in correct column sequence, a row at a time. Each row when typed must consist of the required number of observations, in this case four, each separated by a space. If too few observations are input the computer will request the line to typed again. If too many observations are typed those in excess of the required number are ignored. The input is concluded by typing **END**, upon which you are informed how many rows have been read and the **MTB>** prompt will reappear on the screen.

Our next instruction would normally be to **SAVE** the data in a file for later use. The **SAVE** instruction requires only the specification of a unique filename of up to eight characters, enclosed within single quotes (see Figure 2.6). When the file has been saved confirmation will appear on the screen followed by a prompt for further instructions. This file can be recalled to the system at another stage using the **RETRIEVE** instruction, again followed by the filename within single quotes. The system automatically attaches the delimiter **.MTW** to the filename when the **SAVE** instruction is used after which the file is stored in the system for later use. The **RETRIEVE** command assumes this delimiter to be present and it need not be added by the user. Successful retrieval of the file is acknowledged on the screen and followed by the **MTB**> prompt for further instructions.

As Figure 2.6 illustrates, the **SET** instruction is similar to **READ** but entry is by individual columns. In this example we have recalled the file and added a further fifth column (in this case denoted by **C5**) using **SET**. The contents of the file can be checked by using the **PRINT** command (Figure 2.6) followed by the columns whose contents are to be listed on the screen. Any new file created by modification of an existing file has to be saved using either the same, or a different, filename in order that the changes can be retained. If the changes are not 'saved' they will be lost upon closing the session. The file that has been retrieved, together with any modifications made to it during that session, becomes the *active file* and all analysis will be performed on the data in that file even if it has yet to be saved. Any other **.MTW** file can be made active by using the **RETRIEVE** instruction. This replaces the active file but any changes to the earlier file that have not been saved will be lost and only the original version retained for subsequent use. Only one file can be active at a time. All sessions are concluded by typing **STOP** at the **MTB**> prompt upon which you are taken out of MINITAB but, once again, will lose all files, etc. that have not been saved.

When using simple, generally one-line, instructions such as **READ**, **SAVE** or **RETRIEVE** the line must have no terminator such as a full stop or semi-colon. In later examples we may be required to enter two or more lines of *subcommands* on successive lines in order to initiate the option. To indicate that you wish to enter a subcommand (which usually contains additional information to help the system run) the introductory line should be concluded with a semi-colon. If this is done the prompt **SUBC**> will appear and replace **MTB**>. All subsequent subcommand lines must conclude with a semi-colon except the final one where a full stop informs the computer that the subcommands are complete. The semi-colon terminator should only be used where subcommands are to follow. Later examples will indicate those options to which subcommands are applicable.

In addition to creating **.MTW** files within the system, MINITAB can also read any ASCII code file provided that it contains only numerical data. Such a file, however, can be made active only by using the **READ** and not the **RETRIEVE** command. For example, to read an ASCII code data file of two columns (stored under the name **INFO.DAT**) from a diskette placed in floppy drive A: of the computer we would type:

READ 'A:INFO.DAT' C1 C2

and 'return' the instruction. It could then be saved within MINITAB in the manner described above. In line with MS-DOS practice, the drive specification is separated from the filename by a colon but notice that the drive specification, filename and three-character delimiter must be placed within quotes in this application.

2.9 Data management in SPSS

SPSS is usually accessed from the local system by typing **SPSSPC**, followed by pressing **RETURN**. Setting up data files in SPSS follows a generally similar path to that employed in MINITAB though it is slightly more involved. There is no screen prompt as with MINITAB and all SPSS commands can be entered by selecting them from the 'menus' which appear in the upper half of the screen display. It is, however, often easier when using small data sets and simple instructions to use only the lower half of the screen – the so-called *scratch pad*. All SPSS instructions and commands can be typed in by this means, the typed items then appearing on the scratch pad, from where they can also be executed. In order to be able to type directly onto the scratch pad the keys **ALT** and **E** must be pressed simultaneously. Pressing the **ESC** key, or the **ALT** and **E** keys again, will return you to the menu-driven mode of SPSS. The command **DATA LIST** is used if data are to be introduced initially through the keyboard or imported as raw data from another file either on the machine's hard disc or on floppy disc. **DATA LIST** is accompanied by information concerned with the layout of the data which is to follow. In this simple introduction we can use *freefield* format which requires only that the data appear in rows, each observation being separated from the next by a space. This is the easiest means of setting up the data file and is used in all the following examples. The code word **FREE** is used to indicate this form of data layout and must be followed by a slash (/) which is, in turn, followed by the names of the variables (up to eight characters each, starting with a letter and separated by either a space or a comma). A typical instruction is shown in Figure 2.7. As usual each line is input by pressing **RETURN**. In accordance with the conventions we have already established, each column represents data for one variable, and each row represents a 'case'. The order in which each row of data is input must always be the same and must agree with the order they appear in the variable list. If categorical (nominal) data are being studied it is often useful to denote classes not by numbers but by letters. Should this be done the variable has to be marked as one which contains alphabetical characters and not numbers. This is achieved by using the 'A' format in which the letter **A** is followed by the number of characters used to describe the categories. For example in a study of voting habits in the male and female population we might want to code the sex as the single character M or F, and the voting preference as the

```
DATA LIST FREE / VAR1 vAR2 VAR3 VAR4.
BEGIN DATA.
3.8 5.6 1.2 5.1
4.5 7.9 0.5 0.8
2.9 7.0 4.1 5.6
9.4 7.3 2.1 9.5
END DATA.
SAVE /OUTFILE 'TESTDATA'.DAT'.
GET /FILE 'TESTDATA.DAT'.
LIST.
FINISH.
```

```
VAR1    VAR2    VAR3    VAR4

3.80    5.60    1.20    5.10
4.50    7.90     .50     .80
2.90    7.00    4.10    5.60
9.40    7.30    2.10    9.50

Number of cases read = 4        Number of cases listed = 4
```

Figure 2.7 SPSS screen display illustrating use of DATA LIST, SAVE, LIST and GET instructions. The section above the break line shows the 'scratch pad' instructions; that below the break shows the results screen display. This presentational convention is used in all figures depicting work in SPSS.

two character CO for Conservative and LA for Labour. The command line would read:

DATA LIST FREE / SEX (A1) VOTE (A2)

The next command is **BEGIN DATA**. The data can now be entered row by row as a single and a pair of characters. At its conclusion the **END DATA** command is typed, followed by the **SAVE** instruction. The latter, as we see from Figure 2.7 requires additional information. At this stage we would want to save everything we have typed so we will use the option **OUTFILE** (there are others described in the SPSS manual) followed by the name we are to use to identify the file, the latter enclosed within single quotes. As this file contains data it would be appropriate to use a three-character terminator such as **DAT** in order that it can be recognised as such within the user-directory. This saved file would then contain all the SPSS instructions cited above, together with the raw data, all in the order in which they were typed. In this form the file is known as a *system file*.

Should we then wish to conclude the SPSS session we do so with the instruction **FINISH**. All SPSS sessions conclude in this way, but data files that have not been saved beforehand will be lost.

The typing in of instructions at the keyboard does not cause them to be automatically carried out as they are in MINITAB. To execute the commands up to and including the **FINISH** instruction, use the arrow keys to return the screen cursor (flashing light) to any point along the first line that was typed, or indeed to any line that requires execution. The F10 key should now be pressed. A message will appear at the foot of the screen stating 'run from cursor', press **RETURN** and all instructions from the cursor's current line onwards will then be executed. The scratch pad/menu screen will be temporarily suspended to display messages associated with these instructions. If there have been any errors they will be indicated on the screen. Equally, if the instructions can be properly executed information will appear showing that this has been accomplished. The SPSS output includes also a great deal of additional diagnostic information on the system, how much information has been read and stored, etc. In the following figures illustrating the use of SPSS we have edited out such information which for our immediate purposes serves little purpose – though it is useful when things go wrong as it usually indicates the source of your error.

It is often necessary to run each major instruction as it is input as subsequent commands often require, for example, that the data file has already been made active. In the figures that are used to illustrate the use of SPSS break lines are used to indicate the switch between 'scratch pad' and the 'results screen' displays.

In Figure 2.7 a small data file has been read in. After typing the line **END DATA** the cursor should be moved to the line **DATA LIST** and the tasks executed as outlined above. Then the **SAVE** instruction can be added and run, again by moving the cursor to the **SAVE** instruction line and pressing **F10** and **RETURN**. The file should now be saved in the SPSS system under the name specified in the command line. Its contents can be checked on the screen using the **LIST** command. It must also be noted that a full stop appears at the conclusion of each major command such as **LIST**, **DATA LIST** or **FINISH**. Where the command has secondary lines, examples of which will appear in later chapters, the stop only appears at the end of the final secondary line. These stops are mandatory; if they are omitted an error massage may appear on the results screen following the **F10/RETURN** message. In 'scratch pad' mode the cursor keys can be used to move about the screen to the point where any errors are located. The errors can then be corrected by deleting and retyping. The tasks can be easily rerun after editing by repeating the **F10/RETURN** instruction.

To recall stored system files at a later stage the **GET /FILE** command must be used, which is followed by the name of the file and its terminator, all within single quotes as shown in Figure 2.7. To check the data the **LIST** instruction can be used. Upon execution of this instruction the file's contents, together with information on the numbers of rows read and listed, are displayed on the results screen. The system file may do nothing more than set up the data for later analysis within SPSS, but it might include other processing or analytical instructions. To run the system file use the arrow keys to move the cursor to anywhere on the line containing the **GET /FILE** instruction. Then follow the standard procedure of pressing **F10** and then **RETURN** to execute the file and its commands. The scratch pad and menus are again temporarily suspended to display information on errors, if there are any, or to indicate that the tasks have been successfully completed. When listing is completed the screen messages will prompt you to return to the scratch pad/menu screen to carry out further work (you may need to use the **ALT** and **E** keys to get back onto the scratch pad). Only system files, i.e. those containing not only data but also some SPSS instructions can be recalled using the **GET /FILE** instruction. The file then becomes the active file until replaced by a later file retrieval or **DATA LIST** instruction. However, one of the assets that SPSS possesses is its ability to read any ASCII data file even if not set up within the system and files from a variety of sources can be used in analyses. But the means by which such files are made active differs from that of system files which already contain the ancillary information that SPSS requires. ASCII code data files can only be read by SPSS if they contain data (numerical or categorical) devoid of any column headings, comment lines or additional material. As with any other such data, the information must be stored by columns (for the variables) and by rows (for the cases). To activate a non-system ASCII data file the **DATA LIST** command can again be used but instead of being followed by the specification **FREE** (for the data format) as above the keyword **FILE** is used to indicate the data is to be found in a disc file on one the machine's drives. The filename then follows within single quotes. The filename must include its three-character delimiter (if it has one) and be preceded by the drive letter if it is not on the hard disc of the machine. The single line instruction to activate the file would read:

DATA LIST FILE 'A:INFO.DAT' FREE/ VARX VARY.

In this hypothetical example a data file **INFO.DAT** is stored on a floppy disc in drive A: of the machine. Following this specification the system needs to know the data format, which we will keep as the simple freefield form (keyword **FREE**) then, following the slash (/) the names attributed to each column in which the data are stored. Our example has only two variables, **VARX** and **VARY**. They must be listed in the order in which they are to be read. Should the file be a large one and contain more variable names than can be accommodated on the one line they can be continued on a following line or lines. Data files set up in this way should not be saved as system files with the **SAVE** instruction. At this stage it is simpler to use a DOS command to save the data file in the SPSS directory for retrieval at any later stages using again the **DATA LIST FILE** instruction. No operations can be carried out on non-system data files of this type until the **DATA LIST FILE** line has been run using the **F10/RETURN** instruction.

2.10 General points for using MINITAB and SPSS

Where large data sets are used care must be taken to avoid undue haste and risk of error when entering data at the keyboard. It is easy to type 10.0 instead of 100.0 or 1.0. Data verification at this early stage is vital. Both systems allow for data files to be edited and these methods are given in detail in the manuals that accompany the programs. SPSS requires that all variables have names which are indicated in the **DATA LIST** instruction. At the **READ** or **SET** stage in MINITAB the system needs to know only the number of columns (variables) that are being entered and attaches the indicators C1, C2, etc. to them. Variable names, if required, are given later using the **NAME** command at the **MTB>** prompt. For example:

MTB> NAME C1 = 'rainfall' C2 = 'altitude'

These names can also be added to or changed at later stages if necessary.

Screen displays of statistical analysis are unquestionably useful but it is, nevertheless, often helpful to have a 'hard copy' of the results of any analysis. The precise means by which you can print your results will depend upon your local network. But MINITAB and SPSS both produce files that can be printed and contain information from the screen. SPSS does this automatically setting up a file stored under the name **SPSS.LIS** which contains all instructions and results carried out at a session. You may be advised to edit it before printing the whole document (again local advice should be sought) as it will contain a large amount of unnecessary information and blank pages. Upon initiation of a new SPSS session this file will be automatically over-written, so it is important to edit and print it immediately upon completion of your work. MINITAB, on the other hand, needs to be prompted to produce the same information. This can be done by typing **OUTFILE** followed by a file name, for example **RESULTS**, at the **MTB>** prompt. The full instruction would simply read:

MTB> OUTFILE 'RESULTS'

The system will automatically attach the three-character terminator .LIS to the file which will then contain all screen information, instructions and results that appear after that point. This file can be closed at any time (and automatically saved) by typing **NOOUTFILE** at any **MTB>** prompt. This concluding instruction does not require the file's name. Following the conclusion of the session any LIS file can be edited if necessary and listed at the printer. The MINITAB **OUTFILE** file will not be over-written in subsequent sessions unless you select the same filename under which the later work is to be saved. Although of limited application at this early stage in our studies, the ability to store and print the results of lengthy and complex analyses such as those we will encounter in later chapters is important. Without such a capacity valuable information would otherwise be lost at the session's conclusion.

With data stored in files within the systems, or accessible through 3.5- or 5.25-inch discs, we are in a position to analyse them in whatever way we decide. The various commands by which this can be done and the options that are available will be introduced in the following chapters.

2.11 Conclusions

This chapter is not an exhaustive course in mathematics or computing, and neither is it intended to be so. Major areas have been omitted, partly because of their complexity but principally because they are unnecessary for the understanding of the statistical methods to be described in the following chapters. If the limited contents of this chapter are clearly understood the reader is left free to consider the interpretational and geographical issues surrounding the application of those methods. In the case of the MINITAB and SPSS packages we have done little more than provide the barest essentials of data input and retrieval, but they should be sufficient for students to get started on either system. Familiarity with the basic syntax of their command structures will allow students to experiment for themselves and with the guidance of the exhaustive manuals that accompany the packages will soon develop their skills beyond the grounding provided here.

References

Bryman, A. and Cramer, D. (1990) *Quantitative Data Analysis for Social Scientists*, Routledge, London.

Davis, J.C. (1973) *Statistics and Data Analysis in Geology*, Wiley, London.

Kirkby, M.J., Naden, P.S., Burt, T.P. and Butcher, D.P. (1987) *Computer Simulation in Physical Geography*, Wiley, Chichester.

MINITAB (1989) *MINITAB Reference Manual: release 7*, MINITAB, Pennsylvania.

Monkhouse, F.J. and Wilkinson, H.R. (1971) *Maps and Diagrams*, 3rd edn. Methuen, London.

Norusis, M.J. (1990a) *SPSS/PC+: statistics 4.0 for the IBM PC/XT/AT and PS/2*, SPSS, Chicago.

Chapter 3

Geographical Sources, Data Collection and Data Handling

3.1 Introduction

Data, in various forms, are the raw material with which geographers pursue their research. Yet collection, acquisition and, even, availability of data can pose major problems in themselves. The purpose of this chapter is to explore the basic problems associated with data sources in geographical analysis. As we shall see, the quality, and in some circumstances the quantity, of our data determines not only which statistical tests should be used, but also the relevance of our results. In this respect it is important that we understand the strengths and weaknesses of our data before we embark on the application of quantitative techniques. This section is not, however, intended to be a comprehensive review of geographical sources, but rather an introduction to the major types of data that are available, and a guide to the literature concerned with these sources.

The two main types of data source used by geographers are published or 'archival' material and field observations. Obviously, the use made of each of these will depend on the type of study being undertaken. For example, Haggett (1965) estimated that about 95 per cent of the articles published in human geography between 1960 and 1965 were based on secondary data sources. We may suppose, considering the nature of physical geography, that such a figure could well be reversed on the side of field observations.

Secondary, or 'archival', sources can be classified in terms of whether they are spatial or non-spatial in form. In terms of the former type, aerial data traditionally refers to maps, aerial photographs and, more recently, remote sensing information from an increasing variety of satellites. A general review of remote sensing is presented in Curran (1985) and Sabbins (1987). In both human and physical geography, information from satellites, mainly in the form of photographs, is an important source. The application of such techniques touches most aspects of geography from climatology and geomorphology (Millington and Townshend, 1987) to urban geography (Dickinson, 1979; Lo, 1986).

The changed nature of geography has also transformed the role of maps in many branches of the subject. Traditionally, the map was very often the geographer's single means of description, as well as being used to measure the locational properties of objects. While these facets are still important, maps are just one of a number of data stores for locational information that are available to the geographer. Within this context perhaps the most important advances have taken place with the development of geographical information systems (GISs). Much of the early demand for such systems arose in the USA with an emphasis on land management and the refinement of GISs has been very strongly related to technological developments in computer hardware and software (Maguire, 1989). Indeed, geographical information systems refer to nothing more than integrated systems of geographical data, together with hardware and software: these enable large amounts of collected data to be stored, manipulated and presented – often in cartographic form. Such systems may be classified into two main types, based on how they handle geographical data: namely, vector-based systems and raster-based ones. Vector-based GISs handle data encoded as vectors using Cartesian co-ordinates and are much used within the context of thematic and topographical information. By contrast, the raster-based type use data encoded in a grid cell format, and these are often utilised within remote sensing.

The use of GISs is becoming increasingly widespread as computer systems become more flexible and sophisticated. For example, UK-based students can now gain access to one such system, that of the National On-line Manpower Information System (NOMIS) based at Durham University, England. This stores employment and population census data for the UK, and makes them quickly available at a variety of spatial scales as defined by the user (Townshend *et al.*, 1987).

In terms of the use of maps as information stores, a number of texts explore the problems of cartographic design (Monkhouse and Wilkinson, 1971; Robinson, 1952; Board, 1967). The use of maps as actual data sources is less well-documented, although a few specialist contributions do exist. Thus, Clark (1966) discusses the use of maps as sources of morphometric analysis, while Harley (1980) has reviewed the data contents of Ordnance Survey maps, especially within the context of historical geography.

The non-aerial data used by geographers, as we shall see, covers a wide variety of sources, but most share the characteristics of being readily transferred to a map or converted into some type of spatial index. In most cases, however, these information sources have not been compiled for geographical purposes and therefore a number of problems arise in their application. First, many of the units within which the data are collected are unsatisfactory for some types of geographical analysis, usually because such areas vary in size and shape. Second, the release and publication of official data is often only at specific time intervals, such as every ten years for the full UK census, which may not be frequent enough for those studies interested in changes over time. Finally, the overall accuracy of these sources is outside our direct control and many do suffer from problems of inaccuracy.

Geography has traditionally used fieldwork as a major source of information, and field techniques have become increasingly more sophisticated. Thus, much of the early fieldwork merely involved visual observation and description, based on the geographer's powers of perception. Such observations can be recorded by field sketches, written accounts or on field maps. Unfortunately, this type of data collection is very subjective because it depends on the perception of individual geographers. It should be noted, however, that direct field observation is still an important process in the development of ideas, and is often the forerunner to field measurements. It is in the area of field measurement that we can recognise the greatest changes in the processes of data collection. Such measurements can range from the simple enumeration of objects – for example counting numbers of shops in a study of service provision – through to the setting up of large-scale systematic surveys. In human geography these surveys are often associated with collecting attitudes and responses from people, and questionnaire design is now an important part of the geographer's training. Similarly, in many branches of physical geography much emphasis is given to the systematic monitoring and collection of data.

Finally, mention should be made of the differences between the total enumeration and measurement of particular objects, and the use of samples. The traditional prejudices against sample data described by Haggett (1965) are now no longer such a problem, since the acceptance of sampling methods is widespread throughout all branches of the subject. Sampling techniques are discussed more fully in section 3.4.

3.2 Sources in human geography

As we have already seen, a considerable amount of data in human geography is derived from secondary or 'archival' sources, whose great diversity precludes complete coverage. Only the most significant ones are examined here. These sources can be subdivided into a number of different groups, in terms of their origin, purpose and content. Thus, at the most general level we can distinguish between those issued for official purposes, and non-official sources. Furthermore, within each of these broad categories a distinction can be made between primary statistical sources and those which merely summarise other, more detailed, published statistics. As Figure 3.1 shows, each of these categories can be further subdivided according to the geographical scale at which the data are issued and, finally, the topic covered. This simple classification can be used as a framework for our review of 'archival' sources.

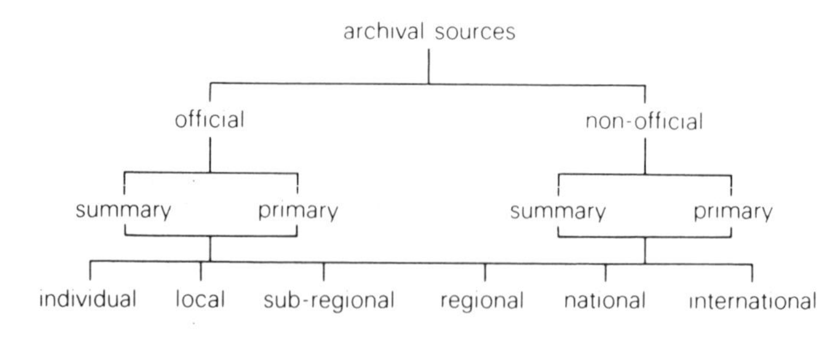

Figure 3.1 Typology of archival sources for geographers.

In terms of official sources we can start our review by comparing the availability of statistics issued by international bodies with those produced by individual governments. For example, Table 3.1 illustrates the variety of sources available at a world scale, produced by the UN and OECD. In most instances these are essentially summary statistics, collated from the official statistics of each individual country. Similar, though more detailed, statistical publications are also available for particular organisations of countries, such as those produced by the European Community and EFTA. Such sources are extremely valuable for comparative work, although in the use of summary statistics consideration needs to be given to the type of classification system used. Different individual countries may have used different methods of classification and the summary statistics may not resolve such differences, thereby hindering any comparison. An outstanding example of this problem is the definition of 'urban', and the variety of criteria used by national censuses for delimiting urban populations. The attempts to overcome this problem are discussed in a number of United Nations publications, but particular attention is drawn to *Growth of the World's Urban and Rural Population, 1920–2000*, published by the UN Department of Economic and Social Affairs in 1969. Similar types of problems associated with variations in the classification of industrial activities are discussed by Smith (1975) in a comparison of the Standard Industrial Classification used in the USA and UK.

In the case of the European Community the problems of comparative statistics are being solved by applying standard definitions and in some instances by instituting new surveys for all member states. In this way a comparable set of statistics is becoming available, creating a stronger database for the geographer.

In any country the government is the main compiler of statistics, which are published in a variety of forms. The variety can be illustrated in the case of the official statistics published in the UK (Table 3.2). Initially such statistics were usually collected as a by-product of routine administration. However, in advanced economies the collection of statistics becomes more important in the planning of the economy and greater emphasis is placed on the integration of data sources. Despite the

Table 3.1 Selected data sources available at an international level

Topics	Worldwide	European Community
General	*UN Statistical Yearbook* (1948–)	*Eurostat* (1968–) General statistics (11 issues/year) data on short-term economic trends Basic statistics (1960–)
Population/ social trends	*UN Demographic Yearbook* (1948–)[1]	*Demographic Statistics* (1977–)[4] *Regional Statistics (1971–)*
Agriculture	*UN Yearbook of Food and Agricultural Statistics* (1947–)[2]	*Yearbook of Agricultural Statistics* (1970–) *Agricultural Statistics* 6 issues/year, and annual volume, (1959–)
Trade and industry	*UN Yearbook of International Statistics* (1950–)	*Industrial Statistics,* quarterly (1959–); also a yearbook (1962–)
	UN Yearbook of National Account Statistics (1957–)[3]	*External Trade Statistics* (1962–)

( ) = date of first publication. 1. Contains introduction about previous statistics published by League of Nations, 1922–42. 2. From 1950 onwards publication issued in two volumes, one covering production, and the second, trade and commerce in agriculture. 3. Gives details of accounts by country, e.g. GNP and distribution of national income. 4. First edition in 1977 covers statistics for the period 1960–76, and annually thereafter.

Table 3.2 Selected official published statistics available for the UK

Topic and source	Frequency of publication	Nationally	Availability	Settlement
1. **Population/social**				
Census reports	10-yearly	*	*	*
Housing Statistics (GB)	Quarterly	*	*	*
Local Housing Statistics (E & W)	Quarterly	*		
Digest of Health Statistics	Annually	*		
2. **Production**				
Agricultural Statistics	Annually	*		
Census of Production	10-yearly	*		
Business Monitor	Periodically	*		
3. **Distribution**				
Census of Distribution (1951–71)	10-yearly	*	*	*
Business Monitor	Periodically	*		
4. **Transport**				
Passenger Transport in GB	Annually	*		
Highway Statistics	Annually	*		

* Indicates availability

recognition of this fact many governments' statistics are still not available at either a comparable spatial scale or for the same time period. One useful UK Government publication that presents data on a variety of topics at a common scale is *Regional Trends* (Central Statistical Office, 1971–), which gives information for counties and standard statistical regions. These regions correspond to the Registrar General's Economic Planning Regions, although their definitions did change slightly in 1967. Other examples of standard classifications adopted by official sources in the UK are the Standard Industrial Classification and the Registrar General's socio-economic groupings.

A further problem concerns the changing nature of official statistics over time. Unfortunately, for a variety of administrative reasons, the way in which many of these official statistics are collected may have changed, thus making statistical comparisons between one time period and another more difficult. In order to understand the scale and importance of such changes it is worthwhile consulting some of the guides to official statistics. A number of these exist, the most notable being the Central Statistical Office's *Guide to Official Statistics* (1980) and Edwards' *Sources of Economic and Business Statistics* (1972).

In working with official statistics the geographer is often tempted into believing that such data are always totally accurate; but in fact all sources contain some degree of measurement error. Often in official sources this may take the form of under-enumeration, as is the case with the Census of Distribution. The first of these censuses in 1951 covered 91 per cent of retail establishments, while in 1961 this figure fell to 88 per cent. What we would need to consider is whether such variations were random or biased, and only related to certain types of shops. Sometimes these measurement errors are due to a lack of co-operation between organisations or people, resulting in less than a total response to government and official surveys. In other cases such errors arise owing to a mishandling of data, which is often more difficult to detect.

If we are interested in obtaining detailed official sources for other countries we can conveniently use statistical year-books. One of the most detailed and comprehensive of these is that issued by the United States Government, which contains not only a wealth of statistics, but also a section on their collection. To this can be added the equally wide-ranging and exhaustive publications of the United States Bureau of the Census. Both the UK and the US Governments are responsible for the preparation and the publication of official statistics for their respective countries covering a wide range of demographic, social and economic aspects. For either nation they should be the first point of departure in the search for information. Both Governments supply information on the availability, and the cost, of published statistical information.

So many and varied are the official sources, not only of the USA and UK but also of the UN and its related organisations, that it is often more efficient to consult one of the many directories of data sources that are available. The most important of these publications are cited fully in the paragraphs that follow. In this respect the two-volume publication *Statistical Sources*, edited by J. Wasserman O'Brien and S.R. Wasserman, the tenth edition of which was published by Gale Research Co. of Detroit in 1986, contains information not only for United States sources but also for those world-wide and in other specific nations. To this volume should be added G.T. Kurian's *Sourcebook of Global Statistics* which was published by Longman of London in 1985. This volume contains not only international sources for socio-economic and demographic statistics but also, and unusually for such items, for data in physical geography (mainly climatology and hydrology). A summary of similarly varied statistics is the *Directory of International Statistics* published by the United Nations (New York) in 1982. The series edited by J.M. Harvey and published by CBD Publications of Beckenham, England is also useful. The fifth volumes in the series appeared in 1987. The first of these was *Statistics Europe 1987: sources for social, economic and market research*. But other volumes cover Africa, America and Asia and Australasia. The same publishers are also responsible for F.C. Pieper's *Subject Index to Sources of Comparative International Statistics* which though published in 1978 remains a useful item.

Finally, while the geographer's interest may well focus on present-day conditions and circumstances there are occasions when historical, or time-based, data are required. Historical statistics can be found in two publications by B.R. Mitchell. The first of these is *European Historical Statistics 1750-1970* published by Macmillan of London in 1975. More recently the same author has written *British Historical Statistics*, published by Cambridge University Press in 1988. Equally valuable is the American source material to be found in *Historical Statistics of the United States: colonial times to 1970*, published by the US Bureau of the Census (Washington DC) in 1975.

Given the large amount of official data that are available it is hardly surprising that human geographers use non-official sources in a limited way. Perhaps two of the most commonly used of these non-official sources are town plans and directories. The former include a variety of fire insurance plans giving details of property ownership in urban areas and, since 1967, Goad Plans of British shopping centres (Rowley and Shepherd, 1976). These indicate street numbers, the name of the occupant and the type of retail activity at a scale of 1 : 1056. Both types are useful sources for data on land-use changes in city centres. However, of greater use are directories that are available in a wide variety, ranging from town to international directories (Bull and Bull, 1978). The importance of directories is that many allow the location of data and give information about individuals and organisations which is not often available in aggregated official statistics. However, problems of inaccuracy often hinder their use, and measurement errors are sometimes difficult to determine (Shaw and Tipper, 1989). Nevertheless, unofficial sources can be useful and should not be dismissed out of hand. Many such sources have been collated and listed for the UK in D. Mort's

Sources of Unofficial UK Statistics published by Gower (Aldershot, England) in 1990.

From the geographer's point of view one of the major problems associated with most published statistics is that the spatial units used for compiling them are seldom meaningful, being often irregular in shape and size, and in some instances changing over time. These problems are well illustrated by the enumeration districts used in the UK census of population, which are of variable shape and, depending on population movements, change from one census to another. Furthermore many different area divisions have been used in the census, the aerial base for the 1971 census being the most complex. This provided statistics for local government areas and in total some 22 different types of aerial units were used (Denham, 1980). Systems such as the NOMIS database, as previously discussed, remove some of these difficulties by allowing the user to select a consistent spatial framework.

There are two possible methods of solving the problems of unsuitable aerial units. The first of these is to compile the official data in statistically sensible and consistent spatial units. To some extent this was attempted in the UK with the 1971 census, which made data available on a 100m grid basis. These grid squares were based on the Ordnance Survey National Grid system and had the advantages of being spatially regular and permanent over time when compared with enumeration districts (Table 3.3). However, they also have their disadvantages in that they do not correspond with any physical features, and in some instances grid lines may be drawn through the middle of a dwelling.

Table 3.3 Advantages and disadvantages of grid square and enumeration districts

Strengths and weaknesses	Enumeration districts	Grid squares
Ready availability of maps showing boundaries	Yes	Yes
Permanence of boundaries and spatial regularity of areal base	No	Yes
Correspondence with physical features or administrative areas	Yes	No
Areas recognized in planning	Yes	No
Limited variability of population in areal base	Yes	No

The second method of overcoming these problems is for the geographer to collect his own data from field surveys. One of the most significant ways of collecting information in human geography is through questionnaires. These require careful planning and preparation and their effectiveness is largely determined by their design. In geography, questionnaires are used to measure such things as behaviour, perceptions and attitudes, and characteristics. In most instances a questionnaire may attempt to collect information on all three areas as, for example, in work on consumer studies where data are required on patterns of shopping behaviour and attitudes of various consumer types. Obviously, each one will require different approaches in terms of the structure and design of the overall questionnaire.

Only a brief guide is given here to the techniques of questionnaire design, and for more detailed information the reader is directed to more specialist texts (Berdie and Anderson, (1974), Oppenheim (1966) and Fowler (1993)). The design of questionnaires can be summarised under five major headings as illustrated in Table 3.4. In the first instance the decisions concern the type of information required, the population under consideration and the method of reaching the respondents. These problems relate not only to the construction of the questionnaire, but also to questions of data collection and survey design which are outlined

Table 3.4 Stages in the questionnaire design process

1. **Initial decisions**
   (a) Decide what information is required
   (b) Who are the respondents?
   (c) What method of survey approach will be used?

2. **Content of questionnaire**
   (a) Which questions are essential?
   (b) Is the question sufficient to generate the required information?
   (c) Are there any factors that might bias or negate the response to the question?

3. **Phrasing and format of questions**
   (a) Is the question phrased correctly or could it mislead respondents?
   (b) Can the question be asked best as an open-ended, multiple-choice or dichotomous question?

4. **Layout of the questionnaire**
   (a) Are the questions organized in a logical way?
   (b) Is the questionnaire designed in a manner to avoid confusion?

5. **Decision to carry out pilot questionnaire**

in section 3.4. In terms of the phrasing of individual questions one of the major decisions is whether or not open-ended questions should be used. These leave the respondent free to offer any reply that seems appropriate and the answers are not pre-consolidated by a set of response categories. However, the questions suffer from interviewer effects, whereby interviewers vary in their objectivity, and the responses are also extremely difficult to handle in an objective manner and often preclude the use of statistical techniques (Dohrenwend, 1965). A further critical area is the layout of the questionnaire, which is a frequent source of error. In general the initial questions should be simple and interesting in order to gain the attention of the respondent, while overall one topic should follow on logically from the next. Finally, the questionnaire should be kept as short as possible, although in instances where respondents were asked to recall past events longer questions appear to be most effective (Laurent, 1972).

A critical problem in the design of questionnaires is measurement error, since differently worded questions on the same topic often produce very different response rates. In one study by Noelle-Newman (1970) a group of women were questioned on the same topic using slightly different questions. Thus, in answer to the question 'Would you like to have a job if possible this were possible?', 81 per cent stated that they would like a job, compared with only 32 per cent who made the same reply to the question 'Would you prefer to have a job, or do you prefer to do your housework?'. The second version of the question is more explicit and produced a quite different response. Many of these problems associated with questionnaires often only come to light during their application, and it is for this reason that they should be tested before the full survey is undertaken.

The collection of primary data using field surveys is one important method of solving some of the problems associated with archival sources. However, secondary sources are still used extensively because they are so readily available, whereas primary data are often very costly to collect. In some circumstances this type of accessibility can also be given to data collected in the field, through the creation of data banks. A large number of these now exist, some of the earliest experiments having been pioneered in Sweden during the 1950s, using census data. In the UK one of the most significant developments has been the ESRC Survey Archive based at Essex University, which is concerned with the dissemination of computer-based files containing social and economic data (Tannenbaum, 1980). Furthermore, a growing number of geographical data banks are being initiated, particularly in the field of industrial geography.

3.3 Sources in physical geography

Though the distinction between human and physical geography is a largely artificial one, in respect of the published and archived databases it must be acknowledged that major differences do exist. The impressive array of statistical information available through official UK and USA sources covering a huge variety of human activities and economic matters have no counterpart in the fields of physical geography. On the other hand reliable topographic maps have a long history as a valuable data source in geomorphological studies (Gannett 1901; Wooldridge, 1928; Chorley, 1958; Wheeler, 1979).

These observations notwithstanding, physical geography is not without some important statistical sources though they are largely confined to the areas of climatology and hydrology. The sources cited in the following paragraphs share the virtues of being accurate, accessible and, largely, inexpensive. Depending upon the precise needs of the enquirer, there are few areas of the UK or the USA for which reliable hydrological and climatological data are not available. Such data also cover periods of time ranging from little more than a decade, to well over a century. Nevertheless, many of the problems discussed above in connection with sources in human geography – the difficulties of comparability using different spatial and temporal scales, the accuracy of the data and the problem of frameworks of data collection that change over time – all apply with equal vigour when examining sources in physical geography. Once again, we cannot hope to identify all sources of information but will endeavour to identify some of the principal means by which physical geographers can obtain information for their researches from published or archived sources.

Climatological and hydrological databases for the UK and the USA are found in two forms, either as archived material which requires abstraction 'to order' or in the form of widely-available published items and statistical digests from which geographers can abstract what is required for their immediate purposes. The major organisations responsible for these sources sources are listed in Table 3.5 and their addresses are given in Appendix XIII.

Table 3.5 Custodians of major UK and US archives of hydrological and climatological data

Name of body	Data type
UK organisations	
The Meteorological Office	Climatic data
The Institute of Hydrology	Hydrological and climatic data
The British Geological Survey	Groundwater data
USA organisations	
The National Climatic Data Center	Climatic data
National Oceanographic and Atmospheric Administration	Climatic data
US Geological Survey	Hydrological data
US Department of Agriculture	Soils data

The addresses of all the organisations cited in this table, together with others for sources in human geography, will be found in Appendix XIII

The UK Meteorological Office has an Education Service to deal with enquiries for data from schools, colleges and universities. The Service has at its disposal a large archive of climatological data for England and Wales with separate departments to deal with Scotland and Northern Ireland. The leaflets EI1 to EI4 are published free of charge and explain the work of the Service and what facilities it can provide. Climatological data are catalogued by individual stations and they are prepared in the form of yearly records of monthly data, monthly records of daily data and monthly records of daily fixed-hour (0900GMT) data. Although the Service has access to information from over 400 weather stations the data may be incomplete because not every station records all weather elements. More importantly, the stations have been operational for different lengths of time and it may be necessary when comparing different sites to ensure that the data embrace the same time periods. Nevertheless every county in the UK has at least one station with a complete and long-standing record which includes data for temperatures (maxima and minima), rainfall, sunshine and the frequency of snow, thunder, fog, gales and frosts. Information can be supplied for specific periods of time or as long-term averages based on the 1951 to 1980 data. The Service is always prepared to advise on data availability provided that the requirements are clearly set out in terms of location, period of time for which data are needed and weather elements for which information is required.

A further valuable source, though restricted largely to rainfall data, is found in the annual publication *Rainfall* which contains information on over 3000 stations in the UK. Each entry consists of the monthly and annual rainfall for the year in question, the station's location, altitude and (if available) long-term mean annual total. Stations are grouped regionally and the report includes a brief written account of the year.

Rainfall has been published since 1969 and supersedes the more detailed publication *British Rainfall* which ceased in 1968. Though changing its presentation over the years, *British Rainfall* appeared annually from as early as 1864. Originally published by the British Rainfall Organisation, responsibility for its preparation and publication was assumed by the Meteorological Office in 1919. Though such data have been published for a long period of time a problem is posed by the stations which comprise the observational network and, once again, very few of them have been in constant operation for more than a few decades. The changing composition of the rain gauge network needs always to be borne in mind.

The newly-established UK National Water Archive is maintained by the Institute of Hydrology. It contains river discharge data for some 1400 sites in the UK. The average period of record is a little over 20 years but, once again, the period of operation of the individual measuring sites again varies greatly. In addition to mean and extreme discharge statistics, the archive also holds catchment rainfall data.

The most useful data are probably the daily mean gauged flows. The data can be purchased as 'hard copy' line printer output or, perhaps more usefully in the current context, on computer disks. In addition to mean daily flows, statistics for monthly and annual extremes (maxima and minima) are available, together with rainfall and runoff data aggregated over specific catchments on a monthly basis.

For each station additional information is held describing the means by which the data are obtained, as they vary from site to site. Information is provided on the factors within the catchments which may influence flow characteristics. For example, the presence of storage reservoirs, the regular abstraction of water for domestic and agricultural use or additions from groundwater abstraction or effluent returns will all impose a degree of artificiality on the discharge levels and their variations. Such considerations are important and if not recognised then the statistical information on river levels can be subject to misinterpretation, though in itself it is entirely 'accurate'.

The British Geological Survey simultaneously is responsible for the groundwater section of the Archive which holds data on water levels for 175 boreholes and wells in the UK. Time series of groundwater levels for each site are readily retrievable in hydrograph form if required. Annual or monthly summary statistics (maxima, minima and means) can also be obtained. Data are stored as a series of water levels with the dates of each level. Complete site histories are supplied at costs comparable to those for surface water data. The periods of observations are generally 20 years and very few records are continuous from pre-war times though in one remarkable instance data is held from 1836 to the present day.

At present the National Water Archives Retrieval Service provides 17 standard surface water data and four groundwater retrieval options which allow for the provision of graphical and descriptive as well as statistical information. Table 3.6 lists those options most appropriate for the provision of data in a form that might be useful in statistics teaching. The contents of the National Water Archive are usefully summarised in the annual publication *Hydrological Data UK* which contains daily discharge data (for the appropriate year) for about 50 stations, with monthly summaries of a further 150. Site and catchment details are also given indicating catchment area, station altitude, aggregated monthly rainfall and runoff over catchment during that year, data on flow frequencies and long-term averages and extremes. Each publication also includes written reports and some basic information on borehole records and water quality data.

The United State's counterpart to the UK Meteorological Office's archives is the Climatic Data Center. The service is again based on individual weather stations, some with records going back over 100 years, but now includes much satellite and upper air information obtained through the latest technologies. The present Center was set up in 1952 and became part of the National Oceanic and Atmospheric

Table 3.6 National Water Archive retrieval option*

Option	Title	Description of data
1	Daily mean gauged discharges	Monthly and annual summary statistics
2	Daily mean naturalised discharges	Monthly and annual summary statistics
4	Monthly mean gauge discharges	Monthly and annual summary statistics
6	Yearbook data tabulation	Monthly riverflows and catchment rainfalls with comparative statistics from the past record
9	Monthly catchment rainfall and runoff	Individual monthly rainfall and runoff data with long-term averages and extremes
7	Monthly extreme flows	Highest and lowest daily mean flows
12	Flow duration statistics	1 to 99 percentile flows with graphical output option

* Only those options immediately appropriate for this book are cited
See Appendix XIII for the address of the custodians of this archive

Administration (NOAA) and today receives data not only from 10,000 volunteer observers in the United States but also from around the World. Regional centres assist in the processing of data for what has become the world's largest climatic database. The Climate Service System was established in 1978 and continues to provide support to educational establishments. Climatological data have been published for over a century and now include a wide range of material summarised in Table 3.7.

Table 3.7 Summary of data available from the US National Climatic Data Center

1. Local climatological data. Produced monthly for 270 cities in the USA and includes all principal climatic elements based on three-hour interval readings.
2. Climatological data are also produced monthly for 10,000 sites in the USA. Based on single daily observations only, these data are published monthly for 45 regional groups.
3. Hourly precipitation data. Available for 2400 stations in 45 regional groups. Published monthly.
4. Monthly climatic data. Monthly data for 1500 stations world-wide.

The address of the custodians of this data archive is given in Appendix XIII

In addition to these regularly-published sources individual enquiries of a more specific nature can be made, with data available on printed sheets, floppy disks or magnetic tape (see Appendix XIII).

The US Geological Survey is responsible for the collation and distribution of hydrological data throughout the United States. Annual state by state reports are published by the Survey's individual State Offices (Water Resources Division) which contain the mean discharge, mean concentration of sediment load and total sediment load for each operating station for each day of the hydrological year. The State Annual Reports also contain additional information on the gauging sites, water quality and some groundwater information. If such information is sought, initial contact is best made though the State Office.

The USA also has a valuable source of data for soil studies in the county-based reports prepared by local agencies in co-operation with the US Department of Agriculture. These reports contain information on the soil characteristics such as the USDA texture classification, drainage properties, reaction (on the pH scale), particle-size distribution and data on the exchangeable cation (H, Ca, Mg and K) levels of each soil type. This forms a most valuable, if very locally-based, data resource.

Finally attention should be drawn the the commercial sale of data through agencies such as Earthinfo of Colorado. Because of the inevitable variability of prices in such commercial activities none are quoted here and the interested reader is advised to consult the organisations listed in Appendix XIII. Earthinfo supplies data on Compact Disc Read Only Memory form (CD-ROM). This new form of data storage and retrieval is one of rapidly growing importance. In this case, each disc contains as much information as 2000 floppy discs. Data are regionally-based for both the US and Canada and are brought up-to-date annually. The information is read using software supplied with the CD and is fully indexed. Obviously a CD-ROM reader would be a necessary adjunct to the computer system but subsets of some of the data can be purchased on floppy disc.

The volumes of data available using CD-ROM technology are formidable. Earthinfo's system includes all daily discharge data for the USGS sites over the history of each site's operation. This accumulates to over 600,000 years of streamflow data. The original daily values and summary statistics are available on a four-disc set. They also have the peak values for 25,000 present and past stations (on one CD) together with a vast store of water quality and groundwater data for all official sites (numbering over 200,000) on a set of eight CDs.

Many of the data already described under the heading of the National Climatic Data Center are available on CD-ROM. A four-CD set can be purchased which contains daily observations of precipitation, snowfall, temperatures and evaporation for over 25,000 past and present stations. Further disc sets contain hourly data.

In the UK, the *World Climate Disc* is now available from Chadwyck-Healey (Cambridge, UK) and includes much of the data gathered by the Climatic Research Unit of the University of East Anglia. The five discs contain temperature data for over 3000 stations from 1854 to 1990, five-degree gridded mean temperatures over the same period, rainfall for over 7000 stations from the eighteenth century to 1990 and gridded sea-level air pressure over the northern hemisphere from 1873 to 1990. The software included with this system allows tables, maps and graphs to be plotted.

It is clear that vast amounts of data can be made available through such new technology and that developments are taking place at a rapid rate. For many teachers there may, ironically, be too much information for most teaching purposes, but its research role is of growing importance and should not be overlooked.

3.4 Research design and data collection

Research design is used here as a term to encompass a number of different organisational stages in the gathering of information. Once we have defined our subject of investigation the next step is to collect the necessary data with which to test our ideas. The problem posed is one of research design, which requires decisions on what information to generate, the methods of data collection, the coverage of the data and the way in which the data are to be analysed. In ideal circumstances these stages in the design of any research project can be viewed as a sequential series of logical steps as outlined in Table 3.8. As we have shown in the preceding sections the major methods of data collection are associated with the use of either 'archival' sources or primary fieldwork. Measurement techniques will vary considerably within each of these major categories but both have in common the problem of how much of the 'population' to measure.

Table 3.8 Stages of research design

1. Define the problem and the type of information required
2. Select the method of data collection and determine whether secondary or primary data are required, or some combination
3. Select the technique of measurement – for example, decide whether to use questionnaires, and of what type
4. Decide whether to measure the total population or a sample: if the latter, select the appropriate sample strategy (Tables 3.10 and 3.11)
5. Determine the appropriate means of analysing the data

Geographers have become increasingly involved with this problem of coverage. As Haggett (1965) suggests, there are two ways in which it may be overcome: first, in a direct fashion by increasing the amount of data available through the accumulation of information, which may be achieved by building up a computer bank of, say, survey results.

Table 3.9 Stages in the sampling process

Stage	Process
1. Define the population	Defined in terms of (a) units, (b) elements, (c) area, (d) time period
2. Define sampling frame	How the elements of the population can be described
3. Specify sampling unit	Identify units for sampling, e.g. city, streets or households
4. Determine sampling method	Method by which units are to be sampled. e.g. probability or non-probability schemes
5. Determine size of sample	The number of units/individuals to be selected
6. Specify sampling plan and method of collecting data	The operational procedures necessary for selecting the sample data

Secondly, sampling methods can be used, and it is to these that we now turn.

There are a number of factors that need to be considered in the application and use of samples, as illustrated in Table 3.9. The initial step is to specify the population and the individuals contained within it.

The second stage is to develop a *sampling frame*, which locates the individuals within the population. Typical sampling frames list all the objects with the population; electoral registers or street directories are good examples. In many areas, however, sampling frames are more difficult to construct. For example, sampling from mobile populations such as car traffic or river water pose considerable problems (Harvey, 1969). Another problem is where the statistical population is not known, as for example in some less-developed nations. As Harvey (1969) points out, geographers have spent relatively little time analysing the nature and definition of the statistical populations in which they are interested. To some extent such conceptual definitions are of little direct importance although, as we shall see in later chapters, they may impinge upon the interpretation of the results of statistical analyses.

The third step in the sampling process is to specify the sampling unit or individual, the selection of which depends on the nature of the topic, the sampling frame and the design of the project. Thus, in a study of residential mobility our basic sampling unit would ideally be individual households, although data constraints may force us to work with aggregated census information at the enumeration district level. In geography, spatial sampling from maps is obviously of importance, and a number of geometrical sampling units may be used, including points, lines or transects, and quadrats. Each has its own advantages and disadvantages which are reviewed in Haggett (1965), Berry and Baker (1968) and Harvey (1969).

A critical stage in the sampling process is to choose the method by which the sample units are selected. A range of sampling methods exist but an essential distinction can be made between purposive or non-probability sampling, and probability sampling.

In the case of non-probability sampling, sample units are selected for economy or convenience, while at the same time hopefully representing the characteristics of the population from which they have been drawn. The reliability of these samples depends to a large degree on the skill and knowledge of the researcher. As Table 3.10 shows, there are three main types of non-probability samples, ranging from the simple convenience sample, where the main criterion for selection is ease of collection, through quota samples, where individuals are selected to represent collectively a replica of the population. For example, in a study of consumer behaviour we may select quota samples based on such controls as age, income and geographical location. These methods are used fairly often in geographical field surveys as an alternative to probability sampling schemes. They do, however, suffer from two problems: first they may fail to secure a representative sample; and, second, the method may make fieldwork more difficult. For example, if the selection of the sample was controlled by six age brackets, four income ranges and four geographical areas, there would be 6×4×4 or 96 cells in the sample each of which would require information (Kish, 1965). A common type of purposive sample is the 'case study'; the typical farm, city or river for example. The problem here is its lack of generality, in addition its 'typicality' cannot be guaranteed.

Table 3.10 Types of non-probability sampling schemes

Types	Characteristics
Convenience	May just select first group of units from population e.g. the first 200 consumers to interview in a street
Purposive	Sampling units are selected subjectively by research worker, on the basis of background knowledge
Quota	Selection of sample that is as close as possible to a replica of the population

Because of these various problems it is not surprising that probability sampling is often preferred. A probability sample is one in which the sampling units are selected by chance. The problem facing a geographer is which of the numerous types of sampling design to select (Tables 3.11 and 3.12).

A simple *random sample* is one of *x* units or individuals selected in such a way that each of them had an equal chance of being chosen. Such

Table 3.11 Types of probability sampling designs

Design types	Characteristics
Simple random	Assign to each population element a unique number; select sample units by use of random number tables
Systematic	Determine the systematic interval, e.g. every fifth individual; select the first sample unit/individual randomly, and select remaining units according to the interval
Stratified	Determine strata; select from each stratum a random sample of the size dictated by analytical considerations
Cluster	Determine the number of levels of cluster; from each level of cluster select randomly or stratify sample

Table 3.12 Example of how to select a stratified random sample

The problem is to select a 10 per cent stratified sample of firms from a survey of 100 businesses of different sizes, namely:

Size of firm (number of employees)	Number in stratum	Sample size
Less than 10	40	4
10 – 50	30	3
51 – 100	20	2
101 and over	10	1

In this example, we would than proceed to sample randomly from each of the four strata

a sample could be taken by assigning to each unit in the population a number, use then being made of random number tables (Appendix XI) to select the sample, as is demonstrated by the following example. A simple random sample of 5 students needs to be taken from a group of 50. We must first of all list the students, label the first of them 00, the second 01 and so on. We can then enter the random number tables at any point and read the list, either horizontally or vertically, of successive two-digit numbers. We would have to ignore all out-of-bounds numbers, in this example 51 up to 99, but the first five within-bounds numbers on the list will indicate the five population members to be chosen. It should be noted that random number tables are constructed in such a way that the same procedure can be used to read off three-, four- or even more digit numbers.

The *systematic sample* is, when properly applied, a form of random sampling. Instead of predetermining by random numbers which individual to include in the sample the researcher might decide to include individuals at regular intervals through the population. For example, in a traffic census the researcher might stop every tenth vehicle. Or in a shopping survey they might interview every 20th person to pass along the street. If the population is distributed through space, rather than time, we might overlay a regular grid on the geographic area concerned and then include in the sample all points or individuals at each grid line intersection. This might, for example, be done for a soil survey of a predetermined area, a sample being taken for analysis at each grid intersection. Such methods are efficient and simple to manage. In most instances they will also importantly produce a random sample provided that the regularity of the sample frame does not coincide with some regularity in the population.

In a *stratified sample*, units or individuals do not have equal chances of being selected, since some strata or subgroups of the population may be deliberately over-represented in the sample. This method is frequently used by geographers, some of the early applications having been associated with the study of land use (Wood, 1955). It is particularly useful when the environment under study is of an extremely variable nature, as in surveys of coastal sedimentary environments with varying proportions of dunes and intertidal beaches (Cole and King, 1968) or in studies of residential mobility in cities with a diverse range of housing tenure systems. The method of obtaining a stratified sample is outlined in Table 3.12. Such stratified samples have two advantages: first, they make it possible to sample in proportion to the characteristics of the population; and, second, by doing so they render the sample a more faithful representation of the population.

A final major type of sampling scheme is *cluster* or *area sampling*, in which the sample is selected in stages or groups. For example, if we were undertaking a survey of national trends in consumer behaviour on 5000 households we would in the first stage draw a random sample of five districts. We would then, should we want a sample of 500, draw a further 100 households at random within each of the five districts. Such cluster sampling might also be combined with a stratified scheme. Clustered sampling may be very efficient in terms of time. For example, in a survey of agricultural production we might, at stage one, select a sample of villages; at stage two, select a sample of land owners within each village and finally, stage three, take a sample of plots from each sampled owner to determine agricultural production. In this way considerable time is saved by not having to prepare a sampling frame for a possibly very large number of individual plots worked by all owners in all the villages.

Having decided on the sampling frame, the next step is to determine the size of the sample. This will depend on the degree of certainty required compared with the resources that are available. Haggett (1963) has demonstrated how the accuracy of a sample increases with sample size, but also that such a relationship is neither simple nor linear. Indeed, the question of how large a sample to take is related to the concepts of a sampling distribution and the notions of probability discussed in

Chapter 6. Here it is sufficient to note a few central points related to the practical aspects of sampling. First, it should be stressed that the form of the relationship between sample size and accuracy has been calculated for most probability sampling schemes. Thus, for a simple random sample the relationship is that the sampling error is proportional to the square root of the number of observations (Berry, 1962). Second, because such relationships are known, it is possible to calculate the required sample size by specifying the allowable error, the confidence level and the coefficient of variation (these issues are discussed in Chapters 4, 5 and 6). Tull and Hawkins (1980) illustrate how to determine sample sizes, while Som (1973) discusses the same problem for cluster samples. Little has been written about the size of non-probability samples which are generally determined on the purely practical grounds of the cost and effort involved in collecting the information.

References

Berdie, D.R. and Anderson, J.F. (1974) *Questionnaires: Design and Use*. Scarecrow Publ., New York.

Berry, B.J.L. (1962) *Sampling, Coding and Storing Flood Plain Data*. US Dept. of Agr., Washington DC.

Berry, B.J.L. and Baker, A. (1968) 'Geographic sampling' in B.J.L. Berry and D.F. Marble (eds) *Spatial Analysis: a reader in statistical geography*. Prentice Hall, Englewood Cliffs.

Board, C. (1967) 'Maps as models' in R.J. Chorley and P. Haggett (eds) *Models in Geography*, Methuen, London.

Bull, C.J. and Bull, P.J. (1978) 'Regional directories as a potential source for the study of intra-urban manufacturing industry'. *Geography*, 63, 198–204.

Central Statistical Office (1980) *Guide to Official Statistics*. HMSO, London.

Central Statistical Office (1971–) *Regional Trends*, HMSO, London.

Chorley, R.J. (1958) 'Aspects of morphometry of a polycyclic drainage basin'. *Geogrl. J.*, 124, 370–380.

Clark, J.I. (1966) 'Morphometry from maps' in G.H. Dury (ed.) *Essays in Geomorphology*, Heinemann, London.

Cole, J.P. and King, C.A.M. (1968) *Quantitative Geography*. Wiley, Chichester.

Curran, P. (1985) *Principles of Remote Sensing*. Longman, London.

Denham, C. (1980) 'The geography of the census: 1971 and 1981', *Population Trends*, 19, 6–12.

Dickinson, G.C. (1979) *Maps and Air Photographs*, 2nd edn. Arnold, London.

Dohrenwend, B.S. (1965) 'Some effects of open and closed questions', *Human Organisation*, 24, 175–184.

Edwards, B. (1972) *Sources in Economic and Business Statistics*, Heinemann, London.

Fowler, F.J. (1993) *Survey Research Methods*, Sage, London.

Gannett, H. (1901) 'Profiles of rivers of the United States', US Geol. Surv. Water Supp and Irr. Paper 44, Washington DC.

Haggett, P. (1963) 'Regional and local components in land-use sampling: a case study from the Brazilian Triangulo', *Erdkunde*, 17, 108–114.

Haggett, P. (1965) *Locational Analysis in Human Geography*, Arnold, London.

Harley, J.B. (1980) *The OS and Land-use Mapping*, Hist. Geog. Research Series, 2, Geobooks, Norwich.

Harvey, D. (1969) *Explanation in Geography,* Arnold, London.
Kish, L. (1967) *Survey Sampling*, Wiley, New York.
Laurent, A. (1972) 'Effects of question length on reporting behaviour in the survey interview', *J.Am. Stats. Assoc.*, 67, 298–305.
Lewis, P. (1971) *Maps and Statistics*, Methuen, London.
Lo, C.P. (1986) *Applied Remote Sensing*, Longman, Harlow.
Maguire, D.J. (1989) *Computers in Geography*, Longman, Harlow.
Millington, A.C. and Townshend, J.R.G. (1987) 'The potential of satellite remote sensing for geomorphological investigations – an overview' in V. Gardiner (ed.) *International Geomorphology 1986*, Wiley, Chichester.
Monkhouse, F.J. and Wilkinson, H.R. (1971) *Maps and Diagrams*, 3rd edn, Methuen, London.
Noelle-Newman, E. (1970) 'Wanted: rules for wording structured questionnaires', *Public Opinions Quart.*, 34, 200–210.
Oppenheim, A.N. (1966) *Questionnaire Design and Attiude Assessment*, Heinemann, London.
Robinson, A.K. (1952) *Elements of Cartography*, Wiley, New York.
Rowley,G.and Shepherd, P. (1976) 'A source of elementary spatial data for town centre research', *Area*, 8 201–208.
Sabbins, F.F. (1987) *Remote Sensing: principles and interpretation*, 2nd edn, Freeman, New York.
Shaw, G. and Tipper, A.(1989) *British Directories: a bibliography and guide*, Leicester University Press, London.
Smith, D.M. (1975) *Patterns in Human Geography*, Penguin, Harmondsworth.
Som, R.K. (1973) *A Manual of Sampling Techniques*, Heinemann, London.
Tannenbaum, E. (1980) 'Secondary analysis, data banks and geography', *Area*, 12, 33–35.
Townsend, A., Blakemore, M. and Nelson, R. (1987) 'The NOMIS data base: availability and uses for geographers', *Area*, 19, 43–50.
Tull, D.S. and Hawkins, D.I. (1980) *Market Research*, Macmillan, New York.
Wheeler, D.A. (1979) 'Studies of river long profiles from contour maps', in A. Pitty (ed.) *Geographical Approaches to Fluvial Processes*, Geobooks, Norwich.
Wood, F. (1955) 'Use of stratified random samples in land use study', *Anns. Assoc. Am. Geogrs.*, 48, 350–367.
Wooldridge, S.W. (1928) 'The 200-foot platform in the London basin', *Proc. Geol. Assoc.*, 39, 1–26.

Chapter 4
Measurement and Descriptive Statistics

4.1 Data characteristics and scales of measurement

Measurement is concerned with the assignment of values or quantities to particular objects or events. However, it must be recognised at the outset that measurement exists in a variety of forms owing to the diversity of data used by geographers and the range of phenomena that they study. With this fact in mind geographers, and other scientists, have found it extremely useful to identify different scales along which data can be measured. In his original work on this topic, Stevens (1946) classified four common scales of measurement, each of which identifies and uses different properties.

These four scales are summarised in Table 4.1 and are represented in ascending order of the information that they offer, with the lowest being that of the nominal scale. As one progresses through these scales the data must satisfy more rigorous requirements. For example, the *nominal* scale involves only the classification or naming of observations; numbers, if used at all, are arbitrarily assigned. For example, in a questionnaire survey there may be two groups of people to be studied and they might be described simply as groups 1 and 2. Hence a member of group 1 may have the measurement attribute '1' in order to identify the group of which it is a member, but the number has no arithmetic significance. The simple classification of consumers by sex or occupation provides a more commonplace example of nominal scale data where 'male' or 'female', 'white-collar worker' or 'unemployed' are the attributes. In contrast, the *ordinal* or *rank* scale of measurement involves the ranking of one observation against another. Thus, we may say that one city is larger than another and proceed to place a number of cities in a rank ordered list, even though we may not have any precise information on their populations. The availability of this type of measurement is important for many of the studies concerning perception and the cognitive components of behaviour, when individuals may be asked to rank particular features in terms, for example, of their perceived attractiveness (Downs, 1970).

Table 4.1 Scales of measurement

Scale	Characteristics	Measurement
Nominal	Determination of equality; data can be placed in classes	Discrete
Ordinal	Determination of greater or lesser; data can be ranked	Discrete
Interval	Determination of equality and of intervals or difference	Continuous
Ratio	Determination of equality and of ratios; measurements have a true zero	Continuous

The *interval* scale permits us not only to sort and rank observations but also to establish the magnitude of the differences separating each observation. Temperatures are a good example. Not only can we say that one area has a higher mean annual temperature than another, we can also estimate the degree to which the two differ. For example, mean annual temperatures for two regions will tell us not only that one region is warmer than the other, but that it is warmer by a specified amount. Importantly, such data do not possess an absolute zero. Temperatures on the Celsius scale comprise interval data, yet 0°C does not represent the total absence of heat. In such systems the number zero has been arbitrarily attributed a particular point – in this case the freezing temperature of pure water. On the Fahrenheit scale that same point is denoted by 32°F. In both cases negative quantities (sub-zero temperatures) are possible. Notice also that an object with a temperature of 24°C is not one with twice the heat content of an object at 12°C. In this sense the temperature scales, though measuring precisely the attributes of a region or place, do so using an arbitrary scale.

Finally we can identify the *ratio* scale, which is the highest level of measurement. It shares many features with the *interval* scale but differs in that it has a known and absolute zero. If we return to our earlier example of city size we could, under the ratio scale, and assuming that we had sufficiently accurate and reliable data, say not only by how much one city was larger than another but also the ratio of the two figures. Thus, if one city has a population of 1,000,000 and another has only 250,000 we can state that not only is the former 750,000 greater than the latter but also that it has four times its population. Rainfall or income are other typical ratio scale variables where zero is not arbitrarily defined, but has a real physical basis; there is nothing arbitrary about zero rainfall or zero income for both of which negative quantities are impossible. Most geographers frequently use data measured on the interval or ratio scales, and for most analytical purposes they can be treated in the same way. The two are sometimes collectively known as *parametric* data. Information at the nominal or ordinal scale might be loosely described as *non-parametric*. The distinction between these two forms of data, and their implications for analytical methods, are discussed further in Siegel (1956).

A further distinction can be made between the nominal/ordinal and the interval/ratio scales. In the former the measurements can be termed discrete as observations can be placed only within certain mutually-exclusive classes. This is not the case with interval/ratio scale measurements, where observations can occupy any position along the measurement continuum and hence are assessed on what is termed a continuous scale. Consider, for example, the difference between measuring the nationality and the income of an individual. In the former case the attribute places the individual into a clearly-defined class. In the latter the attribute is a number located at some point along a measurement continuum. This contrast is often helpful in deciding on the methods of analysis to which the data may subsequently be subjected.

In some areas of geographical work it is difficult to assign precise numerical values to particular features, and in such cases the data severely limit the types of statistical analysis that can be undertaken. Under some circumstances this difficulty may reflect the problem of defining and actually measuring the geographical individual. What, for example, do we mean by the terms 'hillside' or 'region'? How can we determine their limits and quantify their numerous attributes? There are also other problems. In countries such as the USA and the UK, we can obtain an accurate measure of population for each urban area from census publications, and we therefore have data that can be assigned to the ratio scale. However, in some of the less developed countries such accurate census data may not exist, and settlements may have to be classified in a more general way using broad groups and thus restricting measurements to the ordinal scale. Such seemingly abstruse problems are all too commonplace when we attempt to gather data and information and they often impose severe restrictions on the forms of analysis that geographers can use.

4.2 Descriptive and inferential statistics

A division needs to be recognised between descriptive and inferential statistics. This allows us to distinguish between those approaches that merely describe, in numerical terms, an event or data set, and those which enable us to infer relationships between variables and to test hypotheses about the nature of phenomena that are under study.

Descriptive statistics, which are introduced later in this chapter, are the simplest way of summarising and presenting what might constitute large volumes of data. They may take a variety of forms, ranging from the use of basic graphs and frequency tables through to a whole range of spatial statistics such as those presented in Chapter 13. In making such descriptive summaries geographers may also be able to speculate more confidently on the character and behaviour of the phenomena being studied. This step takes us to inferential statistics which, though not discussed fully until later chapters, can be introduced at this juncture. Inferential statistics are concerned with mathematical probabilities and are characteristic of scientific investigation, which involves a search for principles that have a degree of generality and allow us to explain why things happen. In this type of study the findings are often applied to a wider setting than the cases or samples that were represented in the data to hand. The making of such generalizations from a sample and extending them to a population is termed statistical inference.

Inferential statistics allow us to make probabilistic statements about the following:

1. Hypothesis testing – whether a particular supposition is true or false;
2. The relationships between two or more variables;
3. The characteristics of the population from which a sample is drawn.

Despite the importance of inferential statistics problems still exist in their application. Specific difficulties have been outlined by Gudgin and Thornes (1974), particularly in relation to geographers' frequent neglect of the data requirements and assumptions in inferential methods. These topics will be returned to in later chapters but, for the moment, attention will be confined to descriptive statistics.

4.3 Data reduction, tables and graphs

In geography we are often faced with the task of collecting and describing large amounts of data before we can reach any firm conclusions about particular issues. To illustrate the difficulties let us take a specific case, and consider the information that may be collected from the UK decadal census. Even if we concentrate on the 10 per cent sample publications, the total amount of data to be handled is vast. If we turn to the complete census reports it becomes proportionally greater. To yield useful results the data must first be summarised and reduced to manageable proportions.

One of the most useful initial tasks is to describe a data set having counted how often each value, or range of values, occurs. This information may then be used to construct either a frequency table or a graph. If we consider the journey-to-work habits of the British population we can illustrate the simple but effective nature of such techniques. Table 4.2 shows the breakdown of methods of travel (a discrete variable at the nominal scale) for the two million or more employed people that represent the 10 per cent sample from the 1981 UK census. Notice that the groups are mutually exclusive and could have been placed in any order without prejudice to the sense of the information.

Table 4.3 illustrates the case using a continuous variable on the ratio

Table 4.2 Journey to work survey

Means of travel	Number of people (in 000s)
Train	80.6
Underground	42.2
Bus	314.2
Motor cycle	61.0
Car – pool driving	82.4
Car – driver	805.5
Car – passenger	165.5
Pedal cycle	79.8
Pedestrian	309.0
Works at home	70.8

Source: Census 1981 (10% sample) – Workplace and Transport to Work, England and Wales, HMSO, London (1984)

scale, in this instance perinatal mortality per thousand births. Although the measurement scale is a continuous one, it is often helpful to aggregate the data into classes, as has been done here. The perinatal mortality within every county in England and Wales can be placed into one, and only one, of the classes defined in Table 4.3. Such summaries are known as frequency tables and they are equally useful for data measured on any of the scales already described. A problem may be posed in deciding on the number of classes to use in the case of a variable measured on a continuous scale. While writers such as Huntsberger (1961) make very specific suggestions, Croxden and Cowden (1968) offer the more sound general advice that frequency tables should ideally have no less than six but no more than 16 classes. Evans (1977) has also written in this problem in a geographical context.

Table 4.3 Perinatal mortality (deaths/1000 births) 1987 – 89 for English and Welsh counties

Class	Frequency	Percentage	Cumulative percentage
5.25 – 5.74	1	1.8	1.8
5.75 – 6.24	0	0.0	1.8
6.25 – 6.74	2	3.7	5.5
6.75 – 7.24	4	7.4	12.9
7.25 – 7.74	3	5.6	18.5
7.75 – 8.24	13	24.1	42.6
8.25 – 8.74	10	18.5	61.1
8.75 – 9.24	9	16.7	77.8
9.25 – 9.74	6	11.1	88.9
9.75 – 10.24	3	5.6	94.5
10.25 – 10.74	2	3.7	98.2
10.75 – 11.24	1	1.8	100.0

Source: Regional Trends 1991, HMSO, London (1991)

In some cases, such as the 10 per cent sample census reports of the UK, the data have already been processed before publication but geographers and students will often want to obtain their own data sets which, though not as large as that used for the UK census, may nevertheless provide substantial volumes of data. For example, a recent survey of consumer shopping habits in central Exeter questioned 2000 people each of whom responded to 18 questions. This alone gave 36,000 items of data. When large numbers of observations are involved then the production by hand of frequency tables is both laborious and time-consuming. It is for such reasons that geographers often rely on computer programs to carry out these tasks.

In many situations geographers prefer to present information in the form of graphs; especially important are frequency graphs, either as bar charts (used for discrete variables), histograms or ogives (the latter being appropriate for the display of continuous variables). Histograms and bar charts, in which the heights of the columns are proportional to the frequency of observations in their respective classes, are especially useful. Figure 4.1 is a bar chart representation of the data in Table 4.2. Where discrete nominal classes are portrayed the bars are separated in the graph. They should be plotted as 'touching' bars only when they depict continuous data that have been arranged into classes.

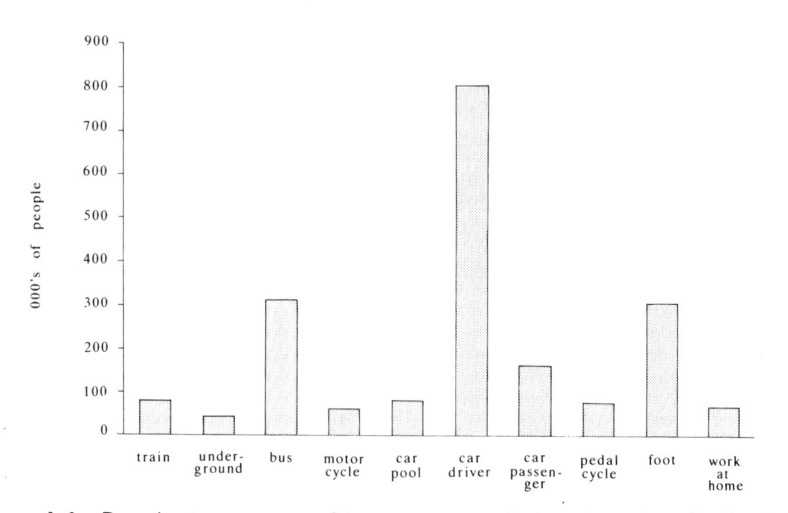

Figure 4.1 Bar chart summary of journey to work data based on Table 4.2.

Such presentations of nominal data can have their columns rearranged without prejudice to the data set. Clearly such reordering cannot be done for data measured on a continuous scale. While continuous variables can be plotted as histograms, more information can sometimes be revealed by plotting a cumulative frequency curve (or ogive) where the slope of the curve is determined by the number of observations in each class. Table

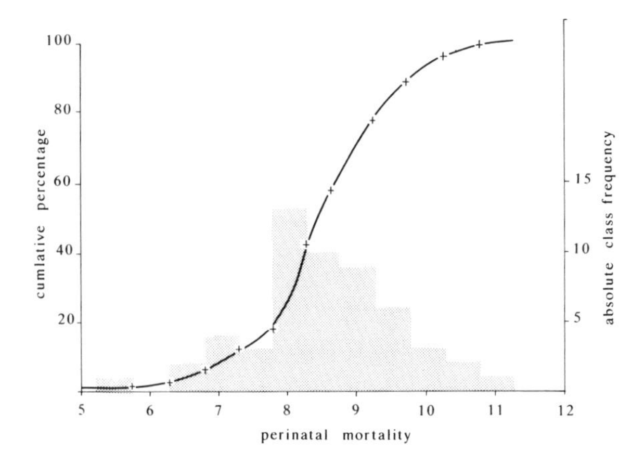

Figure 4.2 Histogram (shaded) and ogive based on Table 4.3.

4.3 also contains the absolute and cumulative percentages of the raw data from which a cumulative frequency curve can be plotted (Figure 4.2). It is common to use cumulative percentages rather than absolute frequencies in such graphs.

Ogives need to be plotted with care. In the case of Figure 4.2 the data are plotted as cumulative percentages 'less than' the upper bound of each class. As a result the cumulative percentage points must be plotted at the upper bound of their respective classes, these points are then connected by a smooth curve which represents the ogive. When passing through classes with greater frequencies of observations the ogive rises more steeply, when passing through classes with fewer observations it is less steep. In general the accuracy of all such sketched curves increases with the number of classes that we can employ in their definition.

One important point to consider in the construction of frequency tables is the selection of class limits and class intervals when dealing with interval/ratio-scale data. Class limits should be assigned to avoid any ambiguity and to ensure that there is only one possible place for each item. Thus, if a continuous variable is being measured then the notation used in Table 4.3 is appropriate since it leaves no gaps and each individual belongs uniquely to a particular class. Class intervals, the statistical distance between the upper and lower limits of a class may, in exceptional cases, vary in size throughout the table in order to preserve or emphasize some aspect of the data. In a study of the age structure of a human population we might group all over 65s together as those of pensionable age, leaving the younger groups in, say, ten-year classes. In terms of purely descriptive statistics such variations may not be a problem but, as will be seen in later chapters, when dealing with inferential statistics class intervals need, whenever possible, to be identical. It is not always easy to determine the number of classes to be used when categorising interval/ratio scale data. Only careful inspection of the data will help and, in general, fewer classes can be drawn up from small data sets. The choice of classes is greater when more observations are available.

4.4 Using MINITAB and SPSS to summarise data

Thus far we have concentrated on the ways in which data can be processed 'by hand'. It has, however, been emphasised that geographers are often confronted with volumes of data so great as to hinder such an approach. The SPSS and MINITAB packages are good examples of the way in which computers can come to our aid. Both systems will execute the tasks described in the preceding sections though inevitably they do so in rather different ways.

Let us return to the county-based data published in *Regional Trends 1991* used to construct Figure 4.2. From that same source we can abstract a range of variables covering the 54 English and Welsh counties. The raw data for birth rates and death rates per thousand population were stored in data files, one for use in MINITAB and one for SPSS and named **COUNTY.MTW** and **COUNTY.DAT** respectively, were prepared using the methods outlined in section 2.8 and 2.9. Instead of tabulating by hand the frequencies within each class the task can now be carried out by using simple computer instructions.

Using MINITAB we can specify the option **HISTOGRAM** (Figure 4.3)

```
MTB  >  RETRIEVE 'COUNTY'
  WORKSHEET SAVED 8/23/1992

Worksheet retrieved from file: COUNTY.MTW
MTB  >HISTOGRAM 'B/1000';
SUBC >  INCREMENT 0.6.

Histogram of B/1000 N = 54

Midpoint       Count
  10.200           1   *
  10.800           0
  11.400           4   ****
  12.000           8   ********
  12.600           8   ********
  13.200          12   ************
  13.800          10   **********
  14.400           8   ********
  15.000           2   **
  15.600           1   *
```

Figure 4.3 MINITAB screen display of **HISTOGRAM** option. In this and all subsequent MINITAB figures, those lines beginning after the prompts **MTB>** or **SUBC>** are typed by the user. All other lines are system responses to instructions.

which is an instruction in much the same way as **SAVE**, **PRINT** and **RETRIEVE** are instructions (section 2.8). The major difference is that it will perform a statistical analysis for us and produces histograms and frequency counts for specified variables. It will also automatically select equal-range classes on which to base the analysis. This latter *default* can be overridden by the use of a subcommand **INCREMENT** specifying user-defined class ranges if desired, though only uniform class intervals can be specified. If such additional commands are used the initial **HISTOGRAM** line must be concluded with a semi-colon in order that the subcommand prompt **SUBC>** can be instigated. The final instruction in the sequence must then be concluded with a stop. This important use of semi-colons to indicate that secondary information is to follow was introduced in section 2.8. We see it here applied for the first time and readers are advised to make note of its use. By then striking **RETURN** at the conclusion of the final line the whole series of instructions will be executed, in this example producing a simple histogram and frequency count. If the line beginning **HISTOGRAM** had no terminator the default settings would have been used and the results produced immediately **RETURN** had been pressed.

From Figure 4.3 readers should also note how the data file was identified for retrieval and the specific variable (column) then selected from the data matrix. This example selects the column of the data file containing birth rates and denoted **B/1000**. Because the variable had already been named its specification in MINITAB needs to be within single quotes. The subcommand specification was made for class intervals of 0.6.

The SPSS subprogram **FREQUENCIES** performs much the same task but contains a number of options of which **HISTOGRAM** is only one, it needs therefore to be specified, together with the variable or variables in the data file which are to studied (Figure 4.4). The **DATA LIST** instruction is used to recall the data file **COUNTY.DAT** which is written in freefield format and input at an earlier stage. Upon its retrieval the same two variables are now denoted as BIRTH and DEATH though SPSS does not require them to be written within single qoutes. Section 2.9 describes the nature of freefield data and demonstrates how such a data file is set up and saved for future use. Options such as **FREQUENCIES** and **HISTOGRAM** can only be executed on an appropriate *active file* from which the columns (variables) to be analysed can be chosen. The key word **FREQUENCIES** must be followed by **/VARIABLES** after which the choice of columns can be entered. In this case we will analyse only the column containing death rate which is denoted by DEATH. In order to indicate that further instructions are to follow this line must have no terminating stop. This following line specifies **HISTOGRAM** and must be introduced by the conventional slash. Had we then terminated the line with a stop mark a histogram using the default settings would have been produced. Instead we

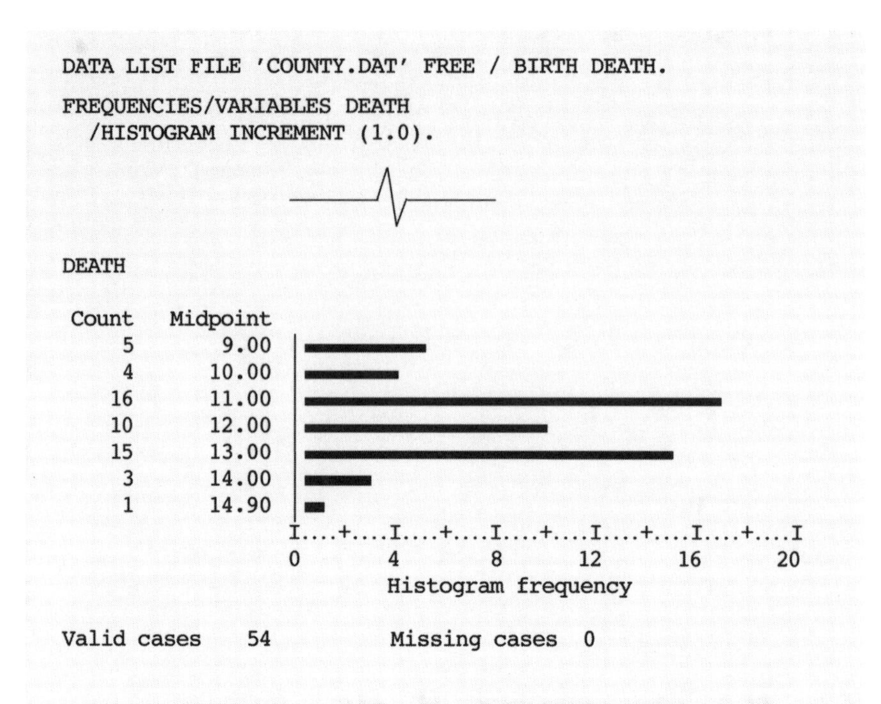

Figure 4.4 SPSS screen display and results of **FREQUENCIES/HISTOGRAM** options.

preferred to override this by invoking the **INCREMENT** command followed, in parenthesis, by the new class interval of 1.0. Only now do we indicate to the computer that all options are concluded by typing a stop. To then execute those three lines of instructions we would, as described in section 2.9, move the screen cursor to the first of them, and press **F10/RETURN**.

If nominal data had to be plotted in which an individual is attached to a class rather than having an attribute measured on the interval or ratio scale, the option **BARCHART** would replace that of **HISTOGRAM**. In such cases, and as indicated in section 2.9 the data can be entered as either alphabetic or as numerical information. This application counts the frequency of observations in each class and plots then in bar chart form. It is otherwise introduced in the same way but, clearly, can have no **INCREMENT** option as the classes are already determined by the nature of nominal data. The **HISTOGRAM** option is applicable only to parametric data where classes need to be set up by either the computer (by default) or the user (with the **INCREMENT** option).

4.5 Measures of central tendency

It is often useful to be able to summarize data for a single variable in one figure. A measure of central tendency provides such a summary and there are a number of such statistics, the choice of which is determined by the measurement scale of the raw data. Attention is here confined to the three most important of these measures; the mode, the median and the mean. The central tendency of nominal scale data can only be measured using the mode. Ordinal scale data can be described by the mode or the median. Interval/ratio scale data are better described by using either the median or the mean.

The mode

This is, strictly speaking, the class in a frequency plot for any one variable that contains the greatest number of individuals or observations. Thus in Table 4.2 and Figure 4.1 the mode, or modal class, is 'car driver', i.e. the most frequently used method of journey-to-work is the car. In Table 4.3 the mode can clearly be seen to be the mortality group in the range 7.75 to 8.24 deaths/1000 births. Notice that in nominal data the redefinition of arbitrary classes can change the frequency of observations that they contain. For example, Table 4.2 could be redefined using only two classes, those for public and private transport giving thereby a different modal class entirely. Equally the class boundaries in Table 4.3 could easily be redrawn. On the other hand attributes such as race or sex are absolute and not subject to arbitrary definition. Especially where classes can be redefined, the mode should be used with caution.

Instability in the use of nominal data can be reduced by using equation 4.1 to determine the mode:

$$\text{Mode} = L + \frac{D_l}{D_1 + D_2} \times i \tag{4.1}$$

Equation 4.1

D_1 = difference between modal frequency and frequency of next lower class
D_2 = difference between modal frequency of next higher class interval
L = lower limit of modal class
i = the class interval

There may also be situations in which two classes are equally dominant. Such a pattern is said to be bimodal. It is also possible to find more than two classes to be dominant. The pattern is then multimodal and little can be said of the central tendency of the data set. Such problems are more commonly associated with small samples and the best solution is to increase the sample size if possible.

The mode is, arguably, the least reliable of the measures of central tendency. Yet it remains the only such measure available to geographers when using nominal scale data. Fortunately, more consistent forms are available for the summary of interval or ratio scale data.

The median

This is the mid-point of a set of ordinal or interval/ratio scale data, above which lie half the data points and below which is found the remaining half. To locate the median the data need first to be ranked into ascending order. The data set will consist of n observations in which the median is located in the middle. For example, if there is an odd number of observations in the data set, say 39, then the median is the 20th in the series as there will be 19 observations above and below this. If there is an even number of observations in the series, say 40, then the median is the average of the 20th and the 21st observations in the series. A simple example will illustrate these points. Table 4.4 lists, in ascending order, the unemployment percentages in 1989 for the 54 counties of England and Wales and the ten regions of Scotland. The table gives the ranked listing of the data in which the median is the average of the observations ranked at 32 and 33.

In this example the 32nd ranked observation is 6.0 per cent unemployment and the 33rd ranked observation has a value of 6.2 per cent. Hence the median unemployment rate in England, Scotland and Wales is 6.1 per cent.

Table 4.4 Unemployment (as % of working population) for the 54 counties in England and Wales and the ten Scottish regions

%	Rank	%	Rank	%	Rank	%	Rank	%	Rank
3.1	1	4.6	14	5.9	27	7.1	40	8.9	53
3.2	2	4.6	15	5.9	28	7.2	41	9.0	54
3.4	3	4.8	16	5.9	29	7.3	42	9.2	55
3.4	4	4.9	17	5.9	30	7.3	43	9.3	56
3.6	5	5.0	18	6.0	31	7.4	44	9.3	57
3.7	6	5.0	19	6.0	32	7.8	45	9.8	58
3.7	7	5.1	20	6.2	33	7.8	46	10.0	59
4.4	8	5.4	21	6.4	34	8.0	47	10.3	60
4.4	9	5.4	22	6.5	35	8.2	48	10.6	61
4.5	10	5.6	23	6.6	36	8.2	49	10.8	62
4.6	11	5.8	24	6.8	37	8.4	50	12.4	63
4.6	12	5.8	25	6.8	38	8.5	51	12.9	64
4.6	13	5.9	26	7.1	39	8.9	52		

Source: Regional Trends 1991, HMSO, London (1991)

Use of the median allows us to introduce a concept that can be employed later in the chapter; that of percentile values. In fact we have already touched on this issue when using ogive or cumulative frequency curves (section 4.3). The top end of such curves include 100 per cent of all the observations, the lower end represents the point below which are 0 per cent of the observations. As we move along the data series from the smallest to the highest value an ever-greater percentage of the observations are included until we get to the highest ranked point, by which stage 100 per cent of the data points have been included. On this basis the median can also be termed to 50th percentile. The data in Table 4.4 could, for example, have been grouped into classes and their cumulative frequency plotted. From the resulting ogive the median could have been graphically interpreted by reading off the point on the horizontal axis (see Figure 4.2) that corresponded to 50th percentile, i.e. the 50 per cent cumulative mark on the graph.

The arithmetic mean

This is derived by adding all the observed values of a data set and dividing this total by the number of such observations. Using the algebraic conventions outlined in Chapter 2, this procedure is summarised in equation 4.2:

$$\bar{X} = \frac{\Sigma X}{N} \qquad (4.2)$$

Equation 4.2
$\bar{X}$ = mean
ΣX = sum of all values
N = number of observations

Each member of the data set contributes towards the value of the mean and any data set can have only one mean. There is no need here to rank order the data. The mean has two further distinctive properties, the real significance of which will be apparent in later chapters. First, the sum of the deviations of each individual data point from its mean is always zero, in other words the total amount by which all observations greater than the mean exceed that mean is the same as the sum of deviations of observations less than the mean. Second, the sum of the squares of all such deviations, above and below the mean, is at a minimum. If the mean is miscalculated in any way then the consequent sum of squared deviations will be greater than that obtained by reference to the correct mean.

For our example we can again consider the unemployment data in Table 4.4. If we follow the arithmetic instructions in equation 4.2 we will add all 64 observations together (to give in this case 425.7) and divide by 64, to give a mean of 6.65 per cent unemployment. It will be seen that this figure is not identical to the median, though they differ by only a little. Such differences are not uncommon and both the mean and the median have given us a general picture of the tendency of this one variable.

The mean is also, in general terms, known as the average. It is the most important and informative of the measures of central tendency but, as we will see, there are situations when the nature of the data sets may preclude its use.

4.6 Criteria for selection of mean, median or mode

From our previous discussion it should be clear that the different measures of central tendency are not directly comparable and that each has very different characteristics. Thus we could not compare the mean of one data set with the median or mode of another because each measures a different aspect of central location. For example, the mode measures the highest frequency, the mean the centrality of the values and the median the middle position of the ranked data. If the distinction between the latter two is not clear it should be remembered that the precise numerical value of all observations contributes towards the mean. In the case of the median it is only the order of the observations, and not their values, that determines its location. The choice of the measure of central tendency will therefore depend on a number of criteria, in particular:

1. The individual properties of the measures of central tendency;
2. The type of question we are asking and the points to be illustrated;
3. The characteristics of the data and its pattern of *distribution*.

The distribution of the data is an aspect that we have already touched upon. The mean reflects the summated contribution of each individual data point and, unlike the median, much depends upon the way in which those points are distributed along the magnitude scale. The median, as we know, considers only the order and not the distribution of the data points. Figure 4.3 shows the distribution of a data set which though here presented as a series of classes could easily be generalised as a continuous curve. In this figure we have a symmetrical distribution and under such circumstances the mean, the median and the mode will tend to give the same numerical value. If however the data's distribution is skewed (lop-sided), as in the hypothetical examples in Figure 4.5, then the equality between the measures disappears. The greater the degree of skew the greater are the differences between mean, median and mode. In right-

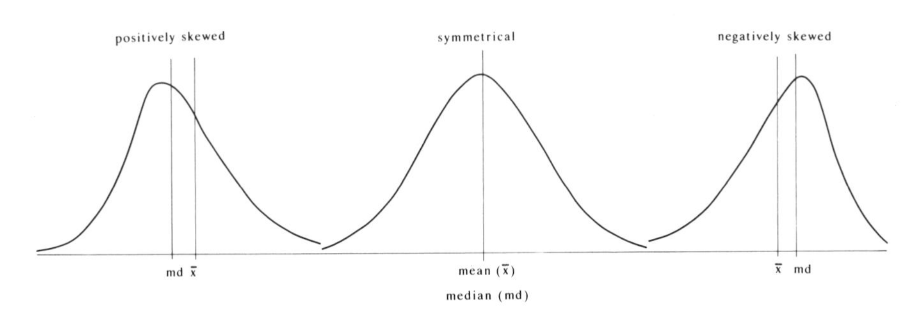

Figure 4.5 Idealised distributions showing positive-, zero- and negative-skewness. The locations of the respective means ($\bar{x}$) and medians (md) are also indicated.

skewed data there is a long 'tail' towards the upper end of the distribution which indicates the presence of a small number of very high values. This is described as a positively-skewed distribution. It is a common feature of geographical data sets. It has the effect of 'pulling' the mean towards that end of the distribution as those high numbers contribute disproportionately to the mean. The median is not influenced by such extreme behaviour because it is the rank and not the value of those observations that are considered. Whether the first-ranked observation is 1000 or only 1 ahead of the second-ranked point is of no importance to the estimation of the median, but it is to the mean. In the case of left- or negatively-skewed data the mean is pulled to the left-side of the median being unduly influenced now by a small number of very low values. In essence the mean is not a good indicator of central tendency where data have a skewed distribution as it no longer represents the most frequent or probable range of values.

Useful though measures of central tendency might be they give us only a partial view of the data. They tell us nothing about the variability of the sample; a feature better described as its dispersion.

4.7 Measures of dispersion

It may be important to describe the variation of data about their average values. We may also want to know how well the average represents the data; if the data vary greatly from one another the average may not be very informative, or at least would have to be used with care. Once again there are a number of measures that can be used. The utility of each depends, as the measures of central tendency did, upon the character of the data and their distribution. As the methods about to be described will make clear, measures of dispersion are generally applicable only to data at the interval and ratio scale.

The range

This is simply the difference between the largest and the smallest observations in the sample. It measures the dispersion of the data, but only in the simplest fashion. It is highly sensitive to individual extremes and completely overlooks the variability of the remainder of the sample.

The inter-quartile range

This overcomes to some extent the disadvantages of the range by excluding the uppermost and lowermost quarters of the distribution and by looking at the overall variability of the remaining central 50 per cent of the data set. It is a direct counterpart to the median. The latter, it will be recalled, can also be described as the 50th percentile. In order to find the inter-quartile range we must now estimate the 25th and the 75th percentiles. These are the points one-quarter and three-quarters of the way along the ranked set of data. They may also be termed the first ($Q1$) and the third quartiles ($Q3$) respectively. By convention the data are ranked from lowest to highest and the first quartile lies towards the lower end of the magnitude range and the third quartile at the upper end (Figure 4.6).

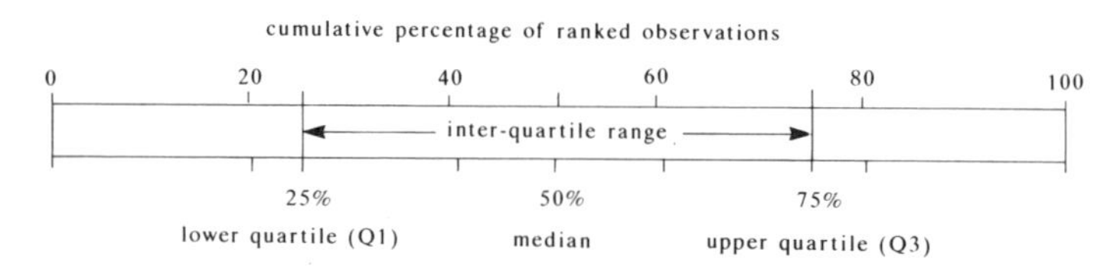

Figure 4.6 Diagram showing the position of median, first- and third-quartiles along the cumulative percentage scale of 0 to 100.

These points can be illustrated if the data in Table 4.3 are again considered. The means by which we estimate $Q1$ and $Q3$ require a little explanation. The convention adopted here is that used in the MINITAB package in which $Q1$ is at position $(N+1)/4$ and $Q3$ is at $3(N+1)/4$ in the ranked data set. The quantity N is the total number of observations that constitute the sample under study. If the answer is an integer (whole number) then the value at that specified rank is used. If the answer is non-integer then the quartile can only be determined by interpolation between the two nearest points. For example, if as in this case, N is 64, $Q1$ is the value at rank $(64+1)/4=16.25$, and $Q3$ is at $3(64+1)/4=48.75$. In the case of $Q1$ the value is one-quarter of the way between the already known 16th and 17th ranked values, whereas for $Q3$ the quantity is three-quarters of the way between the observations ranked 48 and 49. The method of their calculation can be expressed algebraically using the notation principles outlined in Chapter 2. For example, rank 16 of variable X is denoted by X_{16}. This gives the following two expressions:

$$Q1 = X_{16} + 0.25(X_{17} - X_{16})$$

and

$$Q3 = X_{48} + 0.75(X_{49} - X_{48})$$

which, by substitution, yield:

$$Q1 = 4.8 + 0.025 = 4.825\%$$

and

$$Q3 = 8.2 + 0.0 = 8.2\%$$

The inter-quartile range (IQR) is the difference between these two quantities, hence

$$IQR = 3.37\%$$

Hence, the central 50 per cent of the data set by rank cover an unemployment range of 3.37 per cent. A more widely varied data set would give a higher IQR. If less dispersed the IQR would be smaller.

In addition, there is no restriction on the construction of such ranges; for example, an interpercentile range could be used simply by taking the 10th and 90th percentiles. However, all such devices are open to the criticism that they exclude a proportion of the data from consideration. What is required is a measure which, like the mean, includes all observations in the sample.

The variance and the standard deviation

These are measures that do indeed take into account all observations. In statistical terms they are the most comprehensive descriptions of dispersion, since they are given in terms of the average deviation about the mean of all observations in the data set. This will become clear if we examine more closely the properties and uses of these two, closely related, statistics.

The variance is found by calculating the deviation from the mean of each item in a sample, squaring those deviations, adding them and then dividing by the sample size. It is most important in this and all other instances to obey the order of precedence indicated by the equation. In this case the mean is subtracted from each observation, those deviations are individually squared, those squares are then added. Only then is the sum divided by N. Failure to follow such orders of execution will give an incorrect answer. The symbol used to denote variance is s^2, and its derivation is expressed algebraically by equation 4.3.

$$s^2 = \frac{\Sigma(X_i - \bar{X})^2}{N} \tag{4.3}$$

Equation 4.3

s^2 = variance
X_i = ith value of X
$\bar{X}$ = mean
N = number of observations

Although the variance adequately measures dispersion within a sample, it does so in units of squared deviations. The quantities may, thus, be large and to obtain a measure of dispersion in terms of the original data, the square root of the variance is taken. This new quantity is known as the standard deviation. It is denoted by s and is calculated from equation 4.4.

$$s = \sqrt{\left(\frac{\Sigma(X_i - \bar{X})^2}{N}\right)} \tag{4.4}$$

Equation 4.4

s = standard deviation
X_i = ith value of X
$\bar{X}$ = mean
N = number of observations

The example of unemployment may again be used. Table 4.5 is abridged to show how the data might be set out if the calculations are not done on a computer.

Here the standard deviation is 2.25 per cent. It is a measure derived by the inclusion of all data points and by its method of derivation one would not anticipate it to be numerically equivalent to the inter-quartile range. Furthermore it gives the impression of being much smaller than the *IQR* but it must be recalled that the latter, by its very method of calculation, measures the band of dispersion either side of the median (see Figure 4.6). The standard deviation, on the other hand, indicates only the mean deviation to one side of the mean, though it is equally applicable to deviations above as well as below the mean, i.e. it is the mean deviation of values less than the mean, as well as the mean deviation of values greater than the mean. However, when standard deviations are discussed

Table 4.5 Calculation of standard deviation of percentage unemployment for counties in England and Wales and Scottish regions

% unemployed	$(X-\bar{X})$	$(X-\bar{X})^2$
3.1	−3.55	12.60
3.2	−3.45	11.90
3.4	−3.25	10.56
...	.....	.....
...	.....	.....
12.4	5.75	33.06
12.9	6.25	39.06

Mean % unemployment	=	6.65
Total sum of squared deviations from mean	=	325.58
Variance (s^2) – see equation 4.3	=	5.09
Standard deviation (s)	=	2.25

Note: this table does not list all the data, which can be found in Table 4.4. Data need not be ranked for the purposes of estimating the variance.

they are generally expressed in terms of an upper value which is one standard deviation above the mean, and a lower value which is one standard deviation less than the mean. Returning to the unemployment percentage example, the former would be 6.65 per cent + 2.25 per cent, giving an upper value of 8.90 per cent. While the lower value is found by subtracting 2.25 per cent from 6.65 per cent, giving 4.40 per cent.

The standard deviation is extremely useful in describing the general characteristics of data. It does, nevertheless, have its limitations. Obviously it cannot be used to describe nominal or ordinal data. Less obviously, we might be cautious where data have a skewed distribution. For the standard deviation to apply equally on both sides of the mean the distribution should be symmetrical and the positive deviations balanced by the negative. If this is not the case it is better to use the inter-quartile range. Despite its limitations the latter is what can be described as a 'distribution free' statistic. In this example the distribution of data are tolerably symmetrical and the standard deviation is preferred.

The coefficient of variation

The standard deviation is an absolute measure of dispersion which is often of limited use if variables measured on different units are to be compared. The same problem occurs if two samples of the same phenomenon, but with different orders of magnitude, are being studied. Under these circumstances we require a relative measure of dispersion. This can be illustrated by taking the small hypothetical example in Table 4.6 which compares rainfall in two very different areas. Despite the

Table 4.6 Application of the coefficient of variation to hypothetical rainfall data for two stations

Station 1	Station 2
6	56
8	58
10	60
12	62
16	66
18	68
$\bar{X}$ = 11.67	$\bar{X}$ = 61.67
s = 4.23	s = 4.23
CV = 36.2%	CV = 6.86%

differences in magnitude, both data sets have the same standard deviation. Yet clearly a dispersal of 4.23 about a mean of 11.67 is proportionally greater, and geographically more significant, than the same dispersal about a mean of 61.67.

In this situation we can use a relative measure of dispersion such as the coefficient of variation. This is calculated by dividing the standard deviation by its respective mean, as in equation 4.5.

$$CV = \frac{s}{\bar{X}} \times 100\% \qquad (4.5)$$

Equation 4.5

CV = coefficient of variation
s = standard deviation of X
$\bar{X}$ = mean of X

The result of such calculations are expressed as percentages, i.e. the standard deviation as a percentage of the mean. If this is done to the data in Table 4.6, then the coefficient of variation for Station 1 is 36.2 per cent, compared with 6.86 per cent for Station 2. We now have a more sensible view of the relative deviations about the means of the two samples.

When using interval scale data with an arbitrary zero (as in the case of temperatures) there may be difficulties with the coefficient of variation. If the mean of a sample is close to zero or, in the extreme case, is zero, then problems may arise as a result of attempting to divide the standard deviation by such a small quantities. To illustrate this point consider the case of a set of temperature data with a standard deviation of, say, 17°C, and a mean of 0.0°C, the solution to which could not be found as it is impossible to divide by zero. Conversion of such interval scale data to another system, such as degrees Fahrenheit, would be one possible means of overcoming this problem. Finally, if the mean is a negative quantity then the sign of the consequently negative coefficient of variation is ignored.

4.8 Other descriptive measures: skewness and kurtosis

We have already introduced the idea of the distribution of the data. This distribution can be expressed in terms of its central tendency and its spread. Moreover, we have seen that the data may have other properties, in particular it may be skewed. It is useful to have a numerical measure of these other aspects of the distributions under study.

Skewness

This is the degree of asymmetry in the distribution. This can be qualitatively assessed from the histograms of the data. It can be more objectively measured by, for example, studying the differences between the various measures of central tendency. Remember that if the data's distribution is symmetrical there is no skew and the mean, median and mode will be generally coincident. Equations 4.6 and 4.7 make simple use of this property to establish approximate measures of skewness that can be compared from one data set to another.

$$\text{Skewness} = \frac{\text{mean} - \text{mode}}{\text{standard deviation}} \tag{4.6}$$

$$\text{Skewness} = \frac{3\,(\text{mean} - \text{median})}{\text{standard deviation}} \tag{4.7}$$

However, one of the most valuable measures of skewness is momental skewness (equation 4.8). The *moments* of a distribution are important and we have already met two of them. The mean is the first moment and

$$\text{Skewness} = \frac{\Sigma(X_i - \bar{X})^3}{N} \tag{4.8}$$

Equation 4.8

X_i = ith value of X
$\bar{X}$ = mean
N = number of observations

the variance, based on the squared deviations from the mean, is the second moment. Skewness is the third moment and is based on the sum of the cubed deviations from the mean (equation 4.8).

The value for a perfectly symmetrical distribution is zero, negative values indicate negative skewness and positive values positive skewness. Such measures of skewness guide us in assessing which measures of central tendency and dispersion should be used (the means and standard deviation are useful only if the data distribution is not skewed). The measure also allows us to compare different distributions in an objective manner.

Kurtosis

This is the fourth moment of a distribution and indicates the degree of peakedness of a frequency distribution. For example, a distribution with a high degree of peakedness (Figure 4.7) is said to be *leptokurtic* and will have a kurtosis value in excess of 3. A distribution with a low degree of peakedness is *platykurtic* and its kurtosis will be less than 3. The normal condition is *mesokurtic* which is reflected in a value of 3. Equation 4.9 describes the manner of its estimation.

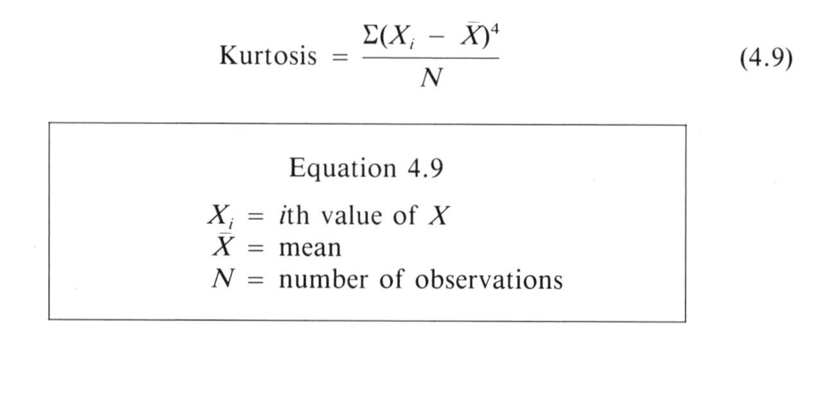

$$\text{Kurtosis} = \frac{\Sigma(X_i - \bar{X})^4}{N} \qquad (4.9)$$

Equation 4.9

X_i = ith value of X
$\bar{X}$ = mean
N = number of observations

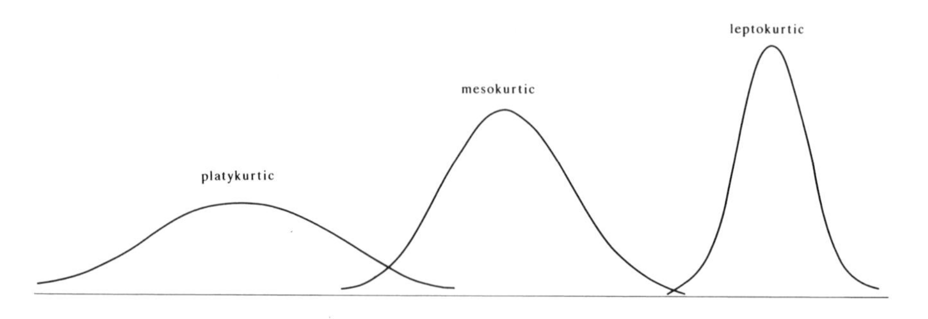

Figure 4.7 Idealised distributions showing contrasting degrees of kurtosis.

Both skewness and kurtosis, though not widely used, are potentially valuable statistics. First, they can identify highly skewed distributions when the use of means and standard deviations can give false values. Secondly, they allow for the objective comparison of different distributions. Finally, they show how far a distribution deviates from the 'normal'. In this latter respect they are important in the application of the statistical tests introduced in later chapters many of which assume the raw data to be 'normally' distributed.

4.9 Descriptive statistics using MINITAB and SPSS

All of the statistical measures described in the preceding sections can be obtained using either MINITAB or SPSS, and where large data sets are being studied such a recourse is scarcely avoidable. In MINITAB the subprogram **DESCRIBE** can be specified to provide summary statistics for any number of variables in the active file. An example of the screen display is shown in Figure 4.8 where data for birth and death rates in 54 English and Welsh counties are held in the file **COUNTY** under the column headings **B/1000** and **D/1000** respectively though their names have each to enclosed in single quotes when specified in MINITAB (see section 4.4). The output is largely self-explanatory and consists of parameters already discussed in this chapter such as the mean, standard deviation (**STDEV**), first quartile (**Q1**) and third quartile (**Q3**). However attention should be drawn to the items **TRMEAN** (which is the sample mean adjusted by trimming away the topmost and lowest 5 per cent of the observations) and the **SEMEAN**. The latter is a quantity known as the standard error of the mean, a full discussion of which is offered in Chapter 6. All of these attributes appear automatically as a result of invoking the **DESCRIBE** command and there are, in MINITAB, no options available to allow you to select only a few of them.

The SPSS subprogram **FREQUENCIES** can be used to provide similar

```
MTB  >  RETRIEVE 'COUNTY'
  WORKSHEET SAVED 8/21/1993
Worksheet retrieved from file: COUNTY.MTW
MTB  >  DESCRIBE 'B/1000' 'D/1000'

                N     MEAN   MEDIAN   TRMEAN    STDEV   SEMEAN
B/1000         54   13.106   13.150   13.100   1.1114    0.152
D/1000         54   11.622   11.700   11.640    1.435    0.195

              MIN      MAX       Q1       Q3
B/1000     10.400   15.700   12.275   13.925
D/1000      8.500   14.900   10.675   12.725
```

Figure 4.8 MINITAB screen display of the **DESCRIBE** option.

information. In this package the attributes to be assessed can be selected by the user but, correspondingly, the command syntax is more demanding than that for MINITAB. In section 4.4 we saw how the **FREQUENCIES** option **HISTOGRAM** could be used to produce a graph of the data. Here we will use two more options, **STATISTICS** and **PERCENTILES**, to produce a summary of the same data.

```
DATA LIST FILE 'COUNTY.DAT' FREE/ BIRTH DEATH.

FREQUENCIES /VARIABLES BIRTH DEATH
  /STATISTICS MEAN MEDIAN MODE STDDEV
  VARIANCE SKEWNESS KURTOSIS RANGE MINIMUM MAXIMUM
  /PERCENTILES 25, 50, 75.



BIRTH

Mean         13.106    Median       13.150       Mode          11.500
Std dev       1.114    Variance      1.242       Kurtosis       -.275
Skewness      -.065    Range         5.300       Minimum       10.400

Percentile  Value      Percentile   Value        Percentile    Value

  25.00     12.275       50.00      13.150         75.00       13.925

Valid cases     54     Missing cases    0


DEATH

Mean         11.622    Median       11.700       Mode          12.100
Std dev       1.435    Variance      2.061       Kurtosis       -.172
Skewness      -.185    Range         6.400       Minimum        8.500

Percentile  Value      Percentile   Value        Percentile    Value

  25.00     10.675       50.00      11.700         75.00       12.725

Valid cases     54     Missing cases    0
```

Figure 4.9 SPSS screen display of the **FREQUENCIES/STATISTICS** options.

In Figure 4.9 the data file **COUNTY.DAT** has been specified which contains the two variables for birth rates (BIRTH) and death rates (DEATH). Statistics are requested for both. The subcommand **STATISTICS** allows the user to select from a number of summary statistics merely by listing the appropriate key words along one or more lines as shown in Figure 4.9. The **PERCENTILES** subcommand prompts the program to estimate any number of specified percentile values for each elected variable; we have chosen those that correspond to the three quartiles, i.e. 25th, 50th and 75th. Each option such as **VARIABLES**,

STATISTICS or **PERCENTILES** must be introduced with the conventional slash (/).

The reader should again note, and become familiar with, the conventions for line terminators. When further instructions are to follow, each of which is introduced by a slash, the line must have no terminator in SPSS. Only when all instructions are fully listed should the stop be typed.

4.10 Conclusions

In this chapter we have studied a range of descriptive statistics. In most geographical analysis these statistics would be used, first, to organize large data sets, and second to summarise such data. It should be stressed that the use of these descriptive statistics must be exercised with care. They may not provide reliable answers if applied to inappropriate data sets. Furthermore the means and standard deviations, because they are based on the whole data set can be distorted by isolated extreme values. This is especially the case when small samples are used and the proportional influence is great. The problem is best overcome by using samples that are as large as are practically possible – sound advice from almost any research point of view.

Table 4.7 summarises the general suitability of descriptive statistics in terms of the scale of measurement and of the character of the data's distribution.

Table 4.7 Scales of measurement and appropriate descriptive statistics

Data scale	Central tendency	Dispersion
Nominal/ordinal	Mode	None
Interval/ratio (skewed)	Median	Inter-quartile range
Interval/ratio (non-skewed)	Mean	Standard deviation

Both the SPSS and the MINITAB systems can be requested to provide valuable summaries of data sets. In some instances analysis may need progress no further than such simple numerical descriptions. More often they will provide a basis for further, more detailed analysis and description, as described in Gardiner and Gardiner (1978), or as a step towards a more thorough analysis of the data and the factors that control the behaviour of the sample variables.

References

Croxton, F.E. and Cowden, D.I. (1968) *Applied General Statistics*, 3rd edn., Pitman, London.

Downs, R. (1970) 'The cognitive structure of an urban shopping centre', *Environ. & Behaviour*, 2, 13–39.

Evans, I.S. (1977) 'The selection of class intervals', *Trans. Inst. Br. Geogrs.*, (NS), 2, 98–124.

Gardiner, V. and Gardiner, G. (1978) *Analysis of frequency distributions*, Catmog 19, Geo Abstracts, Norwich.

Gudgin, G. and Thornes, J.B. (1974) 'Probability in geographic research', *The Statistician*, 123, 157–178.

Huntsberger, D.V. (1961) *Elements of Statistical Inference*, Allyn and Bacon, Boston.

Siegel, S. (1956) *Non-parametric Statistics for the Behavioural Sciences*, McGraw-Hill, New York.

Stevens, S.S. (1946) 'On the theory of scales of measurement', *Science*, 103, 677–680.

Chapter 5
Probability and Probability Distributions

5.1 Introduction

Probability can be defined in both the general and the mathematical senses. We might, for example, say 'it is probable that he will understand this book'. From this we could infer that the reader has a greater chance of understanding the text than of being bewildered by it. But this meaning can be interpreted in only the vaguest terms and scientists prefer to use the word in a more rigorous fashion, attaching some numerical value to the probability of an event. This numerical probability can be expressed in either of two ways – on an absolute scale of zero to one, or on a percentage scale of zero to 100. Both are widely used.

It is possible to think of some events to which numerical probability values can be readily attached. It is, for instance, absolutely certain that, at some unspecifiable time, we will all die. Such an absolute certainty has a probability of 1.0, or 100 per cent. Conversely, there is absolutely no chance of a human being lifting, unaided, a ten-ton weight, and in this case the probability is 0.0 or 0 per cent. However, many events are by no means as clear-cut and their probabilities can be thought of as lying at some location along a probability spectrum between certainty and impossibility. Figure 5.1 illustrates the position for some simple, if non-geographical, events. Other events may be more, or less, probable without necessarily being absolutely certain. But how might we assess those probabilities?

5.2 Assessment of probability values

The numerical probabilities of inevitable, of impossible, and some other events between those extremes can be derived by logical reasoning but alternative methods are needed in the less clear-cut cases. The French mathematician Pierre Simon de Laplace (1749–1827) was the first to define and solve the problem algebraically. If the numerical probability of an event x is denoted by $p(x)$, then:

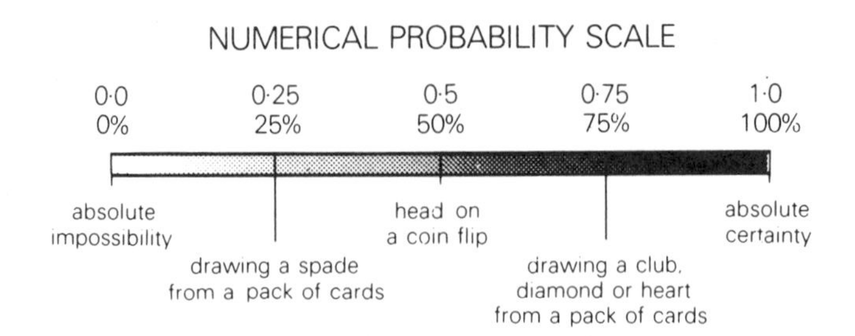

Figure 5.1 Probability scale from 0 to 1 showing the likelihood of some everyday events.

$$p(x) = n/N \qquad (5.1)$$

Equation 5.1

n = number of ways in which a particular event can be realised
N = total number of possible outcomes (the sample space)

To illustrate this important principle, let us consider the problem of drawing one card, at random, from a pack of 52. There are 52 possible outcomes, hence N (sample space) is 52. Suppose, also, that the specified outcome is a spade card. There are 13 such cards in the deck and, hence, 13 ways in which this specified event can be realised. From equation 5.1 the derived probability of obtaining a spade card is:

$$p(\text{spade}) = 13/52 = 0.25$$

Correspondingly, the probability of drawing a card from a suit which is other than a spade – a heart, diamond or club – is given by:

$$p(\text{non-spade}) = 39/52 = 0.75$$

as there are 39 non-spade cards in the deck. We should, however, observe that this derivation of a probability value makes one important assumption – that all 52 cards are equally likely to be drawn. Provided that no duplicate cards are within the deck, this is a realistic assumption. But when looking at more complicated events concerned with, say, human behaviour, this assumption may be violated. A further point to be noted is that of complementarity. Because the result of any draw of a card must be either a spade or a non-spade (no other outcome is possible), the total of the probabilities of the two must be 1.0.

Another example is provided by the flip of a coin. In this case the sample space provides for only two outcomes, a head or a tail. Each outcome can be realised in only one way. Hence, $n = 1$ and $N = 2$, so that:

$$p(\text{head}) = 1/2 = 0.5$$

and

$$p\ (\text{tail}) = 1/2 = 0.5$$

The outcome, if we ignore the vanishingly small probability of the coin landing on its edge, must be one of the above and their sum, as a result, must be 1.0.

We may now introduce two new terms, *mutual exclusiveness* and *independence*. Draws of cards and flips of coins produce outcomes that are mutually exclusive. By this we mean that if a head occurs on the flip of a coin then, for that flip, a tail is impossible. If a spade is drawn from a pack of cards, that same draw cannot simultaneously result in any other suit. This issue, for cards and coins at least, may appear obvious, but it is a valuable illustration of an important concept that will be referred to frequently at later stages in this book.

The other important concept is that of independence – another simple notion at this level, but fundamental, and often less obvious, in much that follows. A flip of a coin or a draw of a card might be described by statisticians as a *trial*. For these two forms of trials the succession of outcomes is a sequence of independent events. If on one trial a head is produced then the next trial is neither more nor less likely to produce another head. This independence applies no matter how many trials might constitute the sequence. Consider, for example, a sequence of ten trials with a coin. If all ten trials yielded a head, unlikely though such a series might be, it does not alter the probability of the eleventh flip also producing a head, which remains 0.5.

Coins and cards provide clear examples of exclusive and independent events. Very many of the numerical techniques and statistical concepts which follow assume the events they treat to be similarly independent. This is an important assumption not always fulfilled by geographical data and one whose validity can be difficult to assess. When collecting data we should always be aware of this principal requirement. Questionnaire responses, soil surveys, meteorological data and many other geographical phenomena provide data in which the constituent observations are not independent of one another and our sampling frame (Chapter 3) should take this possibility into account so that it might be avoided.

5.3 Probability assessments and the geographer

Laplace's method for estimating probability depends on a complete understanding of the circumstances surrounding each event. In particular this means being able to define both the sample space N and the number of ways in which event n can be realised. This thorough understanding is impossible in most geographical studies where both N and n may be difficult, perhaps impossible, to define and to measure.

Consider the notoriously unpredictable mid-latitude climate. Our knowledge of the complex interplay of forces involved is inadequate to allow the Laplace model to be applied to the estimation of the probability of rainfall on any given day. Fortunately it is possible to get around this problem by examining the observable consequences of these complex atmospheric processes and to determine what can best be termed *empirical probabilities*. No rigid numerical framework could answer our question concerning the probability of rain on a given day but, nevertheless, we may be able to obtain a reliable estimate. Let us be more specific and enquire into the probability of rain on any given day in March for a single location, in this instance Durham City in northern England. A study of the records of the Durham University meteorological station shows that over a ten-year period rain had fallen on 196 of the 310 days. From these two figures we can derive the probability of rain on any March day in Durham as:

$$p(\text{rain}) = 196/310 = 0.63$$

and, logically, the complement of this – a day without rain is:

$$p(\text{dry}) = 1 - 0.63 = 0.37$$

The analogies to the Laplace method are clear: the sample space N becomes the total number of available days (310), while the number of realisations of the event is represented by 196. There are, however, some very important contrasts. The results apply, at best only to an area around the meteorological station whose extent depends upon the local variability in rainfall patterns. The result also applies only to March. Seasonal variations are barely apparent with the span of one month, and all days may be considered to be equally probable with respect to rain. Over the whole year, however, this important requirement of equal probability will not be met. Winter days, in England, are more likely to experience rain than are summer days (but note that we make here no reference to the depths of rainfall). Estimates based on data collected over a whole year would, as a result, overestimate the probability of rain during the drier summer period and underestimate it in winter. This necessarily restricted application of the empirical method contrasts with the universality of conclusions gained by the Laplace method where the conclusions apply to all trials at all times in all places.

There are, in addition, further restrictions on the empirical approach to probability estimation. We have assumed that the background circumstances controlling climate and rainfall have not changed during or since the sample period. Whether or not this assumption is correct we might acknowledge that meteorological climates will change only slowly. As a result we might be confident in using the derived probabilities to predict incidences of rain in the near-future. On the other hand, social or economic conditions can certainly change over short periods of time and conclusions based on a data set gathered in one particular time period, even if in the recent past, may give a misleading impression of the probability of future events.

Lastly we may return to the question of independence. For our probabilistic conclusions to be valid each day's weather must be independent of the previous day's. If this is not so the probability of rain on a given day is not a wholly random event but a function of what happened on the previous day. The British climate is certainly variable and any one day may be meteorologically independent of the next, but we should not make such assumptions without further study.

Despite these restrictions, and provided that they are always borne in mind, the empirical method of probability assessment is useful and is often the only one available to us. We can now proceed to apply it in a number of useful ways.

5.4 Estimating the probabilities of multiple events

We have thus far examined only single events – one day's rain or one draw of a card. Frequently, however, we need to examine compound events – events composed of several trials. What, for example, is the probability of three selected days in March all yielding rain in Durham? The probabilities of such compound events remain within the range 0 to 1 but they require careful calculation based on the *multiplication law of probabilities*. Let us term a wet day W and a dry day D. To find the probability of three wet days in succession we can use the multiplication law employing the probabilities for single day events. Thus:

$$p(\mathrm{WWW}) = 0.63 \times 0.63 \times 0.63 = 0.25$$

The probability of the first two days being wet and the third dry is found by the same method:

$$p(\mathrm{WWD}) = 0.63 \times 0.63 \times 0.37 = 0.147$$

Notice that we do not add the probabilities of the individual events. Notice also that in the latter example we are careful to specify the order in which the events take place – two wet, then one dry day. If order is not specified we are presented with a different problem. The arrangement of two wet and one dry day can be accommodated in three ways with the dry day falling either first (DWW), second (WDW) or third (WWD) in the sequence. Expressed correctly we might say that there are three permutations of the combination two wet and one dry day. Now should we wish to know the probability of any one of the three permutations occurring (an essentially different situation to that of any one single permutation occurring) we would add their individual probabilities so that:

$$p(\text{2W1D}) = p(\text{DWW}) + p(\text{WDW}) + p(\text{WWD})$$

which, because the probability of each of the three permutations must be the same (0.147), gives us:

$$p(\text{2W1D}) = 0.147 + 0.147 + 0.147 = 0.441$$

We could apply the same principles to the specification for one wet and two dry days (1W2D). Again there are three ways in which this combination can occur – WDD, DWD or DDW – all of which have the same probability, let us take WDD:

$$p(\text{WDD}) = 0.63 \times 0.37 \times 0.37 = 0.086$$

The probability of any one of the permutations is the sum of each individual's probability:

$$p(\text{1W2D}) = p(\text{WDD}) + p(\text{DWD}) + p(\text{DDW})$$

$$= 0.086 + 0.086 + 0.086 = 0.258$$

To complete the picture we may consider the probability of all three days being dry. As with three wet days, this combination has only one permutation (DDD), so:

$$p(\text{DDD}) = 0.37 \times 0.37 \times 0.37 = 0.051$$

These four combinations of events cover all possible outcomes for three trials. They are themselves mutually exclusive, so the addition of their individual probabilities (Table 5.1) gives a total, allowing for marginal rounding errors in the above calculations, of 1.0.

Table 5.1 Combinations and probabilities for three dichotomous events

Combination		Probability
3W		0.250
2W1D		0.441
1D2W		0.258
3D		0.051
	sum =	1.00

5.5 Histograms and probability

Having dealt with some fundamental concepts in probability, we can now consider their broader application to geographical statistics. There are few more rewarding areas of common ground with which to start than that of the bar chart or histogram. Both are widely used as descriptive devices representing, as we saw in section 4.3, discrete classes of events or arbitrary classes defined along a measurement continuum. In their simplest forms the bar chart and histogram represent the absolute frequency of events in each class. But the relative frequencies, i.e. the absolute frequencies as proportions of the sample size, can also be plotted. Figure 5.2 is based on the same data as Figure 4.1 but plots the relative frequencies.

In the case of Figure 5.2 the data consisted of a 10 per cent sample drawn from the UK census of 1981 and contained 2,056,840 individuals. Of that number 805,535 were, for example, car drivers as far as 'journey to work' transport was concerned. The relative proportion of this class was thus 805535/2056840 = 0.39. Such proportions were used to plot the

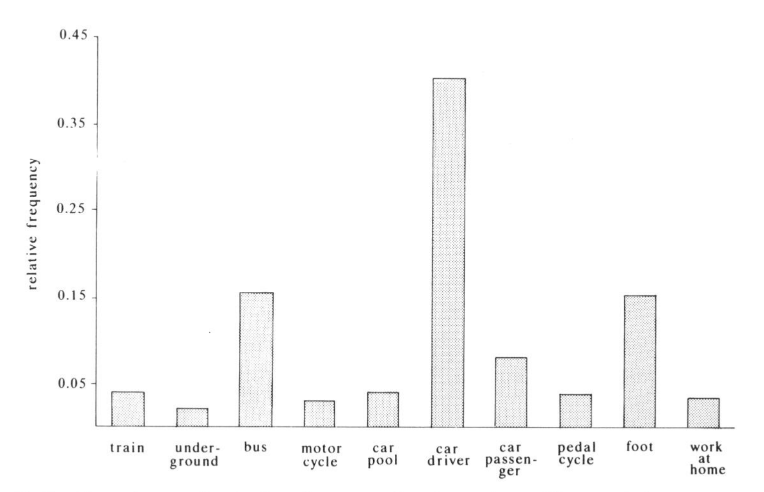

Figure 5.2 Bar chart of relative frequencies based on Figure 4.1.

heights of the bars in Figure 5.2 which, in terms of appearance, is the same as Figure 4.1. But we know from section 5.3 that these ratios are also the empirically-determined probabilities of randomly-selected individuals in the 10 per cent sample falling into those categories. As a consequence there is a probability of 0.39 that such an individual will drive to work. There is also a probability of 0.03 that they will walk, a 0.15 chance that they will go by bus and so on. We can now go one important stage further; if the 10 per cent sample is representative of the whole UK census then we can assume these derived probabilities to hold good for the population. We are, hence, inferring properties of a population from a sample drawn from its members, in this case a large sample, but a sample nonetheless.

These assumptions of representativeness can, for the moment, be considered to apply equally to data based on discrete classes and also to those based on the arbitrary sub-divisions of a continuum. The latter does open other issues, but these are considered in later sections of this chapter. The classes, of either type, represent mutually exclusive, non-overlapping, groups that can be considered to cover all possible outcomes. It is therefore consistent with what we have observed in section 5.4 that the total of the probabilities of each class must be 1.0. Expressed more generally, there is an absolute certainty that any individual must fall into one, but only one, of the specified classes.

In such studies we assume that sampled individuals are free to fall into any of their classes and that there is no bias by which they are excluded from or encouraged into one or other of those classes. In this way any characteristics established from such samples should also be true of the population from which it is drawn. For example, if we interviewed people only at bus and train stations to establish their journey to work patterns our sample would be biased, with little opportunity for car travellers and other groups to get a fair representation. Our sampling frame would have to give equal opportunity for all groups to be registered, perhaps by interviewing people at randomly selected homes.

The unbiased character of the variable's behaviour does not however prevent observed frequencies from varying between the classes and the derived probability estimates show that events are more likely within certain ranges or classes than within others. An important prerequisite in this respect is the requirement that class intervals must be regular if based on a continuous measurement scale such as age, temperature or unemployment rates. If this were not the case then those classes with wider limits would inevitably record unrepresentativly high frequencies of observations compared with those classes within narrower limits. The inferred probabilities would be correspondingly distorted and difficult to interpret.

We can now see that the bar chart or histogram represents more than a graphical device. It demonstrates the partitioning of a sample space with a notional area of 1.0 units. The probabilities of randomly selected individuals falling into each of the classes, or partitioned spaces, is represented by the height of each bar. The sequence of changing probabilities from class to class is known as the *probability distribution*. This concept of changing probabilities along the measurement scale is of particular importance for data measured on a continuous scale and through the vehicle of the histogram we may introduce the idea of the formalised probability distribution – one with recognisable and consistent characteristics.

5.6 The normal distribution: initial considerations

The concept of the statistical or probability distribution is an important one in many areas of research. The bar chart summarising the journey to work habits of UK residents (Figure 5.2) has a decidedly haphazard appearance in which no regularity could be discerned; a problem made all the worse by the use of nominal data which allows the classes to be shuffled without prejudice to the information the chart contains. But when plotting classified data measured on a continuum a different picture emerges. The definition of the class intervals can change, but their order cannot and is determined by the measurement scale – the classes must follow in numerical sequence. Taking Figure 5.3 as an example, a degree of regularity can be discerned in which events close to the mean are more frequent, other classes contain less events in approximate proportion to their distance from the mean. The most extreme cases tend to have lowest frequencies of events. In this particular example the data set consists of annual rainfall totals for the city of Sunderland (England) for the period 1859 to 1990. With 131 observations the histogram presents a near-perfectly symmetrical distribution. Such regularity is generally less apparent in smaller samples in which irregularities from class to class are more likely to occur. Data that approximate to this characteristically bell-shaped curve are said to follow the *normal distribution*. It is a distribution encountered frequently in geographical research and prevails over a remarkably wide variety of phenomena in the

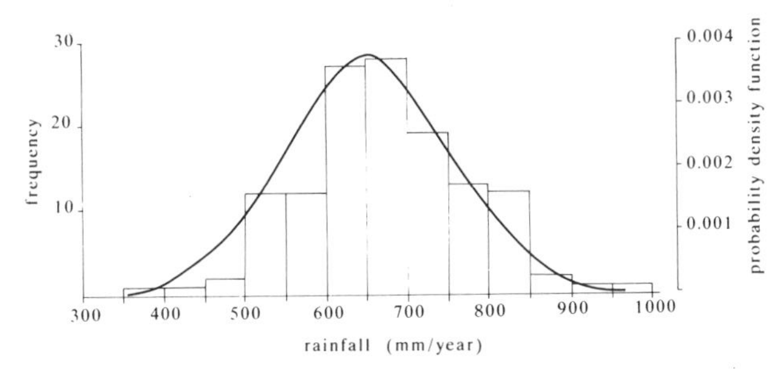

Figure 5.3 Histogram and superimposed normal curve of the distribution of annual rainfall in Sunderland, England, between 1859 and 1990.

physical and social sciences. But the term 'normal' is unfortunate in that there are, as we shall see, other distributions that are no less natural or widespread.

Such a tendency to consistent regularity is no academic abstraction and it reflects intrinsic and definable features of the data set. The normal distribution was discovered by the German mathematician Carl Gauss (1777–1855) after whom it is sometimes named – the Gaussian distribution. Most importantly, however, Gauss was able to express the regularity of the distribution in mathematical terms, and from that definition has sprung a huge area of statistical theory that we can use to advantage.

5.7 The mathematical definition of the normal distribution

The pattern of probability changes in a normal distribution is characteristically 'bell-shaped' with the mean at its peak. It applies to data measured on the continuous (interval and ratio) but not the discrete (nominal and ordinal) scales. Consequently it should be described by a continuous curve rather than by a histogram.

One means by which such a curve could be approximated would be to gather more data and classify them into ever smaller classes as in Figure 5.4. But to do so would often require impossibly large volumes of data to provide sufficient observations in each of the classes. Fortunately we need not resort to this means of determining the precise character of the normal curve and can its mathematical definition to good use. The numerical expression of the normal curve is given by equation 5.2:

$$Y = \frac{1}{\sigma\sqrt{2\pi}} e - ½(X - \mu)^2/\sigma^2 \qquad (5.2)$$

Equation 5.2

Y = height of curve at point X
e = constant 2.7183
π = constant 3.1416
μ = mean of all Xs
σ = standard deviation of all Xs
X = individual value of X

As it shows, we need know only the mean and the standard deviation to be able to calculate its form. Though the equation is scarcely inviting the principles of its application are much the same as the simple expressions used in section 2.5. An unknown quantity, in this case Y (the height of

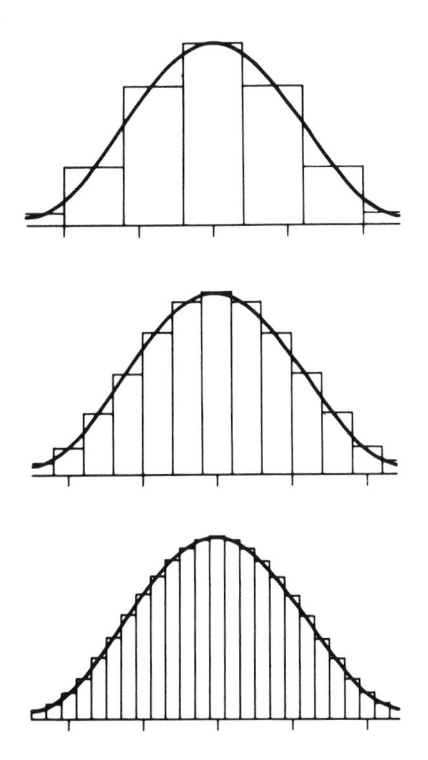

Figure 5.4 The effect of decreasing class intervals on histogram form.

the curve) is estimated from the known terms, which are X (the point along the measurement scale for which Y is required) and the distribution's mean and standard deviation. These items will vary from case to case but the two additional constants of e and π are universal and do not change. Repetitive substitution for one data set whose mean and standard deviation are known, replacing X on every occasion will provide points on the curve that can then be plotted.

Before doing so we should note an important distinction between this continuous distribution and the probability histogram. In the latter the heights and the relative areas of each column were directly proportional to the probability of an event in that class. This is not the case for continuous distributions where the height of the curve is not a similarly direct measure of probability but represents a quantity known as the *probability density function* (PDF). A full discussion of the PDF would take us beyond the scope of this book but some of its features can be understood if it is recalled that the measurement scale consists of an infinite number of points and not a finite number of classes. Such points have no area (Figure 5.5) and probability statements can be applied only to ranges along the continuum and not to individual locations.

How might we set about plotting the form of a normal curve? Clearly we could use equation 5.2 and make a number of repetitive calculations

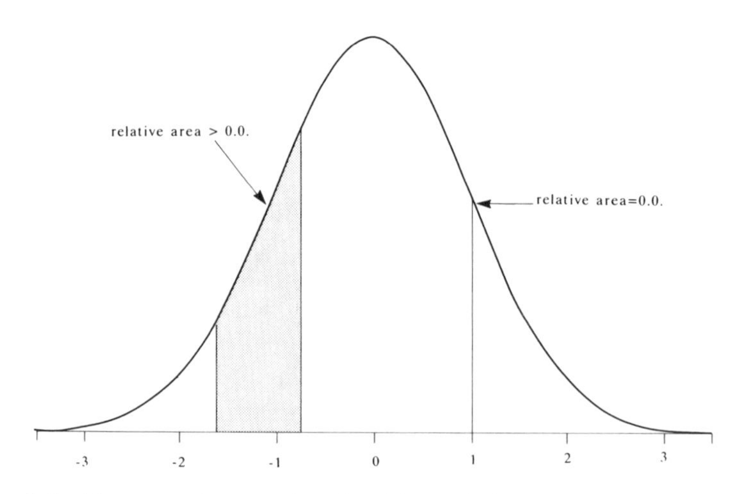

Figure 5.5 The normal curve showing (shaded) the area covered between two points on the horizontal scale. No area can be attached to a single point.

to estimate the PDF at points (X) along the measurement scale. Fortunately there is a far easier means of achieving this. The MINITAB package gives us the option of simulating a 'perfect' normal distribution given a sample mean and standard deviation. We have already seen that these moments can be derived from samples whose plotted distribution may not accord perfectly with the distribution of the population from which they are thought to have been drawn. Notice in Figure 5.3 how the changing pattern of frequencies shown in the histogram is slightly irregular. As we have already explained, such variations about the ideal theme must be expected in any sampling exercise. But if the sample moments are representative of the population, then we can use them to reconstruct the latter's distribution which we assume to be perfectly normal.

Figure 5.6 shows how to use the **PDF** (probability density function) subprogram in MINITAB. Using the Sunderland annual rainfall figures we have a range of observed values from 353mm to 936mm. We could estimate the PDF for each of the 131 observations, instead we will do so only for points at regular intervals along the observed range. For this purpose there is no need to **RETRIEVE** the file containing the original data and as there is no active file the required values can be typed into the empty column (**C1**) following the **SET** command as indicated in Figure 5.6 and discussed in section 2.8. The results, the PDF at each of the points in **C1**, are stored in next available free column, **C2**, by specifying **PUT INTO C2**. Using the subcommand **NORMAL** followed by the observed mean (646.53) and standard deviation (100.49) of the sample the system will produce points on the perfect normal distribution with those defined properties. Because a subcommand is required for the key word **NORMAL** (other distributions, to be described later in this chapter, could have been used) the PDF instruction line must be

```
MTB  >   SET C1
DATA >   300 350 400 450 500 550 600 650 700 750
DATA >   800 850 900 950 1000 1050
DATA >   END
MTB  >   PDF C1, PUT INTO C2;
SUBC >   NORMAL 646.53, 100.49.
MTB  >   PRINT C1 C2

   ROW       C1          C2

     1      300   0.0000104
     2      350   0.0000510
     3      400   0.0001958
     4      450   0.0005865
     5      500   0.0013712
     6      550   0.0025027
     7      600   0.0035664
     8      650   0.0039676
     9      700   0.0034459
    10      750   0.0023365
    11      800   0.0012368
    12      850   0.0005111
    13      900   0.0001649
    14      950   0.0000415
    15     1000   0.0000082
    16     1050   0.0000013
```

Figure 5.6 MINITAB screen display of PDF (NORMAL) commands with a listing of the results.

concluded with a semi-colon as outlined in section 2.8. The conclusion of the subcommands, only one line in this example, is indicated with a stop.

The derived PDFs can be stored and printed out or plotted as necessary. In Figure 5.6 a separate **PRINT** instruction was used to list the contents of columns **C1** and **C2**. It was from these synthesised PDFs that Figure 5.3 was plotted. By doing so we can make at least a qualitative assessment of how closely our observed and theoretical distributions agree. In this case there is close agreement.

The total area beneath such a normal curve has a notional value of 1.0, i.e. all possible events must fall within its total range. It is, however, important to notice that unlike the probability histogram the normal curve in theory can extend indefinitely in either direction and is not constrained within the range of observed values. We could, for example, have estimated the PDF for a rainfall of 1500mm, well beyond the observed upper limit. The resulting PDF would be small, but not zero.

Furthermore, the proportional areas, i.e. of between zero and 1.0, within any specified range can be calculated. The areas represent the probability of random events falling within those limits. But before any such probabilities can be determined the distribution should be *standardised*.

5.8 The standardised normal curve

Individually the form of the normal curve differs widely in response to the mean and standard deviation of the distribution (see Chapter 4 for a full account of these measures). For example, in Figure 5.7a we see three idealised distributions with different means, and in Figure 5.7b another three distributions with different standard deviations.

To take advantage of the potential offered by our knowledge of the normal distribution these differences must be removed by standardisation of the data. This is achieved by reference to each distribution's standard deviation. Individual observations can be converted to their equivalent *z-score* by expressing their departure from the mean as a proportion of one standard deviation. Equation 5.3 is used to achieve this object. When all values have been converted the resulting distribution will have a mean of zero and a standard deviation of 1.0.

$$z = \frac{X - \mu}{\sigma} \qquad (5.3)$$

Equation 5.3

z = required z value
μ = mean of variable X
σ = standard deviation of variable X
X = raw value of X to be converted

A few examples will illustrate the point and we can see how the rainfall data used in Figure 5.3 can be converted to equivalent z-scores. Remembering that the distribution's mean and standard deviation are 646.53 and 100.49 respectively, the following conversions are easily performed.

EXAMPLE 1 The z-score for a rainfall of 800mm.

$$\frac{800 - 646.53}{100.49} = \frac{153.47}{100.49} = 1.53$$

EXAMPLE 2 The z-score for a rainfall of 500mm.

$$\frac{500 - 646.53}{100.49} = \frac{-146.53}{100.49} = -1.46$$

EXAMPLE 3 The z-score for a rainfall of 1050mm.

$$\frac{1050 - 646.53}{100.49} = \frac{403.47}{100.49} = 4.02$$

Notice that both positive and negative z-scores are possible depending upon whether the selected value is greater or less than the mean. We can also calculate z-scores (Example 3) of values of X that fall well beyond the observed, but not the impossible, range of values. In all cases they assess the departure of the selected value from the mean in terms of the standard deviation. In Example 1, 800mm is 1.53 standard deviations above the mean. In Example 2, 500mm is 1.46 standard deviations below the mean. This idea of measuring values in terms of their standard deviation equivalents is an important one that we shall apply in the next section.

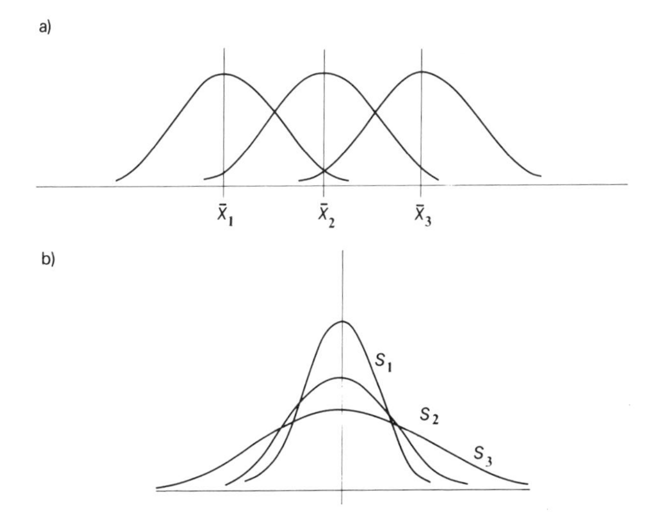

Figure 5.7 Generalised normal distributions with (a) different means and (b) different standard deviations.

It may not have escaped the reader's attention that this standardisation process can be equally applied to the equation of the normal curve. The terms μ(mu) and σ(sigma) on the right-hand side of equation 5.2 are replaced by zero and 1.0 respectively. Simultaneously, as shown in equation 5.3, the term $(X - \mu)/\sigma$ becomes z, giving the following expression:

$$Y = \frac{1}{\sqrt{(2\pi)}} e^{\frac{1}{2}z^2} \qquad (5.4)$$

Equation 5.4

z = z value (see also equation 5.3)
e = the constant 2.7183
π = the constant 3.1416
Y = height of curve at point z

This normal expression retains the characteristic bell shape but the individual effects of differing means and standard deviations have been removed and the curve is standardised to one with zero mean and unit standard deviation.

5.9 The normal curve and probability

In general terms the character of the normal distribution is one in which events close to the mean are the most likely, becoming more improbable as one moves away from that central point. In theory at least the normal curve extends infinitely in both directions. Thus exceedingly large and extremely small values are possible, but become increasingly improbable. Referring back to Figure 5.3, an annual rainfall as great as 1500mm is possible, but highly unlikely and has not occurred in the 131 years of record. In reality the inherent nature of much geographical data imposes limitations on the idealised properties of the normal curve. Phenomena as varied as rainfall, crop yields and incomes all have an absolute base of zero beyond which there can be no observations. On the other hand upper limits are less definite and can be defined only the vaguest of terms.

Accepting these qualifications, we can nevertheless make confident statements about the probabilities of events within specified ranges of our defined normal distribution but we can do so only by referring to z-scores and not to the original data. We have already stated that the area beneath the standardised normal curve has a notional value of 1.0 and must include all events. We know also that the distribution is perfectly symmetrical and half the area lies above and the other half below the mean. Hence there is a 0.5 chance of an event greater than the mean, and 0.5 chance of an event being less than the mean. Usefully, our understanding of the normal curve goes far beyond this and we can estimate the relative areas between any specified limits we choose. These relative areas are also the probabilities of events within those ranges. Such areas could be laboriously calculated but there is no need to do so as the information is available in published table form. Appendix I contains a z-table from which areas between the mean and any selected

z-score can be read off directly. A few examples should be sufficient to clarify their application.

Example 1

What is the probability of an observation (event) with a z-value of between 0.0 and +0.75?

In effect we need here to know the relative area beneath the standardized normal curve between its central point ($z = 0.0$) and $z = +0.75$ (Figure 5.8). Appendix I is arranged to provide the relative area between $z = 0.0$ and any other specified z value. The required probability, accurate to five decimal places, may be read off directly from the table and appears adjacent to the selected z value. In this example $z = 0.75$ which, from the tables, delimits a relative area of 0.2734, this being also the probability of an event within the range $z = 0.0$ to $z = 0.75$.

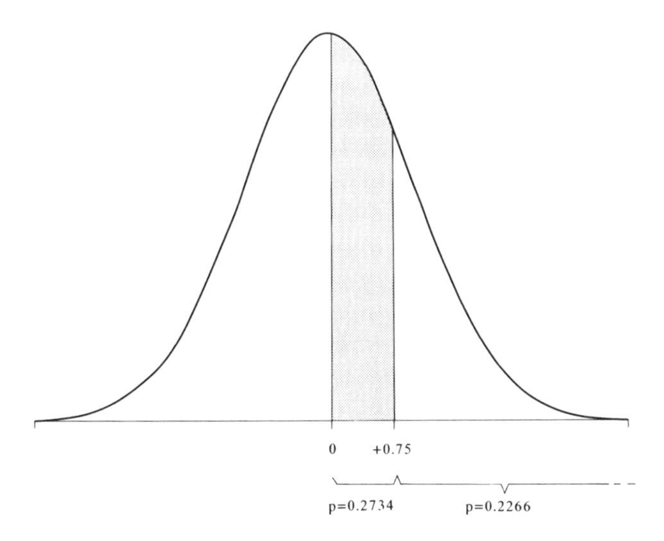

Figure 5.8 Probability of events under a normal distribution between z = 0.0 and z = +0.75.

Example 2

What is the probability of an event with an equivalent z-score greater than +0.75?

The required relative area is now that between +0.75 and infinity. But, as the tabled figure is that between zero and 0.75, some simple secondary calculations are necessary. We know that the relative area under the standardised normal curve between zero and infinity is 0.5 therefore the probability of an event with a z-score in excess of 0.75 must be given by:

$$0.5 - 0.2734 = 0.2266$$

Example 3

What is the probability of an event with a z-score between 0.0 and -1.2?

The normal curve is perfectly symmetrical and Appendix I applies equally to positive and negative values. We can, effectively, ignore the sign and read the probability tabled at $z=1.2$, which is 0.3849 (Figure 5.9).

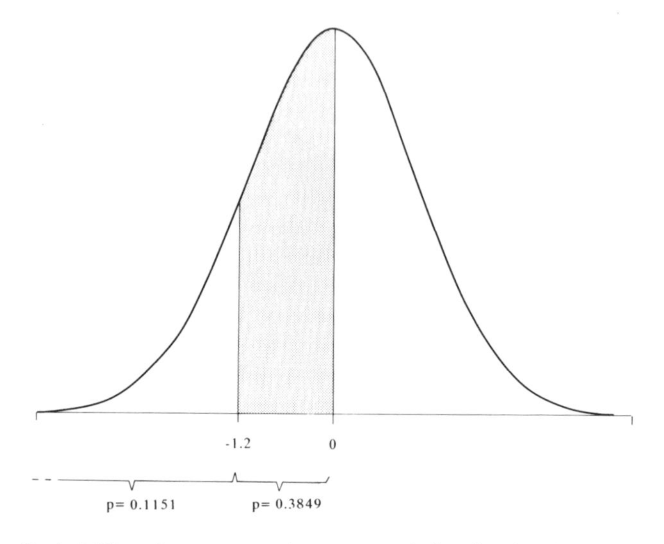

Figure 5.9 Probability of events under a normal distribution between $z = 0.0$ and $z = -1.2$.

For z-values less than -1.2 the probability can found in exactly the same manner as outlined in Example 2. Thus, probability of an event further down the scale than $z=-1.2$ is found from:

$$0.5-0.3849=0.1151$$

Example 4

What is the probability of an event within the range

$$z=+1.5 \text{ and } z=-0.8$$

The same principles apply here, but the problem can only be solved in two parts, one dealing with the probability of events within the range z 0.0 to $+1.5$ and the other within the range 0.0 to -0.8 (Figure 5.10). The two probabilities can be taken from the tables and, finally, added to give the answer:

$$p(z<1.5)=0.4332 \text{ and } p\ (z>-0.8)=0.2882$$

thus

$$p\ (1.5>z>-0.8)=0.4332+0.2882=0.7214$$

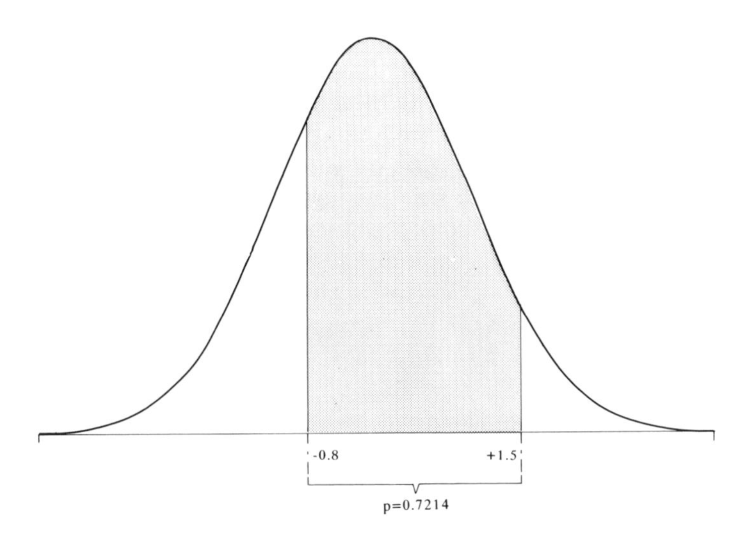

Figure 5.10 Probility of events under a normal distribution between $z = 1.5$ and $z = -0.08$.

Example 5

From the annual rainfall information for Sunderland, England, we can determine the probability of a single year's total being between 600mm and 700mm. The principles outlined in Examples 1 to 4 still apply but the data must first be converted to their equivalent z-scores before the tables can be consulted. With a mean of 646.53mm and a standard deviation of 100.49mm the derived z-scores are found from:

$$z_{600} = \frac{600 - 646.53}{100.49} = -0.46$$

and

$$z_{700} = \frac{700 - 646.53}{100.49} = 0.53$$

Notice that our specified range spans the mean and hence two probabilities must be derived and then added to give the final figure. From the table of z-scores we find that the probability of an observation being between the mean and $z = -0.46$ is 0.1772. Similarly the probability of the observation being between the mean and $z = +0.53$ is 0.2019. The sum of these two probabilities is 0.3791 and we conclude that this is the probability that the annual rainfall in Sunderland will be between 600 and 700mm. Though it must be emphasised that this conclusion applies only to the site from which the data have been derived. Different locations will have equally different probabilities of annual totals between those two limits.

5.10 Further applications of the normal distribution

Any normally distributed variable has predetermined probabilities of events falling within any range that the investigator selects. These ranges are, however, best specified as z-values. The z-tables indicate, for example, that there is a 0.6826 probability (0.3413 either side of $z=0.0$) of an event within the range of $z=-1.0$ and $z=+1.0$, i.e. within one standard deviation either side of the mean. On the same basis there is a 0.9545 probability that the event will be within two standard deviations of the mean and 0.9973 probability of it being within three standard deviations of the mean (Figure 5.11). Thus we would expect that in the long run 99.73 per cent of all events to be within the range $\mu \pm 3\sigma$, 95.45 per cent would be within the range $\mu \pm \sigma$, and so on. Most importantly, however, these expectations apply to all normally distributed variables, be they hydrologic, social, economic, etc., without exception.

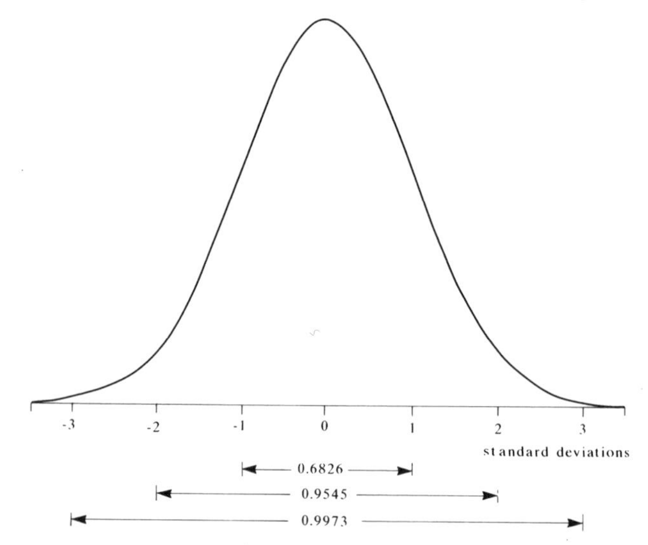

Figure 5.11 Standard deviations and probabilities on the normal distribution.

It is often convenient to express such probabilities and expectations within less awkward ranges. For this reason the 95 per cent, or 0.95, probability limits are often quoted in preference to the 95.45, or 0.9545, limits. These differ slightly from each other but, fortunately, it is not difficult to establish the corresponding z-values. The published tables may again be used, but in the reverse manner. To find the z-value associated with the 0.95 limits we must remember that the tables in Appendix I deal with only one half of a symmetrical distribution and we must first look for the probability 0.475, i.e. 0.95/2, in the body of the table, from which a corresponding z-value of $(\pm)1.96$ is read off.

Expressed in another way, there is a 0.95 probability of an event falling within 1.96 standard deviations of the mean of a normal population. Correspondingly, events so extreme as to fall beyond those limits do so with a probability of only $1-0.95=0.05$. Figure 5.12 demonstrates how these extreme events may be depicted on the distribution. Notice that the relative area of 0.05 is distributed equally at the two ends of the distribution.

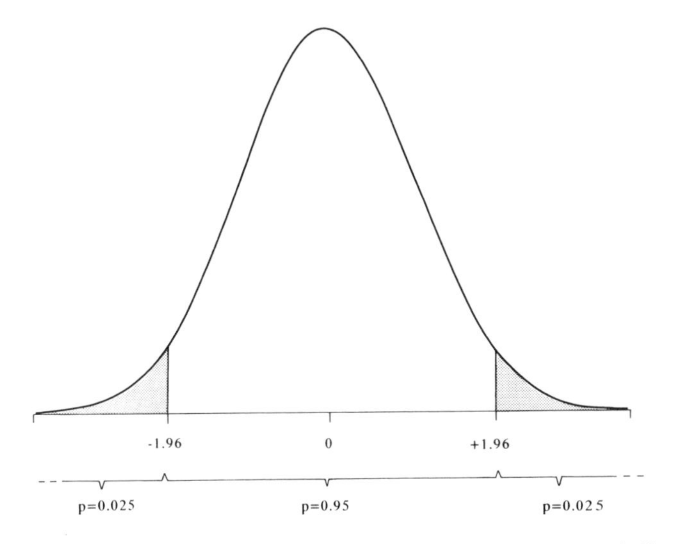

Figure 5.12 Probabilities of events beyond z ±1.96 on the normal distribution.

For any normally distributed variable these z scores of ±1.96 are readily transformed into raw data equivalent values. Consider the annual rainfall example again. By substitution and transposition of equation 5.3 we get for the upper 0.95 probability limit:

$$1.96 = \frac{X - 646.5}{100.49}$$

Therefore, by transposition, the upper 0.95 limit has a value of:

$$X = 646.5 + 1.96 \times 100.49 = 843.5\text{mm}$$

By the same process the lower 0.95 limit is given by:

$$X = 646.5 - 1.96 \times 100.49 = 449.5\text{mm}$$

We can conclude that there is a 0.95 (or 95 per cent) probability that the annual rainfall in Sunderland will be between 843.5mm and 449.5mm.

More generally speaking the required limits, for any probability range, can be expressed as:

$$X = \bar{X} \pm z_c \sigma \tag{5.5}$$

Equation 5.5

$\bar{X}$ = mean of variable X
z_c = critical z value for selected limits
σ = standard deviation of X
X = required raw observation

Nevertheless our interest need not necessarily be with the 0.95 limits and might be equally concerned to identify the wider 0.99 or the narrower 0.90 limits. The critical z-values for these, and other limits, are given in Table 5.2 which also indicates the complementary probability of events falling beyond those limits. These latter areas are, again, divided equally over the two extreme ends of the distribution. The table also gives the commonly-used alternative expressions for the probability of events falling beyond each of the limits. An event with a probability of, say, 0.05, has only a 1 in 20 chance of being realised; a 0.01 event has only a 1 in 100 chance of occurring, and so on.

Table 5.2 Extreme probabilities and associated critical z values

Probability of events		
Within the range	Beyond the range	Critical z value
0.90	0.10 or 1 in 10	± 1.647
0.95	0.05 or 1 in 20	± 1.960
0.99	0.01 or 1 in 100	± 2.586
0.999	0.001 or 1 in 1000	± 3.290

5.11 What to do with non-normal data

Despite the frequency with which the normal distribution is encountered there remain many instances when observed distributions are manifestly not normal. The most common problem in this respect is provided by high degrees of skewness in the distribution. This may appear in one of two forms; either with a long tail towards the high end of the distribution (positive skewness) or towards the low end (negative skewness). The former is far the more common in geographical studies.

The example used here is based on a sample of 123 daily river discharge observations for the River Clyde (Scotland). When plotted in histogram

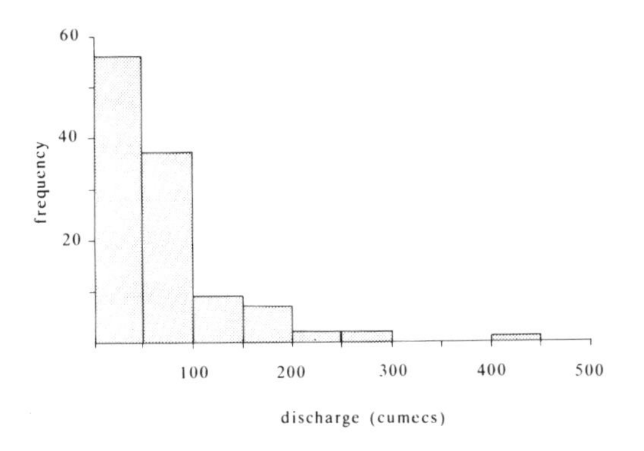

Table 5.13 Histogram of daily discharge data for the River Clyde, Scotland (*source: Hydrogical Data UK 1990*, Inst. of Hydrology, Wallingford (1991)).

form (Figure 5.13) the degree of skewness is readily apparent. Such a distribution forbids the reliable use of measures such as the mean and standard deviation. Table 5.3 indicates the degree of difference between the mean and median of the raw data; usually a clear indication of skewness. As we will see in later chapters, severely skewed data are also limited in the range of statistical tests and analyses to which they can be subjected as many procedures assume the data to be normal or at least near-normal in their distribution.

Table 5.3 Comparison of means, medians and standard deviations for raw and logged data for daily discharges on the River Clyde (Scotland).

	n	Mean	Median	Standard deviation
Raw data	123	60.49	27.35	75.99
Logged data	123	1.539	1.437	0.438

Units: cubic metres per second (cumecs) of discharge
Source: *Hydrological Data UK 1990*, Inst. Hydrology, Wallingford (1991)

Fortunately many positively skewed distributions can have a measure of normality imposed on them by the process of data transformation. Such transformations can assume a number of forms, the most common of which is the taking of logarithms. If the common logs of all the observations of the current example are taken and those logs are plotted in histogram form, then Figure 5.14 results. Its most immediate effect is the removal of much of the skewness that characterised the raw data. In log form the mean and standard deviation of the data set can both be estimated. Table 5.3 shows how the mean and median of the transformed data are now in closer agreement suggesting a lower degree of skewness though, of course, the quantities are now expressed in logarithmic form. Such transformed data might now be confidently used in the forms of

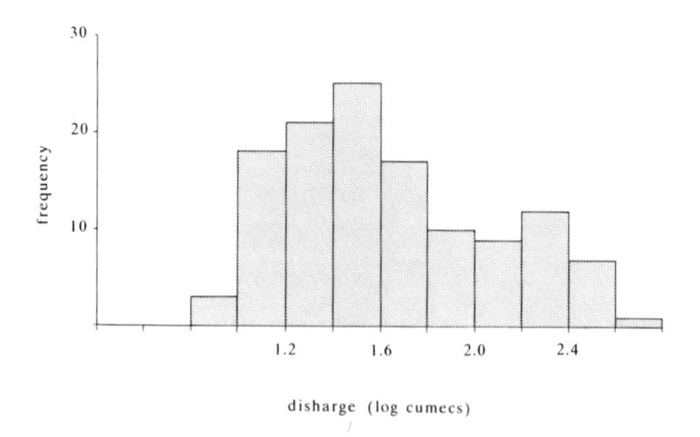

Figure 5.14 Histogram of logged daily discharge data for the River Clyde, Scotland (see Figure 5.13).

analysis to be described in some of the following chapters many of which assume the data to be normally distributed.

Transformations of this character 'work' because smaller numbers as they approach zero become ever greater negative quantities (see section 2.3), thus spreading the tail of the distribution outwards along its lower end. Conversely the long tail at the upper end of the distribution is reduced in logarithmic form. In this and many other examples negative quantities in the raw data are impossible, but may arise when expressed in logarithmic form.

Log transformations of data will not always resolve the problem of skewness some degrees of which may be too great even for this step to help. In other cases the inclusion of zero or negative values in a data set will preclude this course of action as such quantities have no logarithmic equivalents (see section 2.3).

Negatively skewed data may also occur in geographical work. This problem might be eased by taking the squares, or higher powers, of the raw data. This transformation has the effect of drawing out the tail of the distribution towards its upper end by giving greater emphasis to high values. For example $3^2=9$ but 4^2 is 16 while 5^2 rises to 25. Thus the series 3,4,5 can be transformed into 9,16,25, the latter having a quite different distribution.

Nevertheless a word of warning must be sounded. While normally distributed data lends itself to a wide variety of analyses the temptation to juggle with skewed data to attain a high degree of normality should be resisted. Geographers should rightly be concerned with the use of appropriate data, but equally important is the need, once analysis has been completed, to understand and interpret the results. With transformed data this task may not be easy and normalisation procedures should not be adopted without a thought for the interpretational difficulties that it may later create.

5.12 Data transformations and treatment with MINITAB and SPSS

The logging of data is an easy task if small samples are used, but where samples are larger or contain observations on many variables the process is more demanding and subject to error. Fortunately, both MINITAB and SPSS can be instructed to carry out a wide variety of transformations of specified variables. Both do so by setting up new variables that contain the transformed data, the original data still being available after completion of the task. In effect new columns of (transformed) data are added to those already held in the file. If these transformed variables are to be retained for use in later sessions the appropriate **SAVE** instruction must be used or the new data will be lost at the conclusion of the work.

Transformations will generally employ standard functions available within the package, for example logarithms or square roots. Such standard functions are specified by the use of key terms within user-written equations. The manner of their specification is summarised in Figures 5.15 and 5.16 in which files are called up to provide the data to be transformed and into which the new variables will be added.

In the MINITAB example illustrated in Figure 5.15 the **LET** instruction is followed by the equation that specifies the name of the new variable (left-hand side of the equation) and the nature of the

```
MTB >   RETRIEVE 'DISCHARG'

Worksheet retrieved from file: DISCHARG.MTW

MTB >   LET C2 = LOGT(C1)
MTB >   LET C3 = SQRT(C1)
MTB >   PRINT C1,C2,C3

  ROW      C1         C2         C3
    1   27.72    1.44279    1.20116
    2   57.80    1.76193    1.32738
    3   21.56    1.33365    1.15484
    .   .....    .......    .......
    .   .....    .......    .......
  121    8.32    0.92012    0.95923
  122   37.56    1.57473    1.25488
  123  138.50    2.14145    1.46337

MTB >   SAVE 'DISCHARG'
```

Figure 5.15 MINITAB screen display showing LET commands used to transform data using standard functions. A new data file containing the transformed variables was saved under the original name using the SAVE command. Only part of the data file listing is shown here.

```
DATA LIST FILE 'B:COUNTY.DAT' FREE / BIRTH DEATH.

COMPUTE LOGB = LG10(BIRTH).
COMPUTE ROOTB = SQRT(BIRTH).
LIST /VARIABLES = BIRTH, LOGB, ROOTB.
SAVE /OUTFILE = 'B:COUNTY.DAT'.

  BIRTH     LOGB     ROOTB
  14.40     1.16      3.79
  11.80     1.07      3.44
  12.30     1.09      3.51
  .....     ....      ....
  .....     ....      ....
  11.90     1.08      3.45
  14.50     1.16      3.81
  12.60     1.10      3.55

Number of cases read =   54  Number of cases listed =   54

The SPSS/PC+ system file is written to
  file B:COUNTY.DAT
   7 variables (including system variables) will be saved.
   0 variables have been dropped.

  54 out of        54 cases have been saved.
```

Figure 5.16 SPSS screen display showing COMPUTE commands to transform data using standard functions. The data file was recalled from drive B: using the DATA LIST/FILE command and saved on floppy disc in the same drive with the SAVE OUTFILE command. Only part of the data file listing has been reproduced here.

transformation and the existing variable to which it is to be applied (right-hand side of the equation). Common logs, for example, are selected by using the term **LOGT** in MINITAB and **LG10** in SPSS. The square root is denoted by **SQRT** in both cases. The transformed variable is then stored in the specified columns or variables. In Figure 5.15 one variable is logged to produce the new variable **C2**. A further new variable (**C3**) is created by taking the square root (**SQRT**) of **C1**. Figure 5.15 also shows how the file containing the data (**DISCHARG**) is retrieved and, after transformation, the data are listed and saved. The name given to the new file may be the same as the original, in which case the latter is overwritten, or an entirely new designation can be made.

Transformations in SPSS are carried out using the **COMPUTE** subprogram. The keyword is immediately followed by the transformation equation containing the names of the new and existing variables together with the equation that allows the former to be derived from the latter (Figure 5.16). These single line instructions should not be preceded by a slash (/) which are required only for subcommands within options. However each of the lines must conclude with a stop. The equations are otherwise written in much the same way as in MINITAB. In this example the United Kingdom county-based statistics for birth and death rates were retrieved from the file **COUNTY.DAT** written in freefield format and held on floppy disc in drive B of the computer. **DATA LIST /FILE** and its ancillary information activates the file upon which transformations can then proceed. The variable BIRTH was selected and the data were logged to form the new variable LOGB and also square-rooted, the latter transformations forming the variable ROOTB. The transformed equivalent of each original observation is then located on the corresponding row of the next two free columns. In both MINITAB and SPSS such transformations are held in the active file but will be lost at the conclusion of the session unless the appropriate **SAVE** instruction is implemented. In Figure 5.16 the new file containing the raw and transformed variables is given the same name as the original data file and is over-written on the latter stored on the floppy disc in the computer's drive B.

There is no automatic output from either system. MINITAB will simply move to a new **MTB>** prompt if the transformation has been successful. SPSS passes to the results screen on which appears the instruction from the **COMPUTE** line. Further information only appears if errors have been encountered, for example if the user has tried to log zero or a negative number. The nature of the new variables can be quickly assessed using the **PRINT** option (in MINITAB) or the **LIST** option (in SPSS). The screen will then display the list of all specified variables.

Following successful transformation the new variables and their distributions can be checked more thoroughly using the **DESCRIBE** and the **HISTOGRAM** subprograms in MINITAB or the **FREQUENCIES** subprogram (with the **HISTOGRAM** and **STATISTICS** options) in SPSS. Both have been described in Chapter 4 and will quickly convey an impression of the new variable's principal characteristics. It was from such a check that Table 5.3 and Figures 5.13 and 5.14 were prepared.

Non-standard transformations can be also specified by the user. It is possible, for example, to convert a variable to its *z*-scores and equation 5.3 can be used within the two packages to perform this task (Table 5.4). New variables can also be created by treatment of existing data, for example by taking two variables and estimating their average. A summary of some typical instructions for standard and non-standard transformations is given in Table 5.4. The transformed variables are specified in the same way as indicated in Figures 5.14 and 5.15 using standard arithmetic notation but it should be noted that in MINITAB direct use can be made of the means and standard deviations of any

Table 5.4 Examples of data transformations using MINITAB and SPSS

Transformation	MINITAB command	SPSS command
Common logarithm	LET C2 = LOGT(C1)	VAR002 = LG10(VAR001)
Natural logarithm	LET C2 = 2.303(LOGT(C1))	VAR002 = LN(VAR001)
Square root	LET C2 = SQRT(C1)	VAR002 = SQRT(VAR001)
z-score*	LET C2 = (C1 − 10.5)/6.2	VAR002 = (VAR001 − 10.5)/6.2
z-score**	LET C2 = (C1 − MEAN(C1))/ STDEV (C1)	
Sum of two variables	LET C3 = C1 + C2	VAR003 = (VAR001 + VAR002)

* This example assumes the mean and the standard deviation to be known beforehand (10.5 and 6.2 respectively)

** This option is only available in MINITAB and allows the mean and standard deviation for any variable to be calculated by the appropriate specification

variable without those quantities having first to be known. They can be included directly by specifying **MEAN(C**x) and **STDEV(C**x) as items within a transformation equation these are then automatically evaluated by MINITAB. Other items from the **DESCRIBE** subprogram could be specified in the same way.

5.13 The binomial distribution

Thus far only the normal distribution has been examined, but other formal probability distributions exist to which geographical data may conform. Two of the more important are the binomial and the Poisson distributions. Both apply to a range of natural events, but the manner of their application and their specific data requirements are widely different to that of the normal distribution. Nevertheless, there are some links between all three as we will demonstrate and their importance lies in the possibility of them extending the geographer's ability to measure and to understand the World about him.

Both the binomial and Poisson distributions can be defined in mathematical terms. They can also be summarised through their respective measures of central tendency, dispersion and skewness. But these points are better introduced through detailed studies and examples. We will be begin with the binomial distribution.

The properties of the binomial distribution were first defined by James Bernoulli in the late seventeenth century and is sometimes referred to as the Bernoulli distribution. It is the distribution used to describe the probability of events for which only two outcomes are possible. For example, a head or tail in the case of coin flipping, or a wet or dry day in the case of a day's weather. Unlike the standardised normal distribution it has no characteristic form and its shape will vary according the the sequence length, i.e how many times a coin is flipped, or how many days' weather are studied, and the probability of a *success*. Of the two possible outcomes of any trial one is denoted, perhaps on purely arbitrary grounds, as a success. In coin flipping we might decide that a head is the success and a tail, hence, is the failure. The binomial distribution is concerned only with the probability distribution of the different number of successes possible over the specified trial length. If, as in the case of coin flipping, the two outcomes have the same probability (0.5) then the distribution of successes must be the same as that for failures. But this is the only situation where such equality exists. As soon as the probability of a success departs from 0.5 the equality of the distribution of successes and of failures disappears.

Because binomial variables consist of specific and countable events the distribution is discrete and not continuous. In contrast to the rainfall example used to illustrate the normal distribution, fractional values are impossible; the outcome of a coin flipping trial can be only be one or zero heads, and not half or two-thirds of a head. The character of the binomial distribution and the way in which it is plotted takes this into account.

Although the distribution is discrete its form can still be described in mathematical terms. The equation that we use to accomplish this provides a series (determined by trial length) of individual probabilities from which a probability histogram can be constructed to represent the distribution. If the probability of X successes in a sequence of N trials is denoted by $p(X)$, then the binomial distribution is defined by:

$$p(X) = \binom{N}{X}(j^X)\,(1 - j)^{N-X} \tag{5.6}$$

Equation 5.6

$\binom{N}{X}$ = the combinatorial expression
j = probability of a 'success'
X = specified number of successes
N = sequence length (number of trials)

To understand how we might determine the nature of any binomial distribution we must proceed by simple steps. Of the quantities indicated in equation 5.6 only the 'combinatorial' expression requires any elaboration. In so far as the binomial distribution is concerned it is a method, using factorials (see section 2.6), for determining the number of combinations of X items that can be abstracted out of a total of N such items. In general terms:

$$\binom{N}{X} = \frac{N!}{X!(N - X)!} \tag{5.7}$$

If we wish to know how many different combinations there are of three successes in five trials we would proceed as follows: $N=5$, $X=3$ and so

$$\binom{N}{X} = \binom{5}{3} = \frac{5!}{3!(5-3)!} = \frac{120}{6(2)} = 10$$

From which we get the mildly surprising result that there are 10 possible combinations of 3 successes in 5 trials.

These principles may now be included within a simple example. What is the probability of obtaining 4 heads ($p(4)$) from a sequence of 6 flips of a coin? By substituting into equation 5.6 we get:

$$p(4) = \binom{6}{4} (0.5)^4 (1 - 0.5)^{6-4}$$

$$= \frac{720}{24(2)} (0.5)^4 (0.5)^2 = 15 \times 0.0625 \times 0.25$$

$$= 0.2344$$

Thus we find there to be a 0.2344 probability that in any sequence of 6 coin flips 4 will be heads. This, however, is only part of the picture. To complete the binomial distribution for a sequence of 6 and with a success probability of 0.5 we must obtain the probabilities for each of the possible number of heads from zero to six inclusive. Table 5.5 shows how we might evaluate this distribution using the methods outlined above. Figure 5.17 depicts the probability histogram of that same distribution.

Table 5.5 Derivation of probabilities on the binomial distribution

X	$\binom{N}{X}$	j^X	$(1-j)^{N-X}$	$p(X)$
0	1	1	0.0156	0.0156
1	6	0.5	0.0313	0.0938
2	15	0.25	0.0625	0.2344
3	20	0.125	0.125	0.3125
4	15	0.0625	0.25	0.2344
5	6	0.0313	0.5	0.0938
6	1	0.0156	1	0.0156

Note: 0! and anything raised to the power of 0 are both, by convention, 1.0.

5.14 Moments of the binomial distribution

Provided that the probability of a success remains 0.5, then the binomial distribution will retain the symmetry indicated in Figure 5.17 irrespective

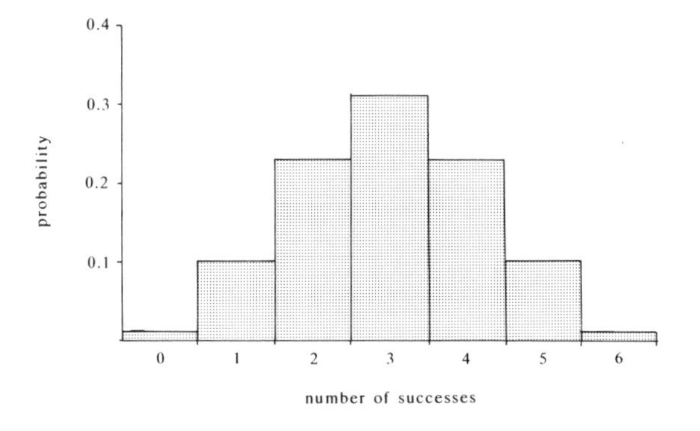

Figure 5.17 Histogram of probabilities on the binomial distribution (see Table 5.5).

of the trial length. Inequalities between the probabilities of a success (j) and a failure ($1-j$) will manifest themselves as distributional skewness. An example is provided by the probability of a rainfall day in Durham, England (see section 5.3). If a dry, rainless, day is defined as the success then j is 0.37. The probability of a failure is 0.63. Let us consider a trial length of three days. In that three-day period only four outcomes are possible, either 0, 1, 2 or 3 successes. Table 5.6 shows how the probability distribution can be calculated from equation 5.6. Figure 5.18 is a plot of

Table 5.6 Derivation of probabilities on the binomial distribution when the two components of the dichotomous variable are not equally probable

X	$\binom{N}{X}$	j^X	$(1-j)^{N-X}$	$p(X)$
0	1	1	0.2500	0.2500
1	3	0.37	0.3969	0.4406
2	3	0.1369	0.63	0.2587
3	1	0.0507	1	0.0507

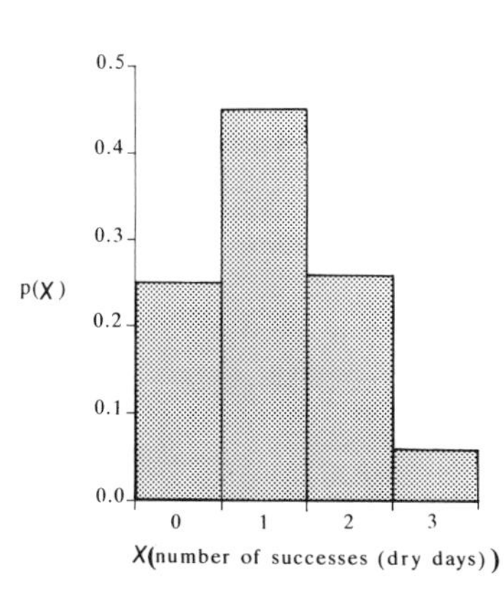

Figure 5.18 Histogram of rain day probabilities under the binomial distribution (see Table 5.6).

the probability histogram showing the degree of skewness that results from the inequality between the probabilities of a success and a failure.

These observations bring us conveniently to the next point, that of the moments of the binomial distribution. Just as the normal distribution can be summarised in terms of its central tendency (mean), statistical scatter (standard deviation) and symmetry (skewness), so too can the binomial. Their meaning and interpretation are much the same as in the normal case, but their derivation differs owing to the obvious implications of the different measurement scales.

The mean of a binomial distribution is best described as the long-term average number of successes in repeated sequences of trial length N. The position is analogous to that of the normal distribution in which the most probable events cluster about the mean. The derivation of the binomial mean is a simple matter and is found, of:

$$\mu = Nj \tag{5.8}$$

where N is the sequence length and j the probability of a success. A sequence of 6 flips of a coin has a distribution mean, and hence most probable outcome, of 3 successes, in this case heads from:

$$\mu = 6 \times 0.5 = 3$$

The situation, however, is not always so simple. Using the Durham rainfall example we find that the mean of the distribution for three-day sequences (taking a 'dry' day as a success) is:

$$\mu = 3 \times 0.37 = 1.11$$

How can we reconcile a fractional mean with a distribution composed of discrete, integer, numbers? To accommodate this seeming paradox it must be recalled that the mean is the theoretical average number of dry days for all three-day sequences. It is the point to which all events will tend to converge though, clearly, that point cannot be realised.

The variance, standard deviation and skewness of a binomial distribution are obtained with similar ease though, again, fractional indices may result. The equations are:

$$\text{Variance} = \sigma^2 = Nj(1 - j) \tag{5.9}$$

$$\text{Standard deviation} = \sigma = \sqrt{[Nj(1 - j)]} \tag{5.10}$$

$$\text{Skewness} = \alpha_3 = \frac{j - (1 - j)}{Nj(1 - j)} \tag{5.11}$$

The characteristics implicit in these measures of the binomial distribution are similar to their normal distribution counterparts. It would not, for example, be expected that every N-trial sequence would provide the same or, indeed, the most probable outcome – some degree of variation is inevitable. More specifically, any 6 flips of a coin will not always yield 3 heads. On different occasions, 2, 4, 1, 5 or even 0 or 6 heads might be registered. The standard deviation measures the events' tendency to disperse about their means. Skewness measures the degree of asymmetry in the distribution. In the case of binomial events it is highly sensitive to changes in the probability of a success (j). When $j = 0.5$, as in Figure 5.17, the distribution is symmetrical and the skewness is zero, but as soon as j departs from 0.5 asymmetry results.

As an alternative to calculating the probability distribution as we did in Table 5.6 we could again use the **PDF** subprogram described in section 5.7. But the subcommand **BINOMIAL** now replaces **NORMAL** and instead of mean and standard deviation the user must now specify the trial length (N) and the probability of a success (j). Figure 5.19 shows how to calculate the binomial distribution for ten flips ($N = 10$) of a coin, with a 'head' being taken as a success ($p = 0.5$). The subcommand prompt appears because we have concluded the previous line with a semi-colon. contrary to the algebraic conventions adopted in this text, MINITAB requires that we denote the probability of a success by **P**, though sequence length remains **N**. This subcommand line, the only one required by this option, must be terminated by a stop. Following its **RETURN** the results appear automatically. Notice that the pattern of probabilities is symmetrical and is beginning to reveal an approximation to the bell shape of the normal distribution; a point discussed in the following section.

```
MTB  >   PDF;
SUBC >   BINOMINAL N = 10, P = 0.5.

    BINOMINAL WITH N = 10 P = 0.500000
         K          P( X = K)
         0             0.0010
         1             0.0098
         2             0.0439
         3             0.1172
         4             0.2051
         5             0.2461
         6             0.2051
         7             0.1172
         8             0.0439
         9             0.0098
        10             0.0010
```

Figure 5.19 MINITAB screen display showing results using the PDF (BINOMIAL) commands.

5.15 The binomial distribution and continuous scale data

The binomial distribution is applicable only to dichotomous variables. Coin flipping undeniably falls into this category, as might the division of the human population into male and female categories. But dichotomous variables may also be derived from the twofold subdivision of continuous variables, particularly where such variables have within them threshold values either side of which different geographical consequences may be realised. Human behaviour, for example, may change abruptly at the threshold age of 65 when retirement occurs, and social geographers might regard age as dichotomous for some purposes although it is itself a continuous variable, frequently demonstrating a normal distribution. But firmer links exist with the continuous normal distribution. Consider the case for large N as it influences a binomial distribution. As the trial sequence increases in length the number of possible outcomes grows proportionally and the columnar histogram begins to resemble the probability curves of Figure 5.4. Provided that the probability of a success (j) does not differ greatly from 0.5, the histogram will tend towards a normal curve with surprising rapidity. As a working rule, if both Nj and $N(1-j)$ are greater than 5.0, then approximation to the normal form can be assumed. Hence, if $j=0.5$ all sequences of events in excess of 10 will be approximately normal in their distribution. For $j=0.3$ the sequence length must be 16 before normality can be assumed. Under these circumstances it is possible to express outcomes on the binomial distribution in terms of z-values by using the moments of the latter and equating them with those of the normal distribution. Thus:

$$z = \frac{\text{difference between individual and mean}}{\text{standard deviation}}$$

$$z = \frac{X - \mu}{\sigma} = \frac{X - Nj}{\sqrt{[Nj(1 - j)]}} \tag{5.12}$$

Equation 5.12

μ = normal mean
σ = normal standard deviation
Nj = binomial mean
$\sqrt{[Nj(1 - j)]}$ = binomial standard deviation
X = individual observation

This conversion allows us to express deviations from the binomial mean as a z-value with, importantly, all the attendant probabilistic properties outlined earlier in this chapter.

5.16 The Poisson distribution

This is the last of the three distributions to be introduced. Named after its French discoverer, it has proved to be a fruitful area for the study of geographical and spatially-organised data. We will come across it again in the concluding chapter of this book which deals with spatial statistics. It may, however, be introduced at this point as it has much in common with the binomial and normal distributions.

The Poisson distribution is most commonly applied to problems concerned with discrete events in space or time. While the binomial distribution can be used when probabilities of a success and of a failure are not vastly dissimilar, problems soon arise when we are dealing with unusual events for which the probabilities of successes (the event occurring) and failure (it not occurring) are enormously different. It is under these circumstances that the Poisson distribution might be used. This is nowhere more clearly the case than when dealing with 'point' events in space or in time, when the number of events can be counted but it may not be possible to contemplate the numbers of points where, or when, an event did not occur. For example, a shop location is a point event in space and it is clearly possible to count the number of such points in a given area. But, because that area is composed of an infinite number

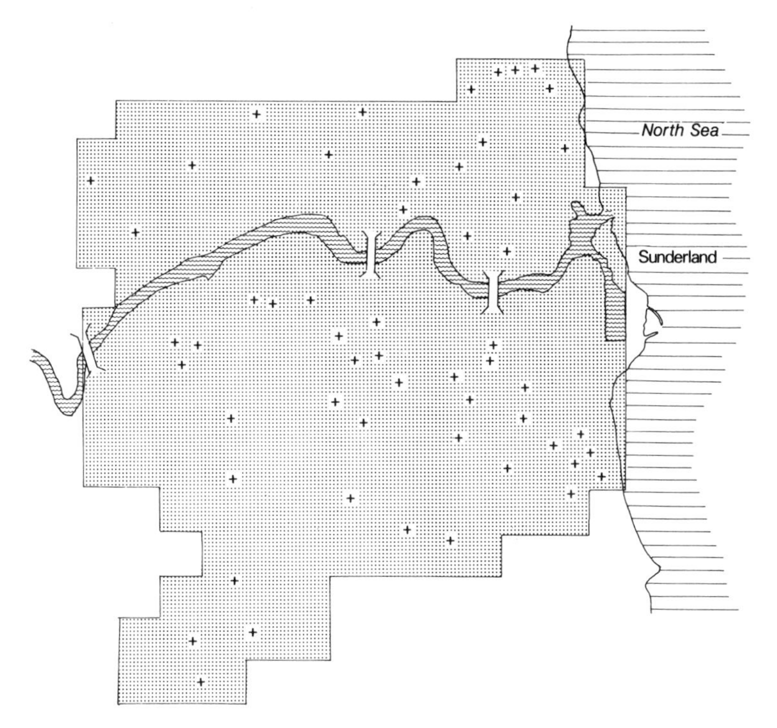

Figure 5.20 Distribution of grocers' shops in Sunderland, England. The shaded area shows the extent of the grid cover used for data abstraction.

of points it is impossible to estimate the number of locations where a shop did not occur. This eliminates any possibility of attaching empirical probabilities to the presence (success) and absence (failure) categories and, consequently, of making use of the binomial distribution. Stream junctions, factory locations, crime locations and a host of other geographical phenomena are of this type.

As an example we can study the location of grocers' shops in Sunderland, England. The first task is to delimit the study area. In this case the built-up area of the city was taken, but the issue is not always so conveniently resolved and it is important to note that the definition of the study area can influence the final outcome of the analysis. Figure 5.20 depicts both the study area, defined on an arbitrary grid square basis, and the location of grocers' shops within it. From this map the average density of points (grocers' shops) per grid square can be estimated by counting. This quantity, denoted by λ(lambda), is vital to the mathematical description of the Poisson distribution and regular use will be made of it in the calculations that follow. Equation 5.13 describes the Poisson distribution, and if it is remembered that each shop has now become a statistical event the distribution's form allows us to estimate the probability p of X events per grid square:

$$p(X) = \frac{\lambda^X}{X!} e^{-\lambda} \qquad (5.13)$$

Equation 5.13

λ = average density of events
e = the constant 2.7183
X = specific number of events

The number of events (grocers' shops) per grid square may vary between zero (a square with no shops) and a theoretical limit of infinity. In practice the distributions are evaluated only within realistic limits, which in this case is 6. But because only integer quantities of event X are possible the distribution is discrete and should be represented in histogram form and not as a continuous curve. The probabilities of 0, 1 and 2, etc. events have to be estimated separately using equation 5.13 but, once calculated, apply equally to all squares in the study area. The importance of λ can now be seen as all the derived probabilities depend upon it but, in turn, λ also depends upon the extent of the study area and the size and number of grid squares – all of which are at the discretion of the researcher.

In the current example there are 55 shops scattered over 135 grid squares, giving a λ value of exactly 0.4. With this information we can now calculate the probability for any desired number of shops in a single square. The probability of a square having two shops is:

$$p(2) = \frac{0.4^2}{2!} \times 2.7183^{-0.4}$$

$$= \frac{0.16}{2} \times 0.6703 = 0.0536$$

A more complete impression is obtained by evaluating the distribution up to $X=6$. The results are shown in Table 5.7 and in histogram form in Figure 5.21.

The very low probability of any square having 6 shops, with the even smaller probabilities for 7 or more, demonstrates why, in this example, no further calculations are needed. However, the range $X=0$ to X of infinity embraces the whole Poisson distribution, therefore $p(0)+p(1)+p(2) \ldots . p(\infty)=1.0$. As the sum of probabilities up to $p(6)$ is 0.99998 we deduce that the probability of any square having 7 or more shops is only $1-0.99998=0.00002$.

Table 5.7 Probabilities on the Poisson distribution for Sunderland shops per grid square

Shops per square X	$\frac{\lambda X}{X!}$	$e^{-\lambda}$	Probability $p(X)$
0	1	0.6703	0.6703
1	0.4	0.6703	0.2681
2	0.08	0.6703	0.0536
3	0.0106	0.6703	0.0072
4	0.00107	0.6703	0.00072
5	0.00009	0.6703	0.00006
6	0.0000007	0.6703	0.0000005

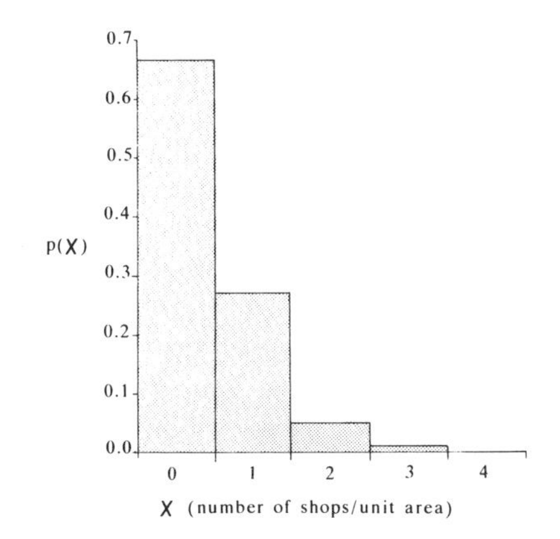

Figure 5.21 Probability histogram of Sunderland shops under the Poisson distribution (see Table 5.7).

5.17 Moments of the Poisson distribution

The Poisson distribution has its own measures of central tendency, dispersion and skewness. Their calculation is simple and based largely on the observed density of events (λ). In regard to such calculations a point, well-illustrated in the example used above, needs to made. The observed density of points depends very much on how we define the area of study and select the size of grid square. Sound geographical reasoning may govern the former, and in the current example observations were taken only within the city's built-up area. The choice of grid square will always be more arbitrary, yet upon it will depend the average density of events. With a finer grid the density of events will decrease; with a wider grid it will increase. Hence the choice should be made with care, though no firm rules can be applied to support this warning.

The moments of a Poisson distribution are found as follows:

$$\text{Mean} \quad (\mu) = \lambda \tag{5.14}$$

$$\text{Variance} \quad (\sigma^2) = \lambda \tag{5.15}$$

$$\text{Standard deviation} \quad (\sigma) = \sqrt{\lambda} \tag{5.16}$$

$$\text{Skewness} \quad (\alpha_3) = \frac{1}{\sqrt{\lambda}} \tag{5.17}$$

One of the most remarkable qualities of the Poisson distribution, almost its hallmark, is the equality of variance and mean. The remaining moments are simple transformations of this quantity. The current example uses points per unit area, but events may also be studied on the basis of points in time e.g. per year, day or hour. The mean represents the average density of events in either space or time and, as with the binomial distribution it may be a fractional and unrealisable quantity but one towards which events tend in the long-term. The average density of grocers' shops in Sunderland was 0.4 per square, clearly no single square can record this quantity which represents the overall trend. However the observed data set represented squares some of which contained no shops, some with one, two and up to four shops. The data are, in this sense, spread about the mean and the variance of the distribution measures this 'spread' of the data.

The inherent properties of the Poisson distribution dictate that measurable skewness is a common feature. Negative frequencies of events (X) are impossible, while probabilities usually tend to tail away with increasing X. These two characteristics combine to create distributional asymmetry. The skewness of the shops example is estimated from equation 5.17 as 1.58, and the asymmetric character of its distribution has already been shown in Figure 5.21.

This discussion of the skewness brings us conveniently to the links

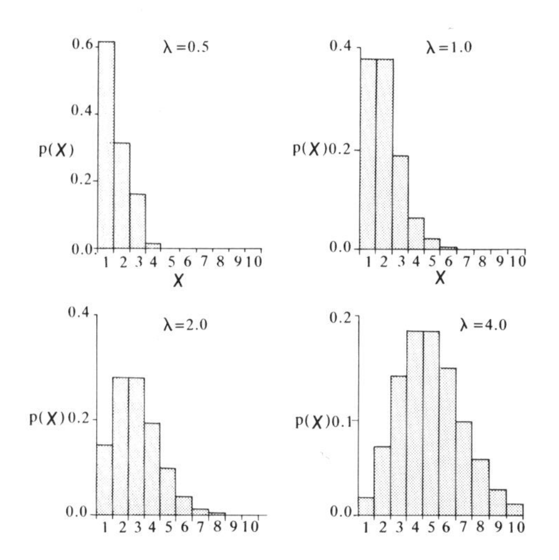

Figure 5.22 The effect of increasing means (λ) on the Poisson distribution.

which exist between the Poisson and normal distributions. As the density of events (λ) increases the Poisson distribution tends to become more symmetrical and to approximate to the normal distribution. This transformation is illustrated in Figure 5.22 and when it occurs deviations about the Poisson mean can be expressed as z-values. Thus, if the mean and standard deviation of the Poisson distribution are known, equation 5.3 can be rewritten as follows:

$$z = \frac{X - \mu}{\sigma} = \frac{X - \lambda}{\sqrt{\lambda}} \tag{5.18}$$

Equation 5.18		
μ	=	normal mean
σ	=	normal standard deviation
λ	=	Poisson mean
$\sqrt{\lambda}$	=	Poisson standard deviation
X	=	individual observation

The calculation of a Poisson distribution can be a time-consuming task, but one which is made easier if we take advantage of a further subcommand in MINITAB's **PDF** subprogram (see also sections 5.7 and 5.14). Because of the characteristic equality of mean and variance this MINITAB option requires only the former in order to complete the task. Taking the example of Sunderland shops, with a mean of 0.4 shops/square, Figure 5.23 shows how the commands are arranged. The values **SET** into the first column **(C1)** determine how far along the distribution, starting at zero, we wish to estimate the probabilities. In this

```
MTB  >   SET C1

DATA >   0 1 2 3 4 5 6
DATA >   END
MTB  >   PDF C1;
SUBC >   POISSON 0.4.
         K           P( X = K)
      0.00              0.6703
      1.00              0.2681
      2.00              0.0536
      3.00              0.0072
      4.00              0.0007
      5.00              0.0001
      6.00              0.0000
```

Figure 5.23 MINITAB screen display of the PDF (POISSON) command.

case we need go only as far as 6 (shops/square), beyond which the probabilities, as we have seen, become remote. The keyword **POISSON** must be followed by the distribution's mean and a stop. The results are listed automatically and are, naturally, the same as those found in Table 5.7 but have been established with far less effort and risk of error.

```
MTB  >   CDF;
SUBC >   POISSON 0.4.

      POISSON WITH MEAN =      0.400
         K  P( X LESS OR = K)
         0              0.6703
         1              0.9384
         2              0.9921
         3              0.9992
         4              0.9999
         5              1.0000
```

Figure 5.24 MINITAB screen display using the CDF (POISSON) command.

The use of another subprogram, that of **CDF** (cumulative density function), can provide the cumulative probabilities along the distribution up to the point where all but the most unlikely of events are included (Figure 5.24). The layout is similar to that for **PDF** and again requires only the Poisson mean. The default listing will give all occurrences up to a cumulative probability of 0.99995 (rounded up to 1.0000 on the print-out).

5.18 Probability and expectation

The three distributions introduced in this chapter, the normal, the binomial and the Poisson, are all clearly defined in quantitative terms. They are therefore quite different to the arbitrary distributions of nominal data described in section 4.3. In the latter instances observed frequencies were used from which probabilities could be inferred. Now, however, we have travelled full circle and can use the theoretical distributions and their properties to estimate 'expected' frequencies. The case of Sunderland shops is a good example. The probabilities of 0, 1, 2, etc. shops per grid square are given in Table 5.7. These values can be easily converted into expected frequencies for a perfect Poisson distribution by multiplying the $p(X)$ values by the number of squares (N) in the study, in this case 135. The result is known as the expectation E of the event X. In general:

$$E(X) = N \times p(X) \tag{5.19}$$

For example, the expected frequency of squares with no shops is the product of the total number of squares and the probability for $X=0$:

$$E(0) = 135 \times 0.6703 = 90.5$$

In this simple example the observed and expected numbers of squares with no shops do not differ by much (96 as opposed to 90.5) and might be expected from random variation from a perfect Poisson distribution.

But why might observed frequencies not follow the theoretical patterns? To answer this question we need to emphasise a point hitherto overlooked. The normal, binomial and the Poisson distributions apply only to *random* variables consisting of independent events scattered randomly either side of the mean. No external, non-random, forces are acting to predetermine that events fall within a specific ranges or are excluded from others. Hence the independence of events is a most important question and researchers assume that one event cannot influence, and hence predetermine, any other event of the same variable. To take the simplest example, because a head results from the flip of a coin it does not dictate that a head is more or less likely with the next flip. All such events are independent and chance can be thought of as having no memory. Similarly, the Poisson distribution assumes that the point events are independent of one another and located randomly in space or in time. Many phenomena are indeed independent in this respect, but some are not. The scatter of residences of those suffering from an outbreak of a contagious disease is likely to be clustered and highly non-random because one event (the home of a person with the disease) is likely to be close to another. As a result, independence of the event is lost and we can no longer assume that all points in the space of the study area are equally available and those closer to existing events are now more probable locations for further events. It is when these conditions of independence and randomness are not fulfilled that differences between the theoretical and the observed distribution arise. Shops locations are a good example of this tendency and some retail functions, though not it appears grocers, tend to cluster together more than others. Those that are more clustered will have distributions that depart more significantly from the perfect Poisson form. Chapters 7 and 13 will examine some of the ways in which we can objectively compare these differences and determine when spatial distributions are random and when they are clustered.

Chapter 6

Samples and Populations

6.1 Introduction

The reader should now be aware of the important distinctions between statistical populations and samples drawn from them. Chapter 3 introduced the theory of sampling and how samples can be designed to accommodate the needs of the geographer. Chapter 5 has drawn attention to the fact that samples can be used to interpret the properties of the original populations. Samples and sampling are an important part of geographic research and study and much has been written on them, for example, Berry and Baker (1968), Court (1972) and Dixon and Leach (1975) demonstrate the range of interests and views.

A wide variety of symbols have been adopted for use in statistics and a distinction must be made between those used to indicate population parameters and those used to describe the properties of the samples. When discussing population parameters it is common practice to use Greek symbols. Thus the mean is denoted by μ and the standard deviation by σ. Latin symbols are often preferred when sample data are used. In this latter case the mean is given by X and the standard deviation by s. Table 6.1 summarises the most commonly used of these symbols. It should be noted that the size of finite populations is often referred to by N but sample sizes by n, though it might be observed that this is the least consistently applied of the algebraic conventions.

6.2 Estimates of reliability: standard errors

Samples do not necessarily reflect the precise characteristics of the populations from which they are drawn and even the most well-designed of sampling frames cannot guarantee to do so. Any sample is capable of misrepresenting the nature of its population. Although our knowledge of a population may be vague and limited by the necessity of examining it only through the medium of a sample we need not be equally vague about the reliability of sample estimates and methods exist with which to determine their accuracy.

Our initial interest is with samples containing 60 or more observations

Table 6.1 Symbols used for describing the characteristics of samples and populations

Parameter	Population symbol	Sample symbol
Mean	μ	$\bar{X}$
Standard deviation	σ	s
Variance	σ^2	s^2
Skewness	α_3	α_3
Number of observations	N	n

per variable. Such samples are said to be *large* (though some texts take smaller sample sizes as the threshold, even down to 30 in some cases) and the so-called *central limit theorem* can be applied to them. The principle of the central limit theorem is best explained as follows: if all possible samples of size n were drawn from a population the distribution of the derived sample means would be normal. Suppose, for example, we were examining rainfall variations across a geographic region. The population of possible point annual rainfalls would be infinite, but we could sample at, say, 100 points from that population to obtain a sample mean which we hope would be close to the population mean. If we could then sample from a further hundred points, and then from yet other sets of 100 points in the region the sample means would not all be the same, some would be larger and some would be smaller. If we plotted the distribution of the sample means (not the original observations from which they were derived) we would find that they plot as a normal distribution. Furthermore this normality in the *sampling distribution* of the sample means can be assumed to prevail irrespective of the distribution of the population. Hence a population with a skewed distribution would still provide a normally distributed set of sample means. In all cases the mean of the sampling distribution ($\mu_{\bar{X}}$) is coincident with the mean of the population (μ), expressed algebraically:

$$\mu = \mu_{\bar{X}} \tag{6.1}$$

For infinite populations we could never know either of the quantities with absolute certainty.

These principles can be applied if we turn to the example of daily discharge data from the River Clyde used in section 5.11. We know the data to be skewed and its population distribution is shown in Figure 6.1. The figure also shows the distribution of the means of samples of 60 observations but these are normally distributed.

Under all but the most unusual of circumstances we would take only one sample from a population. We might therefore imagine ourselves to be in a poor position to examine the sampling distribution of means. This

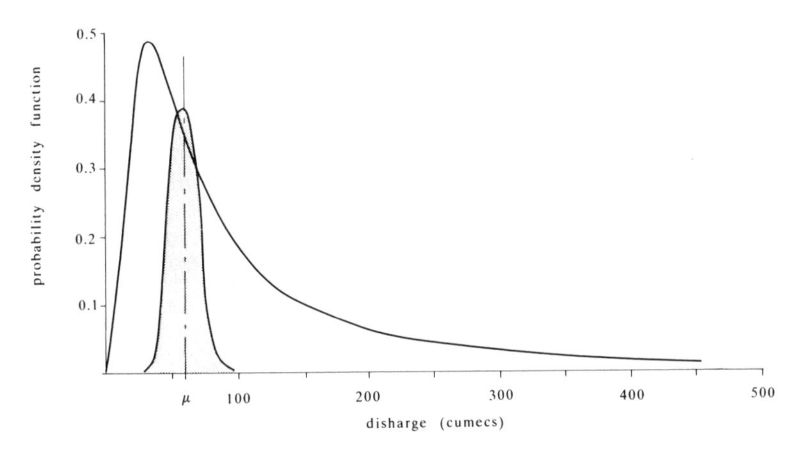

Figure 6.1 Distribution of daily discharge data for the River Clyde, Scotland. The superimposed (shaded) curve shows the sampling distribution of means of sample size $n = 60$.

is not the case, and even with one sample the standard deviation (in the present context more properly termed the *standard error*) of the sampling distribution can be calculated. Clearly such standard errors will vary from case to case but, importantly, they share all those properties of the standard deviation introduced in Chapter 5. The standard error of sample means is easily derived from equation 6.2.

$$\sigma_{\bar{X}} = \frac{s}{\sqrt{n}} \tag{6.2}$$

> Equation 6.2
>
> $\sigma_{\bar{X}}$ = standard error of sample means
> s = standard deviation of sample
> n = sample size

In this way we can determine the degree to which any single sample mean is likely to depart from the mean of the population since sample means are normally distributed about the population mean. Remember that repetitive sampling will not always produce the same sample mean. The degree to which they will vary depends largely on the size of sample – as equation 6.1 makes clear. Estimates are less subject to error and variation as sample sizes increase.

6.3 Confidence limits and standard errors

It follows from what we know of the normal distribution (see section 5.10), and the fact that it applies equally to sampling distributions of large samples, that the probability of a given sample mean lying within specified limits about the population mean can be established from z tables. For example there is a 0.683 probability that the sample mean will be within one standard error (standard deviation) of the population mean, and a 0.954 probability that it will be within two standard errors of the population mean. There is however one major problem – we do not know the value of the population mean. For large, and certainly for infinite populations, it can never be assessed with precision. This failing compels us to readjust our thinking and turn the argument on its head by transposing the roles of sample and population means when making such probability statements. We can sensibly argue that there is now a 0.683 probability of a population mean being within one standard error of the sample mean, and a 0.954 probability of it being within two standard errors. The principle is illustrated in Figure 6.2.

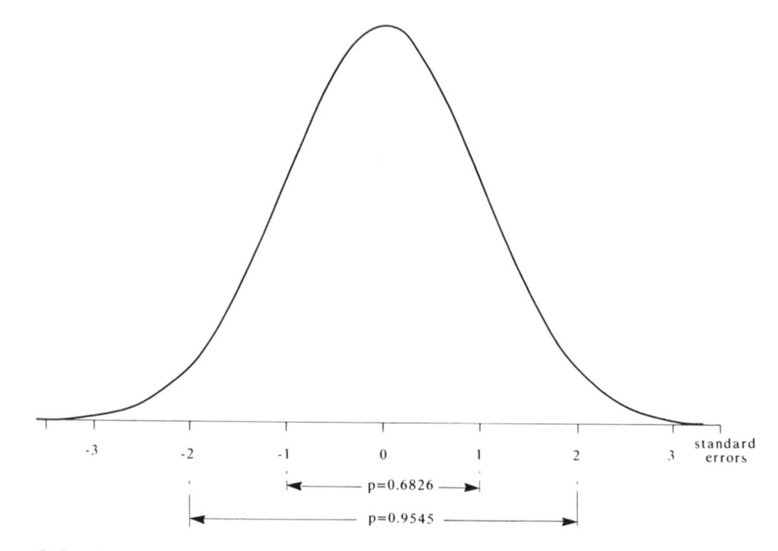

Figure 6.2 Normal distribution with the probability of events within one and within two standard errors of the mean.

The data for the River Clyde show how these procedures might be applied. A random sample of 60 observations gave a mean of 60.99 cumecs and a standard deviation of 9.38 cumecs. How reliable is the estimate of the mean? To answer this question we first evaluate equation 6.2 to give the standard error of the sampling distribution:

$$\sigma_{\bar{X}} = \frac{9.38}{\sqrt{60}} = 1.211$$

We can now conclude that there is a 0.683 probability of the population mean lying within 1.211 cumecs either side of the sample mean. The numerical values associated with those limits can be determined by:

$$\text{upper limit} = 60.99 + 1.211 = 62.20 \text{ cumecs}$$

and

$$\text{lower limit} = 60.99 - 1.211 = 59.78 \text{ cumecs}$$

The limits for the 0.954 (2 standard errors) probability range are given by:

$$60.99 \pm 2 \times 1.211 = 63.41 \text{ and } 58.57 \text{ cumecs}$$

These values are known *confidence limits* and they enclose a range known as the *confidence interval*. Traditionally these limits are described in percentage terms rather than proportions and we might refer to them as the 68.2 per cent and 95.4 per cent confidence limits. But the properties of the normal distribution are well understood and we are by no means restricted to such cumbersome limits. Although we could select any such limits to suit our purposes the 95 per cent and the 99 per cent are frequently used in preference to any others. Whatever limits might be selected they can be expressed algebraically by equation 6.3.

$$\bar{X} \pm z_c \sigma_{\bar{X}} \tag{6.3}$$

Equation 6.3

$\bar{X}$ = sample mean of variable X
z_c = critical z value for selected level
$\sigma_{\bar{X}}$ = standard error of sample mean

It will be recalled that the critical z-scores that enclose 0.95 and the 0.99 probability ranges are ±1.96 and ±2.58 respectively. With this information we can estimate the 95 per cent and the 99 per cent confidence limits for the sample mean discharge of the River Clyde. The 95 per cent limits are given by:

$$60.99 \pm 1.96 \times 1.211 = 63.36 \text{ and } 58.62 \text{ cumecs}$$

and the 99 per cent limits by:

$$60.99 \pm 2.58 \times 1.211 = 64.11 \text{ and } 57.87 \text{ cumecs}$$

From the results we can see that greater confidence of the population mean being within a specified range can be gained only at the expense of widening the confidence intervals. The only way in which the confidence bands can be narrowed without prejudice to the reliability of our conclusions is by increasing the sample size. Here, then, is a substantive justification for the intuitive notion that larger samples may be more representative than smaller samples of their background populations.

6.4 Samples and standard deviations

The mean is frequently the first point of interest in a statistical analysis, but rarely the only one. The standard deviation is also important in providing a clear expression of the behaviour of the study variable. Fortunately assessments can also be made of the degree of uncertainty attached to sample estimates of population standard deviations as the central limit theorem is again applicable. Hence, if all possible samples of size n were drawn from a population the distribution of the sample standard deviations would be normal. The distribution of the background population does not influence the normality of the sampling distribution, the standard error of which is given by:

$$\sigma_s = \frac{s}{\sqrt{(2n)}} \tag{6.4}$$

Equation 6.4	
s =	sample standard deviation
n =	sample size
σ_s =	standard error of sample standard deviation

As the sampling distribution of standard deviations is normal, the confidence limits may again be estimated. The strategy is identical to that used for sample means and can be generalised as:

$$s \pm z_c \sigma_s \tag{6.5}$$

It only remains for the researcher to select the confidence level.

In a study of loess and lacustrine deposits by Campbell (1979) a total of 369 field measurements were made of the percentage of sand in the surface sediments. The variability of sand content across the study area was measured by the standard deviation of sand content, which in this case was 0.78 per cent. But how reliable is this estimate of sand content variability? To answer this question we must first calculate the standard error of the sampling distribution using equation 6.4. By substitution we get:

$$\sigma_s = \frac{0.78}{\sqrt{(2 \times 369)}} = 0.029$$

This standard error can now be used to estimate the 95 per cent confidence limits as follows:

$$0.78 \pm 1.96 \times 0.029 = 0.78 \pm 0.0568$$

$$= 0.723 \text{ to } 0.837$$

Hence we can be 95 per cent certain that the true standard deviation of the sand content percentage is between 0.723 and 0.837. The narrowness of this confidence interval and the accuracy of the sample estimate is principally attributable to the large sample size.

6.5 Samples and non-parametric data

Thus far we have concerned ourselves only with samples measured on the interval and ratio scales. There are, however, many occasions when geographers will be working with samples of nominal data. Fortunately, the central limit theorem can again come to our assistance.

Proportions and non-parametric data generally are widely used in human geography, particularly when analysing questionnaire survey responses. The UK Government's decadal census is a good example. It can take several years before a complete census can be analysed and the results for a preliminary 10 per cent sample are usually published within a few years of the census being completed. Let us consider the case of employment structure. Each individual can be allocated to an employment class. Our interest might be in the proportion of the 10 per cent sample that are classified as unemployed. How reliable is that figure as an estimate of the proportion of the whole population that are unemployed? If we were to take a very large number of such 10 per cent samples we would discover that the sampling distribution of the proportions, like those of sample means and standard deviations, is normal. Furthermore, using equation 6.6 we can estimate the standard error of the sampling distribution of proportions. Let p be the proportion estimated from a sample of size n: then the standard error of the sampling distribution of proportions (σ_p) is given by:

$$\sigma_p = \sqrt{\left[\frac{p(1-p)}{n}\right]} \tag{6.6}$$

This standard error again possesses all the characteristics of the normal distribution and it can be used to determine confidence limits for sample proportions. In similar fashion to equations 6.3 and 6.5 these limits can be written as:

$$p \pm z_c \sigma_p \tag{6.7}$$

Let us take an example from the published 10 per cent sample of the 1981 UK census. The report for Great Britain (HMSO, 1984) showed that of the 136,771 people in Wales that might be economically active 15,596 were unemployed, giving an proportion of 15596/136771 = 0.114. Our problem is to determine the reliability of such an estimated proportion. The first step is to calculate the standard error of the sampling distribution from equation 6.6.

$$\sqrt{\left[\frac{0.114\ (1-0.114)}{136771}\right]} = 0.00086$$

From which we can estimate the 95 per cent confidence limits of the sample proportion. Remembering that the critical values of z for the 95 per cent limits are ±1.96, we use equation 6.7 as follows:

$$0.114 \pm 1.96 \times 0.00086 = 0.114 \pm 0.0017$$

In other words we can be 95 per cent certain that the population proportion of unemployed in Wales was between 0.1123 and 0.1157 of the economically active workforce. Here again the very large sample size gives us good grounds for confidence in the sample's statistical properties.

6.6 Small samples

In the above examples we saw how large samples reduce the error margins in estimating population parameters. But caution is always required when using samples, more so when the sample size is less than 60. Such *small samples* can be examined using procedures which take account of their peculiarities. Large samples (n of over 60) provide sampling distributions that are normal irrespective of the character of the background population. Where n is less than 60 the sampling distributions from normal populations are themselves no longer normal and will instead follow what is known as Student's t-distribution. Where background populations are known or suspected to be non-normal small samples should, if possible, be avoided.

The t-distribution is not unlike the normal. It is continuous and it extends from plus to minus infinity. Although it is also bell-shaped its precise form varies according to sample size and in this important respect it differs from the consistent character of the normal distribution. For very small samples the distribution forms an attenuated bell-shape, but as sample size increases the shape approximates ever more closely to the normal form. Coincidence is usually assumed when sample sizes exceed

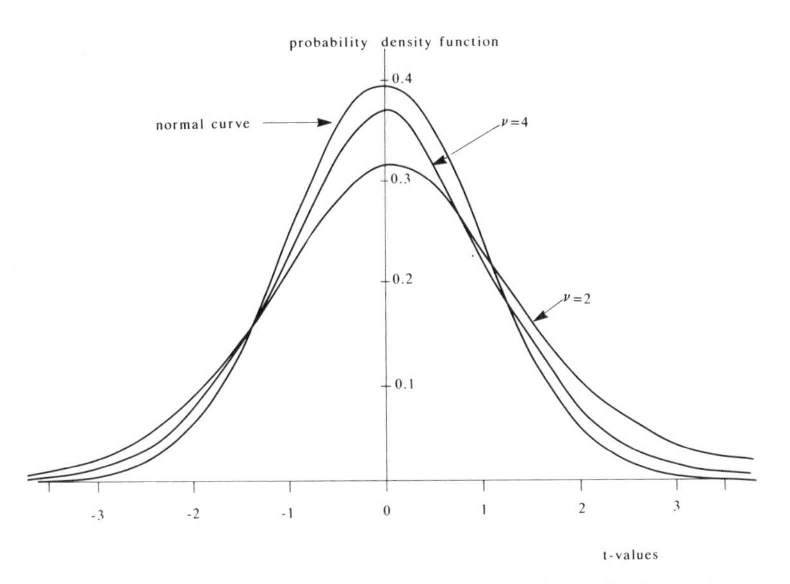

Figure 6.3 The normal and t-distributions showing how the latter becomes increasingly attenuated with decreasing sample size and degrees of freedom (υ).

60. Figure 6.3 summarises the variable character of the t-distribution. This distribution was discovered in the 1930s by an industrial statistician, William Gosset, who published his findings under the pseudonym 'Student'. The variability of the distribution that Gosset discovered presents no difficulties because its form is mathematically definable and t-tables, not altogether dissimilar to z-tables, can be prepared to show the probability of events falling within specified ranges of the distribution. In the case of the normal distribution there is a 0.95 probability of events falling within the range $z = -1.96$ to $z = +1.96$. To find the limits of the t-distribution that delimit the same probability range we can use t-tables (see Appendix II). Now, however, attention must also be given to sample size, this was not the case when using z-tables and the layout of Appendix II requires a little explanation. The critical t-values are arranged by rows according to *degrees of freedom*, a quantity which, in this case, is one less than the sample size and is conventionally denoted by the Greek symbol υ(nu). For the moment we will confine our attention to the question of 'confidence limits' as we have already defined them. The use of these tables for 'significance levels' is discussed in Chapter 7. The columns of the table are arranged according to selected probabilities only. For example, the second column shows how the critical t-values defining the 0.95 (95 per cent) confidence limits change with degrees of freedom. The fourth column gives the critical values for the 0.99 (99 per cent) confidence limits. The critical values for the 0.90 (90 per cent) and the 0.999 (99.9 per cent) limits are also given. The distribution is always symmetrical and, as with the normal curve, the critical limits are defined

by the positive and the negative values of the tabled figures. The z-scores on the normal distribution are given at the foot of each column to show the comparison between the two distributions and how the t-distribution converges towards the former. A simple example will demonstrate the application of the t-distribution.

Lam (1978) studied soil characteristics in the Hong Kong area. A sample of just 20 observations gave a mean soil acidity level of pH 4.75. For such a small sample the t-distribution must be used. The standard error of the sampling distribution can be estimated, as before, from equation 6.2. This gives:

$$= \frac{4.75}{\sqrt{20}} = 1.06$$

With this information we can determine the ranges within which the population mean lies with, say, 95 per cent confidence. The sample size is 20, which gives 19 degrees of freedom in this case. From Appendix II we see that the critical t values that accord with these requirements are ± 2.09. The 95 per cent confidence limits are thus:

$$4.75 + 2.09 \times 1.06 = 6.97$$

and

$$4.75 - 2.09 \times 1.06 = 2.53$$

Expressed in everyday terms, we can be 95 per cent confident that the population mean of soil acidity lies between pH 2.53 and pH 6.97. These are very wide limits and proportionally far wider than in the examples given earlier for some very large samples. Had we selected the 99 per cent confidence limits the band would have opened yet further. The critical t-value is 2.86, which would have given confidence limits of:

$$4.75 \pm 2.86 \times 1.06 = 7.78 \text{ and } 1.72 \text{ pH}$$

In all cases it can be readily appreciated that the critical t-values are larger, creating wider confidence bands, than their normal counterparts.

Some concluding comments on the question of degrees of freedom are necessary. The principles of this concept are demonstrated if we consider a sample of ten numbers which add to, say, 17.5. This total is fixed for that sample of data. Of the ten constituent quantities however nine are free to assume any value whatsoever. But in order for the total of 17.5 to be preserved the tenth quantity is fixed by the other nine and we say that one degree of freedom has been lost. As we will see in the following chapter, there are many distributions that depend upon degrees of freedom and in some cases more than one degree can be lost from any given sample. But we will consider these matters as they arise.

More generally, degrees of freedom are often preferred to sample size as they reduce unwanted bias in sample estimates. Small sample variances, for example, are usually underestimates of the population parameters. This difficulty can be overcome by dividing the sum of the squared deviations (see section 4.7) by the sample size minus one ($n - 1$) instead of n. This adjustment provides what is termed the *best estimate* of the population variance. Notice that division by the smaller 'degrees of freedom' gives a larger variance estimate than would otherwise have been the case. A useful qualitative expression, and one that will be used in later sections, is:

$$\text{variance} = \frac{\text{sum of squares}}{\text{degrees of freedom}}$$

In strict algebraic form this expression is represented by:

$$\hat{\sigma}^2 = \frac{\Sigma(X - X)^2}{n - 1} \qquad (6.8)$$

Equation 6.8

X = individual observation of X
$\bar{X}$ = sample mean of variable X
n = sample size
$\hat{\sigma}^2$ = best estimate of variance

Notice the use of the 'hat' over the variance symbol to denote the *best estimate* aspect of the quantity. Both the MINITAB and the SPSS packages estimate variance on the basis of degrees of freedom rather than sample size. Clearly where n is large there is little difference between the results of the two methods.

6.7 Samples and standard errors on MINITAB and SPSS

Calculations of the type described above are not lengthy unless several variables are being studied but computational assistance can be called upon. Section 4.9 has already described how the MINITAB's **DESCRIBE** and the SPSS's **FREQUENCIES** subprograms can provide estimates of means and standard deviations. The former also lists by default the standard error of the estimate of the mean (**SEMEAN** in Figure 4.8). The SPSS **FREQUENCIES** subprogram has also been described in which the option **STATISTICS** can be invoked to list the

```
MTB  >   SET C1
DATA >   507.8 664.6 586.1 778.8 851.7
DATA >   687.3 771.4 763.8 633.2 907.0
DATA >   END
MTB  >   TINTERVAL 95 PERCENT C1

            N      MEAN    STDEV  SE MEAN      95.0 PERCENT C.I.
C1         10     715.2    122.4     38.7     (   627.6,    802.7)
```

Figure 6.4 MINITAB screen display if the TINTERVAL command. This example used the SET instruction to input ten observations from the keyboard into column one (Cl).

means and standard deviations of all specified variables (Figure 4.9). Additional keywords can be added to the instructional list in order to obtain the standard error of the sample mean (keyword: **SEMEAN**), and the standard errors of the sample skewness **(SESKEW)** and kurtosis **(SEKURT)**.

The MINITAB subprogram **TINTERVAL** goes further and provides confidence limits for sample means. The **RETRIEVE**, **READ** or **SET** commands can be used to make a data file active or to set up the raw data. The **TINTERVAL** command must then be accompanied by the required percentage confidence interval, in the manner indicated in Figure 6.4, to be calculated and the columns or column of data for which the intervals are required. Figure 6.4 illustrates a typical example for a small data set of ten observations input at the time of analysis using the **SET** command. The output lists the sample size, the mean and standard deviation of the sample, the standard error of the mean (derived using equation 6.2) and the confidence limits of the mean at the specified percentage level which in this case is 95 per cent. The confidence limits, listed under **95.0 PERCENT C. I.**, are inevitably wide for such a small sample. The program uses the *t*-distribution to estimate the confidence intervals. Hence it is equally applicable to both large and to small samples as the *t*- and the normal distributions are identical for observations in excess of 60 in number.

References

Berry, B.J.L. and Baker, A.M. (1968) 'Geographic sampling' in B.J.L. Berry and D.F. Marble (eds) *Spatial Analysis*, Prentice-Hall, Englewood Cliffs.

Campbell, J.B. (1979) 'Spatial variability in soils', *Ann. Ass. Am. Geogrs.*, 69, 544–556.

Court, A. (1972) 'All statistical populations are estimated from samples', *Prof. Geogr.*, 24, 160–162.

Dixon, C. and Leach, B. (1975) *Sampling Methods for Geographical Research*, Catmog 17, Geo Abstracts, Norwich.

Lam, K.C. (1978) 'Soil erosion, suspended sediment and sediment production in three Hong Kong catchments', *J. Trop. Geog.*, 47, 51–62.

Chapter 7

Testing Hypotheses with One or More Samples

7.1 Introduction

All geographers form general impressions in the course of their work that are the beginnings of ideas and hypotheses. We might observe that some forms of retail function cluster more readily than others, or that steep slopes are associated with certain types of rock and shallower slopes with others. But whatever the nature of our subjective impressions we would always want to know if they are correct or incorrect. We would wish to 'test' our hypothesis against reality.

The first step in assessing the validity of any hypothesis is that of measurement, and here we find also the first of our problems. For while some phenomena can be measured on one or more scales (nominal, ordinal, interval or ratio), others have no agreed form of measurement or may be measured on only one of the scales. Slopes may be measured by degrees of inclination (ratio scale), rock hardness on the Moh's scale (ordinal scale) and rock type can be allocated to a lithological type (nominal scale). But how might we measure aspects of human behaviour such as environmental perception or class attitude? We are, quite simply, not always able to measure exactly the variables in which we are interested. Even worse, we may be unsure of their definition. Consequently, a distinction has to be drawn between theoretical and operational definitions or concepts. For example, a geomorphologist might be interested in hillside morphology, hence his theoretical concept is that of hillside shape. In operational terms, however, a more precise and pragmatic definition is required. What exactly is meant by hillside morphology? Is it overall slope, the slope of the steepest section, the variability in the slope or its degree of curvature? Indeed, how are we to define and demarcate the length and section of hillside to be studied? Such problems are so commonplace in geography that we often overlook them. But in statistical methods this cannot be done and the difficulties can be resolved only by clearly defining the problem and the terms of reference at the outset.

The substitution of operational for theoretical concepts is similar to translating from one language to another, in which care must be taken not to lose the sense of the statement. But, with a satisfactory operational definition, speculations and hypotheses can be expressed in a testable and examinable form. The term 'test' is here used in the specific context of statistical tests, the study of which occupies this and many of the subsequent chapters.

7.2 The null hypothesis

Statistical tests require numerical data collected within a rigorous sampling framework and drawn from a specified population. During the process of gathering data the researcher must take every precaution to ensure that the samples are unbiased and are not unrepresentative of their respective populations. No matter how reliable or sophisticated a test may be it cannot produce reliable results from unreliable data.

In this chapter attention will be confined largely to *univariate* tests, i.e. those tests that analyse data for one variable only; though there may be more than one sample of such data if, for example, two groups are to be compared. Later chapters will explain how pairs or groups of variables can be tested. These are the *bivariate* and the *multivariate* tests. Though all tests differ in their procedures and characteristics they share the one principal function of testing hypotheses formulated by the researcher.

The hypothesis to be tested however is not a general geographical statement but a *null hypothesis* designated as H_0. The null hypothesis is indispensable in statistical testing and is often one of 'no difference'. For example, that there is no difference between the mean slopes on two contrasting geological strata, or that there is no difference between the degree of clustering of two retail functions. In some cases 'real' data might be compared with a hypothesised set. For example we might hypothesise that there is no difference between an observed and a purely random scatter of retail functions. Such statements of 'no difference' may well run counter to our understanding of the problem but it is important that it is always stated in this manner. The null hypothesis is therefore fundamentally different to the general or working hypothesis on which the research may be based. The statistical test results in the acceptance or rejection of the null hypothesis. If it is rejected then the alternative hypothesis (H_1) must be accepted. The alternative hypothesis is the logical converse of the null hypothesis and asserts that very real differences exist between the two or more samples under study. Hence we may distinguish three forms of hypothesis:

1. The research hypothesis – a general, perhaps lengthy, statement of the factors governing the behaviour of the variable in question;
2. The null hypothesis of no difference (H_0);
3. The alternative hypothesis (H_1).

The latter two are statistical statements only. The null hypothesis is the testable expression of the working hypothesis, and whether we believe it to be true or not is immaterial. It can be thought of as existing only within the setting of the statistical test within which it will be either accepted or rejected.

Statistical tests are invaluable in research but we should see them as a means of helping to clarify our understanding, they are not an end in themselves. It is clear that all three forms of hypothesis are linked and the fortunes of H_0 determine how we view the working hypothesis. Perhaps a cherished working hypothesis fails to stand up to objective scrutiny, or a seemingly weak argument is sustained by close analysis. But the real challenge in research is in explaining and understanding the geographical processes underlying the rejection or acceptance of H_0.

7.3 An introduction to hypothesis testing: the *Z*-test

As stated above the strategy for all statistical tests is the same and requires the formulation and testing of a null hypothesis. Every test produces a final *test statistic*, the numerical value of which determines whether H_0 is accepted or rejected. Sometimes the test statistic is in the form of a z- or t-value the probability distributions of which we have already studied. But many tests produce their own particular statistics and these too have probability distributions that are known and defined. From our knowledge of each distribution we can determine the random probability of any test statistic being equalled or exceeded. Those test statistics that lie towards the extreme ends of the distributions will have only remote probability of occurring by chance. The less extreme test statistics are far more likely to have arisen from random variation within the distribution. We have already shown, for example, that low z-values on the normal distribution are more probable than those found towards its extremities. In general, when a test statistic, of whatever form, is found to be improbable H_0 is rejected, and when it has a greater probability H_0 it is accepted. Clearly we must decide upon a boundary probability beyond which H_0's chances of being correct are so remote that it must be rejected. Equally, much will depend upon our choice of the critical probability that determines our decision to accept or to reject the null hypothesis. This critical probability must always be determined before the test begins. In terms of the distribution of the test statistic this defines and delimits the *rejection region* of H_0.

These principles can be understood from the following example. In section 4.4 we looked at ways of summarising county-based data from England and Wales where the death rate (per 1000 population) had an average of 11.59 and a standard deviation of 1.447. Suppose we now consider the death rates for the ten Scottish regions (administratively equivalent to the English counties). Looking at the raw data we suspect that the English and Scottish death rates do indeed differ; this simple and subjective observation is the basis of our working hypothesis. But to test this proposition it must be expressed in terms of the null hypothesis, i.e. that there is no significant difference between the mean death rates in Scotland and the rest of Great Britain. The alternative hypothesis is that the difference between the two means of the death rates is too large to be explained by random variation within a common statistical population.

The null hypothesis can be tested using an application of the sampling distribution of means introduced in section 6.2. In this case the population mean and standard deviation are known as we have based the figures on all the British counties. The death rates in the ten Scottish regions can be regarded as a sample which can be compared with the British population. The difference between the population mean (11.59 deaths/1000) and the Scottish sample mean (13.04 deaths/1000) can be expressed as a z-value on the normal distribution of county-based death rates. If the difference is great, the z-score will be large and therefore unlikely to be because of random variation; as a consequence H_0 will be rejected and H_1 accepted. But how great must the differences be? What is the critical value beyond which H_0 is rejected?

Such important test parameters must always be determined before the test is executed and we have to decide on the *significance level* to which we will work, i.e. the point or probability beyond which we conclude the test statistic to be so unlikely as to be non-random. Conventionally the choice is between the probabilities 0.05 (5 per cent) and 0.01 (1 per cent), though in reality we are free to select any value. In this example we will use the 0.05 significance level. As our test statistic is a z-value we can use the tables in Appendix I to determine the probability of that value being equalled or exceeded by random variation. If that probability is less than 0.05 then H_0 falls within the defined rejection region of the normal curve. If the probability of the z-value being equalled or exceeded is greater than 0.05, then H_0 is accepted.

So, what are the z-values that delimit the rejection region and beyond which events will occur at random with a probability of less than 0.05? It will be recalled from section 5.10 that the range of z-values within which events will occur with a 0.95 probability are ± 1.96. Equally the probability that a random event will fall beyond those limits, i.e. have a z-score greater than 1.96 or less than -1.96, must be $1.0 - 0.95$ which is of course 0.05. So the *critical test statistic* is $z = \pm 1.96$. Figure 7.1 summarises this argument, and it should be remembered that the rejection region is split equally over the two ends of the distribution. From our discussion in section 5.10 we see that the corresponding critical test statistics for the 0.01 significance level would have been ± 2.58. We have pursued these arguments at some length as they are an important

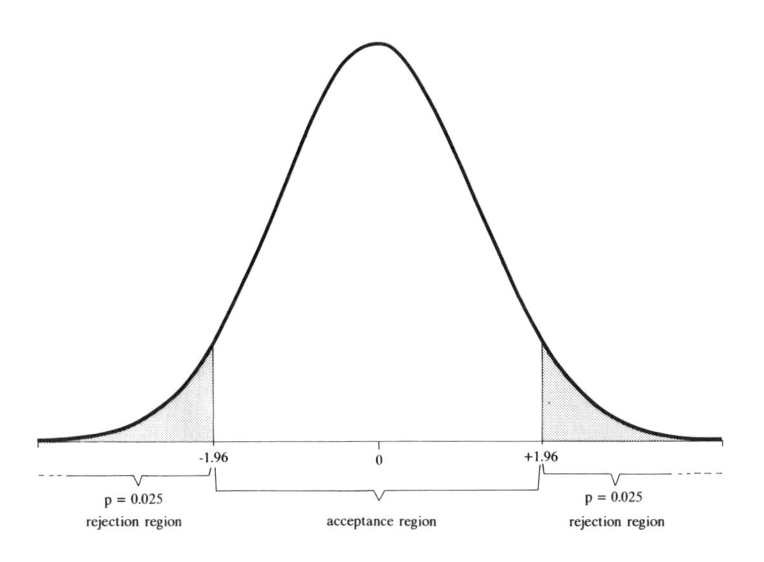

Figure 7.1 Rejection region on the normal distribution for $\alpha = 0.05$.

part of the strategy of all statistical tests, but let us now see how they can be applied in the current example.

The standard error of the sampling distribution of means ($\sigma_{\bar{X}}$) is given by:

$$\sigma_{\bar{X}} = \frac{\sigma}{\sqrt{n}} \tag{7.1}$$

Where σ is the population standard deviation and n is the sample size from which the mean has been estimated. In this example the standard deviation of the English and Welsh counties (1.447) has been used as the population parameter. For the Scottish sample of ten we get:

$$\sigma_{\bar{X}} = \frac{1.447}{\sqrt{10}} = 0.458$$

Readers should notice that this equation for the standard error of the sampling distribution differs slightly from that used in section 6.2 in that the population standard deviation is known and is used in preference to that of the Scottish data. This allows the difference between population and sample mean, given by μ and $\bar{X}$ respectively, to be expressed as an equivalent z- score:

$$z = \frac{\bar{X} - \mu}{\sigma_{\bar{X}}} \tag{7.2}$$

Which, in the present case, gives:

$$z = \frac{13.04 - 11.59}{0.458} = 3.17$$

Such an extreme value lies well into the rejection region for H_0 and we conclude that the difference between the Scottish mean death rate and that for England and Wales is so large that it cannot be attributed to random variation within a common population. Such a finding does not of course tell us why the difference occurs, merely that it confirms our general hypothesis which must now be explored more thoroughly to find the factors controlling Scottish death rates.

The MINITAB package offers a quick and simple means of carrying out this analysis. The **ZTEST** option is illustrated in Figure 7.2 The **SET** command is used to input the sample data (in this case the same death rate example is used). The **ZTEST** instruction must then be followed by the population mean, typed as **MU**, which is 11.59 and standard deviation **(SIGMA)** of 1.447 and the column in which the sample data have been stored **(C1)**. The output includes two lines of summary information followed by the sample parameters including the standard error of the sampling distribution **(SE MEAN)**, the z-value associated with the difference between population and sample mean and, finally, the probability **(P VALUE)** of that z-value being equalled or exceeded. In this example that probability is very remote indeed and far below the critical probability of 0.05. Consequently we reject H_0 and the differences suspected in our working hypothesis are confirmed.

7.4 Errors in statistical testing

The statistical test does not guarantee an accurate result and there is always the possibility of the null hypothesis being rejected when it is

```
MTB  >   SET C1
DATA >   14.4 12.0 15.2 12.4 11.5 12.3 12.2 13.0 13.6 13.8
DATA >   END
MTB  >   ZTEST MU = 11.59, SIGMA = 11.447 C1

TEST OF MU = 11.590 VS MU N.E. 11.590
THE ASSUMED SIGMA = 1.45

          N     MEAN    STDEV  SE MEAN          Z     P VALUE
C1       10   13.040    1.180    0.458       3.17      0.0016
```

Figure 7.2 MINITAB screen display for the ZTEST option.

correct, or accepted when it is false. In this example it was decided that H_0 should be rejected if the test statistic's random probability was less than 0.05. But the significance level (usually denoted as α (alpha)) is the probability of wrongly rejecting H_0 as it is also the probability of random events falling within the rejection region. We have simply decided that this probability is too low to offer an explanation – but it is not impossible for it to do so.

To wrongly reject H_0 is to commit a *type I* error and the likelihood of this happening can be reduced by lowering the significance level to, say, 0.01. To do so, however, is not a perfect solution because it increases the chances of accepting H_0 when it is incorrect. This is known as a *type II* error. The probability of the latter is denoted by β (beta). The *power* of a test is measured by:

$$\text{Power} = 1 - \beta \tag{7.3}$$

The weakness of adhering rigidly to a policy of either a 0.05 or 0.01 significance level should now be clear. The appropriate significance level should be determined by reference to the consequences of committing either of the fundamental errors. For most teaching purposes this problem does not arise, but in research activities, industry and commerce it is often necessary to think carefully about the consequences of type I and II errors and to adopt significance levels that minimise the risk of heavy costs or dangers which might result from either of them.

7.5 One- and two-tailed tests

The Z-test example was termed a *two-tailed* test, so-called because the rejection region is located at both ends of the test statistic's distribution (Figure 7.1). Because the alternative hypothesis did not specify that the sample mean should be either larger or smaller than that of the population, merely that it should be different, both eventualities were allowed for by apportioning the rejection region equally between the two ends of the normal curve.

There is often a need however to indicate in the alternative hypothesis not just that a difference exists but that the difference is in a specified direction. In the death rate example it would have been sensible to stipulate that the sample mean was both different from and larger than the population mean. As a result the rejection region must now be concentrated at one end of the distribution (hence the term *one-tailed*). Care must be taken to allocate the rejection region to the correct end of the distribution and equation 7.2 shows that for $\bar{X}$ to be larger than μ it must lie at the upper, positive, end of the normal distribution (Figure 7.3).

This reorganisation also means that the critical values defining the

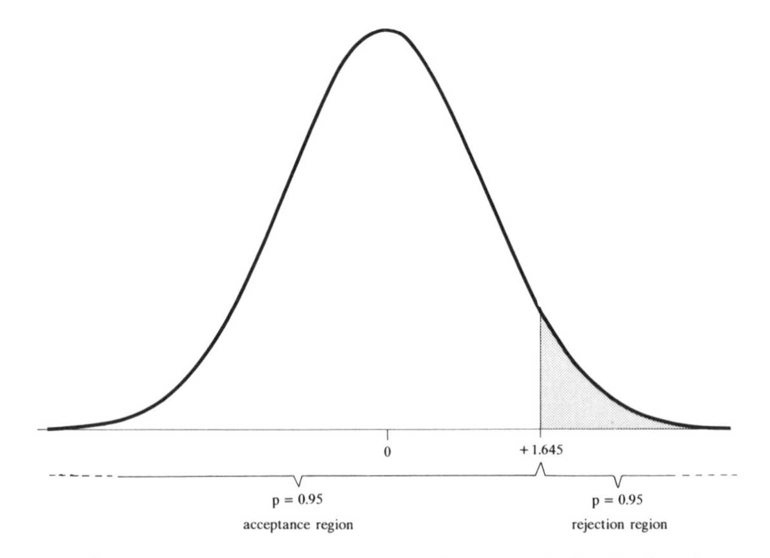

Figure 7.3 One-tailed rejection region on the normal distribution for α = 0.05.

rejection region have to change as it is no longer split over the two extremities of the distribution. The new critical values for $\alpha = 0.05$ and $\alpha = 0.01$ are given in Table 7.1, though z-tables can be used to determine the one-tailed critical values for other significance levels. To do so it must be recalled that the significance levels are proportional areas beneath the standard normal curve and in the case of a one-tailed 0.05 significance level that area must be confined to one end of the curve. Below the critical z-value must lie 0.45 of that half of the distribution, above it must lie the remaining 0.05. From the z-table we find this value to be 1.645. Had H_1 stipulated the sample mean to be smaller than the population mean, the critical value would have been -1.645.

Table 7.1 Critical z values for the 0.01 and 0.05 rejection regions for one- and two-tailed tests

	Critical values	
Tailedness	0.05 level	0.01 level
One-tailed test	-1.645 or $+1.645$	-2.33 or $+2.33$
Two-tailed test	-1.96 and $+1.96$	-2.58 and $+2.58$

These changes to the critical values have important consequences because it is possible for H_0 to be accepted if the test is two-tailed but rejected if it is one-tailed. If, for example, the test statistic is a z-value anywhere between 1.645 and 1.96 and the significance level is 0.05 the implications are clear from Table 7.1: rejection if the test is one-tailed (in the positive direction) but acceptance if it is two-tailed. Consequently, the phrasing and justification of the alternative hypothesis should be formulated with care. In the case of some of the tests which follow their very character, as will be seen, precludes the question of tailedness and the reader's attention will be drawn to instances where this is the case.

The choice of tests represents a small proportion of all that are available to researchers. The selection has not, however, been arbitrary and an attempt has been made to include representatives from each of a number of categories and the tests have been grouped according to two criteria. First we have considered tests appropriate for data measured at different scales and examples are included which are applicable to nominal, ordinal and interval/ratio scale data. Non-parametric tests, those commonly using nominal or ordinal data, are not as powerful as the parametric tests which use interval and ratio scale data. However they enjoy the advantage of being applicable regardless of the distribution of the data; they are said to be *distribution-free*. The parametric tests require, on the other hand, that their data be normally distributed. Secondly, tests have been grouped according to the number of samples on which they operate. In *one-sample* tests a single sample is drawn and then compared with a hypothesised population. But we can also examine tests that compare two or more samples, the *two-sample* and the *k-sample* cases. In the former case two samples of the same variable are included, and in the latter three or more are studied. The structure of the chapter is summarised in Table 7.2. In most cases our interest focuses on the behaviour of one variable. However, especially when using two- and *k*-sample tests our interest may also be with another, second, variable that is used to discriminate the two or more samples. Such considerations provide an introduction to the following chapter which looks specifically at degrees of association between pairs of variables.

Although generally less powerful than the parametric tests, the non-parametric tests nevertheless enjoy wide popularity with researchers.

Table 7.2 Categories and examples of statistical *k*-sample (top row), two-sample (middle row) and one-sample tests

Non-parametric		Parametric
Nominal	Ordinal	Interval/ratio
χ^2 *k*-sample test (section 7.8)	Kruskal – Wallis *H* test (section 7.12)	Analysis of variance (section 7.13)
χ^2two-sample test (section 7.8)	Mann – Whitney *U* test (section 7.10)	*t*-test of difference between means (section 7.11)
χ^2one-sample test (section 7.6)	Kolmogorow – Smirnov test (section 7.7)	Z-test (section 7.3)

Not only are they the only means by which nominal and ordinal scale data can be analysed but their distribution-free qualities are often an advantage. The normality requirement for parametric tests limits their applicability but, fortunately, interval/ratio scale data can always be re-expressed as class-based frequencies or ranks thereby avoiding this problem. In this nominal or ordinal form the data can be tested using non-parametric methods though the conversion will result in loss of information and statistical detail. The county-based unemployment data used in section 4.4 might, for example, be expressed as frequencies within specified classes or as ranks rather than as a collection of individual observations along the measurement continuum. In all cases where parametric tests are to be used it is important to examine the data, using perhaps the MINITAB **DESCRIBE** or SPSS **FREQUENCIES** options, to check for gross irregularities in respect of normality and the data's general distribution.

7.6 The χ^2 test: a one-sample test for nominal data

Although not as powerful as either the parametric tests or many of its non-parametric counterparts the χ^2 (chi-square) test is widely used in research studies. Its lack of power is more than compensated for by its simple data requirements, since it needs observations only at the lowest, nominal, level of measurement. In its one-sample application the χ^2 test compares one set of observed categorical frequencies with an hypothesised set. The example chosen here uses data for days of thunder at a climatological station in Southern England. This example tests for seasonal differences in the incidence of thunder.

Table 7.3 Observed and expected frequencies for seasonal thunder activity

	Spring	Summer	Autumn	Winter
Observed frequency	4	14	7	3
Expected frequency	7	7	7	7
$(O-E)^2/E$	1.28	7.0	0.0	2.29

Degrees of freedom = 3; test statistic = 10.57

The observed frequencies of days (O) with thunder have been aggregated over the four three-month seasons of the year using a two-year sampling period and are given in Table 7.3. The expected frequencies (E) with which they are to be compared are estimated from the null hypothesis which, in this case, is of no difference between the incidence of thunder over the four seasons. Under this H_0 the same frequency of days would be expected in each region. These expected frequencies are found by taking the total number of thunder days and dividing that figure by 4, thus:

$$\frac{4 + 14 + 7 + 3}{4} = \frac{28}{4} = 7.0$$

The complementary alternative hypothesis is that the differences between the seasonal incidences of thunder are greater than might be expected from random variation within a common population. From which we might conclude that seasonal factors, probably in ground level heating, at least partly control the incidence of thunder at this station.

In this example we will use the 0.05 significance level. Our interest is focused on the differences between the observed and expected cell frequencies. If they are small then H_0 may be correct, but if they are large then H_0 is more doubtful. But how great must such differences be before we reject H_0? The normal distribution is not applicable in such cases but we can use one of the non-parametric counterparts – in this case that of the χ^2 statistic. The test statistic is easily derived from equation 7.4 but, most importantly, its distribution is known and definable.

$$\chi^2 = \sum_{i=1}^{k} \left[\frac{(O_i - E_i)^2}{E_i}\right] \qquad (7.4)$$

Equation 7.4

χ^2 = chi-square test statistic
O = observed frequencies
E = expected frequencies
Σ = summation sign

The sub-totals using equation 7.4 are shown in Table 7.3 and sum to produce a χ^2 test statistic of 10.57. Clearly the magnitude of the test statistic varies according to the differences between observed and expected frequencies. The random probability of test statistics exceeding specified values can be established from the properties of the χ^2 distribution which is continuous and also dependent on the degrees of freedom of the data set (υ). In this case υ is determined by the number of cells or categories (k) and is equal to $k - 1$. There are no negative test statistics and the question of tailedness does not apply. Figure 7.4 shows how the χ^2 distribution varies its shape with the degrees of freedom υ.

The rejection region for H_0 is determined from the tabled summary of the distribution (Appendix III). The table is set out by rows and columns

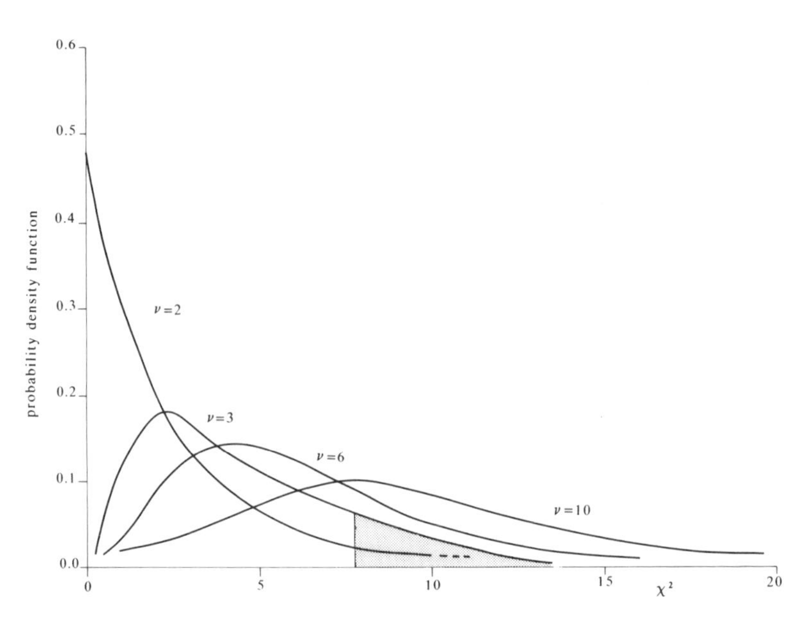

Figure 7.4 The chi-square distribution showing its changing form with increasing degrees of freedom (υ). The rejection region on the distribution for υ = 4 for the test example is shaded.

designated by degrees of freedom and significance levels respectively. After selecting the significance level and calculating χ^2 we can read off the critical value that delimits the lower end of the rejection region. In the present example $\alpha=0.05$ and $\upsilon=4-1=3$, from which we determine a critical value of 7.82. In other words there is a 0.95 probability, under H_0, that the calculated statistic will be less than 7.82 and only a 0.05 probability that it will be equal to or greater than that value. The rejection region always lies at the upper end of the distribution and, for this example, in shown as the small shaded area in Figure 7.4.

In this instance the test statistic is greater than the critical value and falls well within the rejection region as a result of which H_0 is rejected and H_1 is accepted, it being concluded that the observed seasonal contrasts in thunder frequencies are too great to have arisen by chance. It must not, however, be forgotten that there remains the remote (0.05) probability that such differences might indeed have arisen purely by chance.

Useful though the χ^2 test is there are some minimum data requirements that cannot be overlooked:

1. The data categories must be discrete and unambiguous;
2. The frequencies must be absolute counts and not proportions or percentages;
3. The null hypothesis must not permit zero entries in the expected frequencies;

4. There should not be an expected frequency of less than five in more than 20 per cent of the cells.

Provided always that these specifications are met, the χ^2 test can be used, the calculations often being simply executed. Where large data sets are used a considerable amount of time might be spent preparing the table of observed frequencies. Much of this repetitive labour can be avoided by using the SPSS package within which there is a subprogram **NPAR TESTS** which itself contains several statistical applications one of which is the one-sample χ^2 test. The instructions to run this option with SPSS are illustrated in Figure 7.5. All data should be at the nominal scale. If ordinal or interval/ratio data are to be reduced to this level, they must be pre-processed into the classes designed by the user. Data for use in this SPSS application are presented in columns, one column of data for each variable being studied. We can see how SPSS would have treated the data for seasonal frequency of thunder.

```
GET/FILE 'THUNDER'.

VALUES LABELS DAYS 1 'SPRING' 2 'SUMMER' 3 'AUTUMN' 4 'WINTER'.

NPAR TESTS /CHISQUARE DAYS.

- - - - - Chi-square Test

   DAYS
                                Cases
                 Category    Observed    Expected    Residual

   SPRING            1.00           4        7.00       -3.00
   SUMMER            2.00          14        7.00        7.00
   AUTUMN            3.00           7        7.00         .00
   WINTER            4.00           3        7.00       -4.00
                                   --
                    Total          28

        Chi-Square       D.F.           Significance
           10.571          3                .014
```

Figure 7.5 SPSS screen display for the χ^2 one-sample test of seasonal frequencies of thunder.

In the example illustrated in Figure 7.5 the data had already been set up within an SPSS system file (**THUNDER**) which is recalled using the **GET /FILE** instruction (see also section 2.9). The data, placed in a single column denoted by the variable name DAYS, are composed of a string of numbers each coded to represent a day of thunder in each of the regions over the study period. Each observation is represented by a code number, 1 for spring, 2 for summer, 3 for autumn and 4 for winter. It must be remembered that these numbers have no intrinsic numerical weight and are here used merely to allocate an observation to a class. In order to remind ourselves of their meaning we can include the additional **VALUE LABELS** instruction. This keyword is followed by each code number and a 'label' which describes the class and which appears also in the final results screen. Each label appears after its code number and is contained within single quotation marks.

Having defined the value labels the **NPAR TESTS** line can be introduced which also specifies the particular test to be employed (keyword **CHISQUARE** in this case) and the variable name used when setting up the system file which contains the data (see section 2.9 for a description of this process). Both of the latter must follow the slash and be concluded with a stop unless subcommands are to be included.

When executed, the program counts the number of individual observations that fall into each of the classes. The output lists these observed frequencies and their complementary expected counterparts. It includes also the differences between the two, the chi-square test statistic, the degrees of freedom and the probability of the test statistic being exceeded (**SIGNIFICANCE**) in the listing. For H_0 to be rejected the latter must be less than the selected significance level, alternatively the test statistic can be compared with the tabled values.

By default this program assumes the expected frequencies under H_0 to be equal throughout the classes. Under some H_0 the expected frequencies may not be equally distributed through the classes. If this is the case the subcommand **/EXPECTED** can be used followed by the frequency of events anticipated to fall within each class. This instruction would appear after that beginning **NPAR TESTS**. The keyword **EXPECTED** is followed by the list of expected frequencies which must be arranged in the same class order as that for the observed list and separated by commas.

7.7 The Kolmogorov–Smirnov test: a one-sample test for ordinal data

The Kolmogorov–Smirnov test is similar to the χ^2 one-sample in that it compares sample values with hypothetical data prepared by reference to a null hypothesis. It differs in using the cumulative frequencies of the observed and hypothesised observations both of which are expressed as proportions in the range zero to 1.0. Ordinal data, i.e. data which can be rank-ordered is the minimal requirement, though interval/ratio scale data can again be reduced to ranks if necessary. The test produces a

D-statistic given by the largest absolute difference between corresponding points on the two cumulative distributions.

The null hypothesis is of no difference between the two data sets and is rejected when D exceeds its critical value at the selected significance level. The distribution of D when subject to only random variation is defined and the probability distribution is known and summarised in Appendix IV. The distribution is dependent upon sample size.

A popular application of this test is in the comparison of observed data with that representing a known distribution. Reverting to the example of the spatial scatter of grocers' shops (see section 5.16) we can compare the observed distribution of shops with that produced under a perfect Poisson distribution. The latter would be the distribution of points located purely at random. The two data sets consist of the observed and expected frequencies of squares containing 0, 1, 3, 4, etc. shops. Table 7.4 lists the observed data set giving the number of occasions when grid squares contained 0, 1, 2, 3, and 4 shops (none were observed with 5 or more). Table 7.4 also gives the corresponding number for shops distributed completely at random, and therefore conforming perfectly to the Poisson distribution. These latter data were estimated by the method of expectation outlined in section 5.18. The two frequencies are accumulated and converted to proportions of 1.0. The test statistic D is the maximum of the individual differences between the two cumulative distributions F_o and S_n.

$$D = \text{maximum} \mid S_n X - F_o X \mid \tag{7.5}$$

which, by inspection of Table 7.4 we find to be 0.0408. Notice the use of the modulus brackets to indicate that the sign of the difference is ignored, the test being two-tailed only.

Table 7.4 Observed and hypothesised frequencies of grocers' shops for the Kolmogorov-Smirnov test

Shops per square	Observed frequency	Expected frequency	Cumulative observed frequency ($S_n X$)	Cumulative expected frequency ($F_o X$)	Difference
0	96	90.5	0.7111	0.6703	0.0408
1	27	36.2	0.9111	0.9384	−0.0273
2	9	7.2	0.9777	0.9921	−0.0144
3	2	0.97	0.9926	0.9992	−0.0066
4	1	0.10	1.0000	0.9999	0.0001

In this example the 0.05 significance level was used to test H_0 of no difference between the observed and expected frequencies. The critical value of D is easily determined as the table in Appendix IV is arranged in columns according to significance level (α). The rows are arranged by sample size (n) – which in this case is the number of grid squares used in deriving the shop frequency count; sample size is not, it must be observed, the number of shops. Notice also that no correction is made for degrees of freedom in this case. Where n exceeds 35 the tabled data are replaced by critical values obtained by use of the equation appropriate to each significance level. Hence the critical D-statistic for $\alpha = 0.05$ is given by:

$$D = 1.36/\sqrt{n} \qquad (7.6)$$

$$= 1.36/\sqrt{135} = 0.117$$

Test statistics greater than 0.117 lie within the distribution's rejection region and have a random probability of less than 0.05. Only if, as in this case, the test statistic is smaller than the critical value is H_0 accepted and the two distributions concluded not to be significantly different. This being the case here we conclude, in a more general sense, that the observed spatial scatter of shops does not differ significantly from the Poisson and is therefore random with no tendency to clustering.

The SPSS package offers the possibility of using the Kolmogorov–Smirnov test within the **NPAR TESTS** subprogram. The sequence of instructions for running the Kolmogorov–Smirnov (**K – S**) test is shown in Figure 7.6. Data were input as a single series of 135 square by square shop counts. Three hypothetical distributions can be tested using the required keyword; **POISSON** (as above) is one of them but observed cumulative distributions can also be tested against the **NORMAL** distribution and against a **UNIFORM** distribution (one in which the frequencies are the same in all classes). The user may supply the test parameters which should follow immediately after the specification of the hypothesised distribution and all of which must be within parentheses (see Figure 7.6). The Poisson subcommand requires only the mean; the normal requires the mean and the standard deviation, and the uniform distribution requires maximum and minimum values of the original data. If the distribution parameters are not supplied by the user they are estimated from the sample data. Finally the name of the variable which is to be analysed must be supplied, which in this case is SHOPS.

The program will accumulate the observed data set and prepare the observed and the hypothetical cumulative distributions, the latter based on the parameter(s) specified by the user. The output lists the distribution's mean and sample size ('Cases'). The maximum positive and negative class differences are given but, as we indicated above, the signs are ignored for the purposes of defining the test statistic which is the largest absolute difference. The test statistic however is given as a z-value together with its two-tailed probability which, if less than the significance

```
DATA LIST FREE / SHOPS.
BEGIN DATA.
0
1
2
0
.
.
0
1
END DATA.

NPAR TEST /K-S (POISSON, 0.4) SHOPS.

- - - - - Kolmogorov - Smirnov Goodness of Fit Test

     SHOPS

     Test Distribution - Poisson          Mean:   .4000

          Cases: 135

          Most Extreme Differences
     Absolute    Positive    Negative       K-S Z     2-tailed P
      .04079      .04079     -.02734        .474          .978
```

Figure 7.6 SPSS screen display for data input, running and results of the Kolmogorov-Smirnov test. Only part of the data input sequence is shown.

level will result in H_0 being rejected. As the test is always two-tailed there is no need to adjust this probability. Figure 7.6 reveals the z-value of the test statistic to be very small (0.474) and to have, therefore, a strong probability (0.978) of having arisen by chance. A word of warning is necessary concerning the method of calculating the hypothesised distribution. In this, and indeed many other cases, the Poisson distribution is based on the mean of the observed distribution. It should, however, be estimated independently and where this is impossible small errors may result but only where rejection of H_0 is marginal should this difficulty cause concern. Neither the computational nor the hand-worked examples used above correct for this effect.

7.8 The χ^2 test: a two-sample test for nominal data

The basis of this test has been discussed in section 7.6 from which this application differs only in its use of two samples of observed data rather than one. The H_0 remains one of no difference, but now between the two samples rather than between one and a hypothesised set of frequencies. The two samples should have been drawn independently of each other and we are testing to see whether they have been derived from the same background population.

The test statistic is again distributed as χ^2 and the most apparent change is in the method by which the expected frequencies and degrees of freedom are calculated. Beyond that the one- and two-sample tests are similar; large differences between the observed and expected frequencies suggest the H_0 to be doubtful, small differences suggest it to be correct.

Suppose we have carried out a survey of voting habits in a sample of male and of female respondents. The results are given in Table 7.5 from which we can see that there are differences between the two groups in the distribution of votes to each of the parties. But are such differences large enough for us to consider them to be significant? The H_0 states that there is no difference in the pattern of voting between the two samples (other than might occur because of random variation). H_1 states that the two samples differ to such a degree that they represent observations from different populations of voters. The underlying general hypothesis is that we suspect voting habits of the study area to be partly determined by the sex of the voter.

Table 7.5 Sample survey of male and female voting preferences showing (in parentheses) the expected cell frequencies and the test statistic sub-totals

	Party for which vote was cast				
Sex	Conservative	Labour	Liberal Democratic	Others	Total
Male	34 (37.7)	40 (36.7)	8 (9.6)	7 (5.0)	89
Female	41 (37.3)	33 (36.3)	11 (9.4)	3 (4.9)	88
Total	75	73	19	10	177

λ^2 test statistic = 0.363 + 0.297 + 0.267 + 0.800 + 0.367 + 0.300 + 0.272 + 0.737 = 3.40

The expected frequencies under H_0 are provided by the marginal totals of the rows and columns so that:

$$\text{Expected cell frequency } E_{ij} = \frac{\text{row total} \times \text{column total}}{\text{grand total}} \quad (7.7)$$

For example the expected frequency of male Conservative voters would be found from:

$$E_{ij} = \frac{75 \times 89}{177} = 37.7$$

All such expected frequencies are given in parentheses in Table 7.5. The equation used to produce the test statistic is once again based on the squared differences between observed and expected frequencies, all of which are divided by their respective expected frequencies, but must now be summed across a matrix of rows and columns:

$$\chi^2 = \sum_{i=1}^{r} \sum_{j=1}^{k} \left[\frac{(O_{ij} - E_{ij})^2}{E_{ij}} \right] \tag{7.8}$$

In this example we derive a test statistic of 3.40. The degrees of freedom (υ) are determined by reference to the number of rows (r) and columns (k) so that:

$$\nu = (r - 1) \times (k - 1) \tag{7.9}$$

$$= (4-1) \times (2-1) = 3$$

Using the 0.01 significance level Appendix III can be consulted to find the critical χ^2 statistic which is 11.34. Clearly our test statistic does not fall within the rejection region for H_0 and we conclude that, despite our subjective assessment, there is no difference between the habits of male and female voters other than might arise from random sampling variations.

Both MINITAB and SPSS will execute the two-sample test. SPSS does so from the original data, the researcher not having to prepare either observed or expected cell frequencies. The data should be held in two columns coded, in this example, according to sex (**M** or **F**) and to voting choice (**LA** for Labour, **CO** for Conservative, **LD** for Liberal Democratic and **OT** for 'others'). Notice from Figure 7.7 that the groups have been coded as letters and not numbers, this use being indicated by the 'A' format on **DATA LIST** (see also section 2.9). The option **CROSSTABS** then cross-compares the pairs of observations using the variables specified on the same line (**SEX** and **VOTE**) and prepares the observed cell frequencies by entering each male and female voter into their appropriate voting category. **CROSSTABS** allows any two variables in the data file to be tabulated (in this example there were only two). The variable that precedes **BY** defines the rows of the cell matrix, and the variable which follows it defines the columns of the matrix. When the subcommand **STATISTICS CHISQ** is added the expected frequencies and test statistic are calculated on the observed cell frequencies. The additional, and preceding, subcommand **CELLS** allows us to specify that the observed frequencies (**COUNT**) and the calculated expected frequencies (**EXPECTED**) can also be included in the print out.

```
DATA LIST FREE / SEX (A1) VOTE (A2).
BEGIN DATA.
M CO
F CO
M LA
M LA
F OT
. ..
. ..
M LD
END DATA.

CROSSTABS SEX BY VOTE
 /CELLS COUNT EXPECTED
 /STATISTICS CHISQ.
```

```
SEX by VOTE

              VOTE                           Page 1 of 1
          Count
         Exp Val
                                                    Row
                    CO       LA       LD      OT   Total

             F      41       33       11       3      88
                  37.3     36.3      9.4     5.0   49.7%

             M      34       40        8       7      89
                  37.7     36.7      9.6     4.9   50.3%

        Column      75       73       19      10     177
         Total   42.4%    41.2%    10.7%    5.6%  100.0%

       Chi-Square              Value       DF    Significance
--------------------        --------      ----  -------------

Pearson                      3.39271        3          .33495
Likelihood Ratio             3.44223        3          .32833

Minimum Expected Frequency - 4.972
Cells with Expected Frequency < 5 - 1 OF 8 ( 12.5%)

Number of Missing Observations: 0
```

Figure 7.7 SPSS screen display for data input, running and results of the χ^2 two-sample test. Only a part of the data input sequence is shown.

The listing contains two χ^2 estimates, the first (Pearson) is that based on the methods discussed in this section. The Likelihood Ratio is the figure obtained when a correction is applied for loss of information assuming that ordinal or interval/ratio scale data are reduced to the nominal scale. It appears automatically and, as can be seen, tends to differ little from the Pearson χ^2 when samples are large. The degrees of freedom (DF) and the random probability (Significance) of each test statistic are also listed. In both instances the latter are greater than the our selected significance level and H_0 must be accepted, the test statistics being deemed to have arisen by chance alone. This program also checks that the requirements of the test have been met by listing the minimum 'expected frequency' and the number of cells with 5 or less observations.

The χ^2 two-sample test methods are equally applicable, without any change, to three or more sample tests of nominal data. The SPSS **CROSSTABS** subprogram is equally capable of dealing with more than two samples; the use of *k*-sample data would be apparent from the data coding and treated accordingly by the program without further prompting or specification by the user.

7.9 The χ^2 test using 2×2 contingency tables

The most frequent use of the two-sample χ^2 test is when each sample is measured as a dichotomous variable giving giving a two-by-two *contingency table* of frequencies. Where data are in this form equation 7.10 replaces equation 7.8. It has the advantage of providing a more accurate result should interval/ratio scale data have been reduced to just two classes or ranges of values, as in the following example. In all applications of this model there is only one degree of freedom.

An example can be taken from a study of social attitudes undertaken in rural India by Vlassoff and Vlassoff (1980). A sample of adult males were allocated to two groups according to age, 25 to 49 and 50 to 59, and questioned concerning their views of family support for the aged. When asked if they ever considered the problem the younger group more frequently responded 'No'. The older age group revealed this tendency to a less marked extent. The differences, however, are not great (Table 7.6) and we might be justifiably cautious in pressing any arguments based only on visual inspection of these results.

If we test the null hypothesis that there is no difference in the responses between the two age groups, the test statistic can be obtained from:

$$\chi^2 = \frac{n\left(|AD - BC| - \frac{n}{2}\right)^2}{(A + B)(C + D)(A + C)(B + D)} \qquad (7.10)$$

in which the A, B, C and D cell frequencies are as identified in Table 7.6.

The critical value using $\alpha = 0.05$ is 3.84 (Appendix III). By simple substitution into equation 7.10 we get:

$$\chi^2 = \frac{282(|3354 - 4522| - 141)^2}{124 \times 158 \times 205 \times 77} = 0.96$$

As the test statistic falls well short of the critical value we must accept H_0 of no difference between the two groups, and conclude that perceptions of family care of the aged do not differ significantly between the two age classes.

Table 7.6 Data for the χ^2 2 × 2 test (the symbolic notation for use in equation 7.10 is included in parentheses)

Response	Age category 25–49	50–59	Totals
Yes	86 (*A*)	38 (*B*)	124
No	119 (*C*)	39 (*D*)	158
	205	77	n = 282

In the case of this technique neither SPSS nor MINITAB prepare the test statistic using equation 7.10 and would use 2×2 data in the manner outlined in section 7.8. This can lead to slight discrepancies between the two possible results and those based on equation 7.10 should always be preferred.

If we run the MINITAB version of the χ^2 two-sample test the differences appear. In contrast to the SPSS system the cell frequencies may be input directly entering the information row by row using the **READ** instruction as in Figure 7.8, but with care to ensure that they are input in the correct sequence. The test is carried by entering the **CHISQUARE** command, followed by the columns in which the frequencies are stored. The results listing includes each cell's observed (upper) and expected (lower) frequencies, and each cell's contribution to the chi-square total and the final test statistic.

In this case the test statistic of 1.244 is greater than that produced by equation 7.10 and although it makes little difference to our decision to accept H_0, this will not always be the case where decisions are marginal and the critical and test statistics differ by very little.

If the cell frequencies are not known the raw data of individuals listed by their classes can be processed to produce the table of frequencies. In MINITAB this is done using its counterpart to SPSS's **CROSSTABS**, namely the **TABLES** option. Data would be coded as in the SPSS example (see Figure 7.7) and the command **TABLE C1 C2** would then produce the observed cell frequencies with a column for each of the two

```
MTB  >   READ C1 C2
DATA >   86 38
DATA >   119 39
DATA >   END
         2 ROWS READ
MTB  >   CHISQUARE C1-C2

Expected counts are printed below observed counts

              C1        C2    Total
     1        86        38      124
           90.14     33.86

     2       119        39      158
          114.86     43.14

 Total       205        77      282

ChiSq = 0.190 + 0.507 +
        0.149 + 0.398 = 1.244
df = 1
```

Figure 7.8 MINITAB screen display for data input and running of a two-sample (2 × 2) test.

samples. The **TABLES** option also functions with three or more classes and can be used as part of a *k*-sample test. The command **CHISQUARE** would again complete the task, producing expected frequencies using the row and column method from which the test statistic could then be calculated.

7.10 The Mann–Whitney test: a two-sample test for ordinal data

Where data for two samples can be ranked, the Mann–Whitney test is used to examine their differences. One of the advantages of this test is its suitability for very small samples of as few as two observations, though it remains equally suitable for larger data sets.

The following example analyses regional variations in agricultural activity in England. It is commonly thought that the climate and terrain of northern and western England favour pastoral farming, whereas southern and eastern districts are more suited to arable farming. Using data for the proportion of land under pasture, the eight regions of England can be ranked from the most to the least pastoral. Each region is then allocated to either the north/west or the south/east geographical group. The results of both stages are summarised in Table 7.7.

H_0 is of no difference in the mean ranks of the two groups. While H_1

Table 7.7 Data for the Mann – Whitney test of degrees of pastoral agriculture in North-West and South-East England

Region	South West	North-West	West Midlands	North	South-East	Yorkshire	East Midlands	East Anglia
Proportion of pasturage	0.662	0. 654	0.579	0.507	0.368	0.366	0.338	0.132
Rank	1	2	3	4	5	6	7	8
Group	NW	NW	NW	NW	SE	NW	SE	SE

$n_1 = 3$; $n_2 = 5$; $U = 1$.
Source: *Regional Trends*, HMSO, 1981

states that the differences in their ranks are too great to be attributed to random sampling variations and one group tends to rank higher than the other. To test H_0 we must define this *object group* which in this case consists of the north-west group of, supposedly, pastoral regions. If the two groups do indeed differ in their overall ranking some segregation should be apparent in the ranked list with the object group concentrating into the higher ranks. The test statistic (U) measures the degree of segregation and its distribution is known so that we can define a rejection region.

For such a small sample the U-statistic is obtained by counting. Each member of the object group is taken in turn and the number of members of the other group that precede them are counted. The test statistic is the sum of these counts. In the present example we therefore consider each member of the north-west (NW) group in turn. The first four members of this group rank 1 to 4 in the list and are therefore not preceded by any members of the other group and all four count as zero in their contribution to the test statistic. Only Yorkshire of the object group is preceded by any member the SE group, though only by one. Hence the U-statistic is made up as follows:

$$U = 0 + 0 + 0 + 0 + 1 = 1$$

Clearly the greater the degree of segregation the lower will be U and this test is unusual in requiring low test statistics for H_0 to be rejected. Appendix Va gives, for small samples, the probabilities of different U values. The distribution, it should be noted, is discrete and only integer quantities of the test statistic U are possible. If the tabled probability is less than the specified significance level H_0 is rejected as it falls within the rejection region. In order to determine the probability of U attention is focused on the table for n_2 (the larger of the two groups – in this case 5). The required probability values form the body of the table and are located by reference to n_1 (the size of the smaller group) and the already established U statistic. For $n_1 = 3$ and $U = 1$ the probability is

0.036 that this figure will be equalled or exceeded. With a 0.05 significance level such a probability places U within the rejection region for H_0. We accept therefore H_1 and conclude that northern and western regions of England are indeed more pastoral than southern and eastern districts. When used in this manner the test is one-tailed as we specify which of the groups precedes the other in the rankings and the tables are set out with this assumption in mind.

If sample sizes exceed 8 equation 7.11 is an easier means of calculating the test statistic U.

$$U = n_1 n_2 + \frac{n_1(n_1 + 1)}{2} - R_1 \tag{7.11}$$

Equation 7.11

n_1 = size of small group
n_2 = size of larger group
R_1 = sum of ranks in smaller group
U = Mann – Whitney test statistic

The tables of critical values for such larger samples are found in Appendices Vd to Vf. Each table is headed by the significance level, with the critical U-statistics forming the body of the table and being interpreted by reference to the two samples size (n_1 and n_2). But it must be remembered that the test statistic for U must be *less* than the critical value for H_0 to be rejected. It will have been noticed that because either of the two samples may be the object group two U-statistics are possible. It is important to note that only the smaller of the two should be used as the test statistic. When sample sizes are small it is readily apparent which of the two has been derived. But for large samples the position may be less clear. Fortunately a simple check can be carried out. If the two possible test statistics are denoted by U and U_1 then:

$$U = n_1 n_2 - U_1 \tag{7.12}$$

As soon as one of the U statistics has been calculated it can be substituted into equation 7.12 to check that it is indeed the smaller of the two.

Yet further changes take place when one of the samples exceeds 20 in size as the distribution of U then approximates to the normal with a mean of μ_U and a standard deviation of σ_U so that:

$$\mu_U = \frac{n_1 n_2}{2} \tag{7.13}$$

$$\sigma_U = \sqrt{\left[\frac{n_1 n_2 (n_1 + n_2 + 1)}{12}\right]} \tag{7.14}$$

The test statistic is found from equation 7.11 but is then converted to its equivalent z-score using the equations 7.13 and 7.14. In general:

$$z = \frac{\text{difference from mean}}{\text{standard deviation}}$$

hence in this case:

$$z = \frac{U - \mu_U}{\sigma_U} \tag{7.15}$$

which gives:

$$z = \frac{U - \dfrac{n_1 n_2}{2}}{\sqrt{\left[\dfrac{n_1 n_2 (n_1 + n_2 + 1)}{12}\right]}} \tag{7.16}$$

The critical z values are then established in the manner outlined in section 7.3. The probability of z being equalled or exceeded then determines if it falls in the rejection region. The null hypothesis is rejected when the random probability of the test z-statistic is less than the preferred significance level.

Where observations tie on ranks, which may occur when interval/ratio scale data are reduced to ordinal scale measurements, the problem is resolved by reassigning the tied observations over the average of the ranks involved had the ties not occurred. For example, if two observations tie for rank 3 they are each awarded the rank of 3.5, this being the mean of the two ranks (3 and 4) that would otherwise have been taken up. Rank 4 is not then used and the next rank in the sequence is 5.

The SPSS **NPAR TESTS** specification includes a sub-option for the Mann–Whitney test. Where sample sizes are small such a step may not be necessary, but speed and accuracy are gained when larger samples are employed. Figure 7.9 shows how the subprogram **MANN-WHITNEY** is specified and the nature of the output using the same data as above. It should be noted that this test is also known as the Wilcoxson test in which form it yields a test statistic W, which appears in the print-out, and with which we are not here concerned.

The data can be input either at the time or held in a stored file (in this example it is entered immediately before the test using the **DATA LIST**, **BEGIN DATA** and **END DATA** commands). They are set out in two columns one of which contains the numerical value of the variable used to measure the individuals (in this example the proportion of land under pasture), the other indicates to which of the two groups that individual

```
DATA LIST FREE / PASTURE REGION.
BEGIN DATA.
0.507 1
0.662 1
0.132 2
0.654 1
0.338 2
0.368 2
0.579 1
0.366 1
END DATA.

NPAR TEST /MANN-WHITNEY PASTURE BY REGION (1,2).
```

```
- - - - - Mann-Whitney U - Wilcoxon Rank Sum W Test

    PASTURE
  by REGION

   Mean Rank      Cases

        5.80          5   REGION = 1.00
        2.33          3   REGION = 2.00

                      8   Total

                              EXACT          Corrected for Ties
         U          W      2-tailed P       Z        2-Tailed P
        1.0        7.0       .0714      -1.9379       .0526
```

Figure 7.9 SPSS screen display for data input and running of Mann-Whitney test.

belongs. The latter should be coded by a number and its attributed variable name (in this example 'region') appears after the keyword **BY**. The name of the measurement variable ('pasture') precedes **BY**. The **MANN-WHITNEY** instruction also requires the user to specify the first and the second groups by overall ranks as suggested in H_1. This is done by ordering them, in parentheses, at the close of the instruction line, the object group being the first in the list. In this example we state group 1 (north-west) should precede group 2 (south-east). The data need not be ranked beforehand, or entered in any particular order. The output contains the test statistic U and its two-tailed random probability. As this test is mostly used in its one-tailed sense, this probability should be halved to give the correct figure. The output also gives summary data of the two groups, the z-score equivalent of the test statistic and, by default, a correction of the latter for ties even if, as in this case, none are present. Finally it is important, even when using SPSS or other systems to check, using equation 7.12, that the lesser of the two possible Us has been estimated.

7.11 The *t*-test for difference between means: a two-sample test for interval/ratio scale data

This is another test that examines the differences between two samples, but uses parametric data. In contrast to the non-parametric tests there are important prerequisites that the data must fulfil. A general necessity for all parametric tests is that of data normality, though most are tolerant of modest degrees of data skewness. More specifically, this test exists in two forms; the simpler, from the mathematical point of view, requires also that the variances of the two samples are not significantly different. If they are, then a variant form of t-test can be used which is thought by some to be more reliable in both cases.

Nevertheless the two forms are identical in principle and derive their test t-statistics from the division of the difference between the sample means by the standard error of that difference. Expressed algebraically:

$$t = \frac{\text{difference between means}}{\text{standard error of difference}} = \frac{X - \bar{Y}}{\sigma_{X-Y}} \qquad (7.17)$$

The reader will already be familiar with the idea of standard errors for sampling distributions of means and of standard deviations from Chapter 6. In the same manner there is also a sampling distribution of the difference between the means of two samples drawn from a common population. The test statistic which results from the difference is distributed as t. Large differences between sample means combined with low variances of the samples themselves will yield high t-statistics with little inter-group overlap, while small differences produce smaller t values with the two distributions having much more common statistical ground. This distinction is summarised in Figure 7.10.

For our example we will take annual run-off from two geologically and climatologically contrasting areas: South-East England and the Southern Uplands of Scotland. The latter is an area of impermeable Palaeozoic sediments and high rainfall, the former of permeable Cainozoic sediments and lower rainfall. We suspect that geology and climate combine to create important contrasts in the mean annual run-off (measured here in millimetres) of the two areas, with greater run-off from the Scottish region. The null hypothesis, as usual, is of no difference between the two sample means which are, by inference, drawn from the

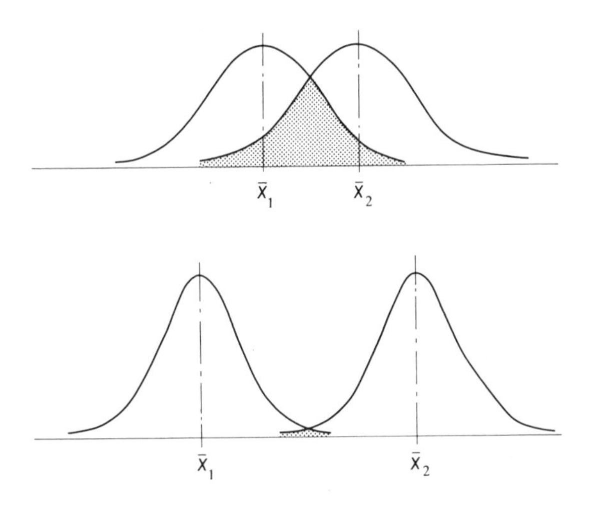

Figure 7.10 Generalised sample normal distributions showing (a) significant overlap and (b) clear distinctions between them.

same run-off population. The alternative hypothesis states that the difference between the sample means is too great to be attributed to random variation within a common sampling distribution. Because we are also specifying that the Scottish sample mean ($\bar{X}$) is greater than that from South-East England ($\bar{Y}$), the test is one-tailed.

The next task is to determine the variances of the two samples. The arithmetic requirements of such parametric tests, often using large volumes of data make computer applications strongly advisable. We can use MINITAB to carry out the processing for us. The **DESCRIBE** option (see section 4.10) gives the sample means and standard deviations and we need only square the latter to obtain the respective variances. The SPSS **FREQUENCIES** option with the **STATISTICS** subcommand will list sample variances without the need for further calculation. Alternatively,the variances can be estimated by hand using the method described in section 4.7.

In this case, with variances of 107950.6 and 10951.4 for the Scottish and English regions respectively, we find that they do indeed differ. We can now test their equality by taking the ratio of the larger to the smaller variance. This produces an *F*-statistic the distribution of which is again known. The tables of critical values appear in Appendix VI. Because of critical *F*'s dependence on the degrees of freedom of both the larger and the smaller variances each significance level requires a separate table; we have included in the appendices only those for the 0.01, 0.05 and 0.10 levels. The degrees of freedom for the greater (v_1) and the lesser (v_2) variance estimates are found from $n-1$, where n are the respective sample sizes (which do not have to be the same). If the observed ratio exceeds the critical value for the chosen significance level the H_0 of variance equality is rejected.

In this example:

$$F = \frac{107950.6}{10951.4} = 9.86$$

The two degrees of freedom here are $\upsilon_1 = 20$ and $\upsilon_2 = 20$. Interpolating the 0.05 significance level table the critical F-ratio is found to be 2.16. We must therefore reject H_0 of variance equality and use the form of t-test in which the two variances are treated separately and not 'pooled' to provide a single figure.

We must now, before embarking on the test, define the rejection region for H_0 of no difference between the sample means. The test statistic is distributed as t and we have already decided that the test is one-tailed and the rejection region is to be concentrated at one, the upper end, of the distribution. If we now take 0.05 as the significance level we can consult Appendix II and determine the critical t-statistic that delimits the rejection region by reference to the degrees of freedom of the test. In this form of the test the degrees of freedom are given by a cumbersome equation based on the two sample sizes (n_X and n_Y) and the sample variances s_X and s_Y.

$$\upsilon = \frac{[(s_X^2/n_X) + (s_Y^2/n_Y)]^2}{\dfrac{(s_X^2/n_X)^2}{(n_X - 1)} + \dfrac{(s_Y^2/n_Y)^2}{(n_Y - 1)}} \tag{7.18}$$

This calculation gives 24 degrees of freedom for which the one-tailed critical t-statistic at the 0.05 significance is found from Appendix II to be $+1.71$. Readers should note that Appendix II is set out to provide one- and two-tailed probabilities.

The formula for the test statistic, loosely described in equation 7.17, can now be expressed in full:

$$t = \frac{\bar{X} - \bar{Y}}{\sqrt{\left(\dfrac{s_X^2}{n_X - 1} + \dfrac{s_Y^2}{n_Y - 1}\right)}} \tag{7.19}$$

Equation 7.19

$\bar{X}$ = mean of variable X
$\bar{Y}$ = mean of variable Y
s_X^2 = variance of X
s_Y^2 = variance of Y
n_X = sample size of variable X
n_Y = sample size of variable Y
t = test statistic

In fact the quantities required for substitution into this equation should present no problem. The two variances we already know from the F-test of variance equality. The means can be easily found from SPSS, MINITAB or by using a simple pocket calculator. But with large volumes of data perhaps already on file or easily input at the keyboard we might prefer to use the computer. MINITAB commands for running the t-test option are shown in Figure 7.11 in which it is necessary only to indicate the subprogram required (**TWOT**) followed by the variables' names or column numbers (**C1** and **C2**) that contain the run-off data and the group, either **1** (for the Scottish) or **2** (for the English), to which each observation belongs. The program then sorts the data into the two groups and completes the analysis. The output not only lists the summary sample statistics (MEAN, STDEV, etc.) but gives also the values that delimit the 95 per cent confidence intervals (95 PCT CI) for the sampling distribution of the difference between the two means – in this case 408 and 719. The final line gives the test statistic (T), its random probability of being equalled or exceeded (P) and the degrees of freedom (DF).

The test statistic ($t = 7.49$) falls well beyond its critical limit. We can also see from Figure 7.11 that its random probability is so remote that it cannot be registered to four decimal places and is far lower than the significance level of 0.05. This program produces a two-tailed probability for the test statistic which should be halved to produce the one-tailed equivalent. From the results H_0 is rejected and the two samples are confirmed as having been drawn from different populations.

Had the requirements of variance equality been met the standard error of the difference between the means would have been found from equation 7.20:

$$\sigma_{\bar{X}-\bar{Y}} = \sqrt{\left(\frac{n_X s_X^2 + n_Y s_Y^2}{n_X + n_Y - 2}\right)}\sqrt{\left(\frac{n_X + n_Y}{n_X n_Y}\right)} \tag{7.20}$$

This uses the 'pooled' or combined variances of the two samples. But is appropriate only when the two variances are similar and the samples larger than those used here. This variant can be used in MINITAB's **TWOT** option by entering **POOLED** on the subcommand prompt that

```
MTB  >   READ C1 C2
DATA >   1324 1
DATA >   879 1
DATA >   401 2
DATA >   583 1
.............
DATA >   871 2
DATA >   421 1
DATA >   END
             42 ROWS READ

MTB  >   TWOT C1 BY C2

TWOSAMPLE T FOR C1
C2      N         MEAN        STDEV     SE MEAN
1      21          803          329          72
2      21          240          105          23

95 PCT CI FOR MU 1 - MU 2: (408, 719)

TTEST MU 1 = MU 2 (VS NE): T= 7.49 P=0.0000 DF= 24
```

Figure 7.11 MINITAB screen display showing the data input sequence and the use of the *t*-test (TWOT) option. In this example the data (only a part of which is shown) are typed in at the time of the test using the READ instruction.

would follow the **TWOT** instruction line, which should, in this specific application, be terminated with a semi-colon (to indicate that a subcommand is wanted on the following line – otherwise the test is carried out as outlined above). In this form of the test the degrees of freedom are given by:

$$v = (n_X - 1) + (n_Y - 1) \tag{7.21}$$

The SPSS option for the *t*-test (**T-TEST**) requires a slightly different set of instructions as Figure 7.12 (using the same data set) indicates. The data, set out as in the MINITAB example, were stored in an SPSS system file **RUNOFF.DAT**. The file contains two columns of data; under the variable **RUNOFF** are the raw data and under the variable **REGION** is the group to which each observation belongs. The **GET /FILE** instruction retrieves the file which is ready to run on SPSS. The request line opens with the keyword for the analysis (**T-TEST**) but has to be followed by **/GROUPS** and the variable name that identifies the column containing the list of the groups to which each individual belongs. In the present example we have identified the Scottish group as **1** and the English group as **2** and this information is in the column denoted by **REGION**. The **VARIABLES** specification requires the identifier for the

```
GET /FILE 'RUNOFF.DAT'.

T-TEST /GROUPS REGION (1, 2)
 /VARIABLES RUNOFF.
```

```
Independent Samples of REGION

Group 1:  REGION  EQ   1.00    Group 2:  REGION  EQ   2.00

t-test for: RUNOFF

            Number        Mean       Standard        Standard
           of Cases                 Deviation          Error
Group 1       21      803.3810       328.525          71.690
Group 2       21      239.6667       104.649          22.836

                  | Pooled Variance Estimate  | Separate Variance Estimate
                  |                           |
    F   2-Tail    |   t   Degrees of  2-Tail  |   t   Degrees of  2-Tail
  Value  Prob.    | Value   Freedom    Prob.  | Value   Freedom    Prob.
                  |                           |
  9.86   .000     | 7.49      40       .000   | 7.49    24.02      .000
```

Figure 7.12 SPSS screen display for file retrieval, running and results of the *t*-test (T-TEST) option.

column in which the run-off data are stored; at the data input stage this column was designated as **RUNOFF**. It should be noted that, as with the MINITAB version, the two-tailed probability of test statistic *t* is given. If our test is one-tailed this probability must again be halved.

The listing provides a summary of the two samples and the standard error of the mean of both of them. The *F*-ratio of the two sample variances and its associated random probability is also given. The *t*-test is executed using both the 'pooled' and the separate variance methods. The results of the *F*-ratio test directs us to the appropriate section of the results listing. It has already been shown that our interest is with the separate variance method. The derived *t*-statistic can be checked against the table of critical values or we can refer to its random probability which is also given. The latter, vanishingly small in this example, is given in its two-tailed form which would need to be divided by two in the case of one-tailed tests.

7.12 The Kruskal–Wallis test: a *k*-sample test for ordinal data

Where individuals can be ranked as well as classified into groups the significance of the inter-group contrasts can be examined by the Kruskal–Wallis test. The number of groups may be three or more and the sample sizes vary from three upwards. The test assists in deciding if the groups could have been drawn from the same population. As with its two-sample counterpart, the Mann–Whitney test, this method is useful not only for intrinsically ordinal data, but also for interval/ratio scale data that have been reduced to ranks to avoid problems of normality or small sample size.

The data are first ranked (by reference to the measurement variable) over all observations irrespective of the groups into which they may subsequently be allocated. The smallest observation is ranked as 1 and so on. Ties are dealt with as in section 7.10. The test proceeds by summing the ranks within each of the k groups to give a set of rank sums (R_i), one for each group. These R values are then included within the equation for the test statistic (H) given by:

$$H = \frac{12}{n_s(n_s + 1)} \sum_{i=1}^{k} \frac{R_i^2}{n_i} - 3(n_s + 1) \qquad (7.22)$$

in which n_s is the total number of observations and n_i the number within each group. Where there are more than five observations in each group the test statistic can be assumed to be distributed as χ^2 with $k - 1$ degrees of freedom, where k is the number of groups. For group samples of less than five the tables of H statistics in Appendix VII should be used in which the critical values are found by reference to the different combination of group numbers and the selected significance level.

In principle, greater inter-group differences provide larger H statistics which, if greater than critical H, cause the null hypothesis of no difference between the groups to be rejected. The test is most often used for small groups: but it is equally applicable to larger samples if its distribution-free qualities are needed. The following example uses data for the incidence of deserted Mediaeval villages (measured per 100sq km) in three English regions, the South-East, the Midlands and the North. Only fragmentary evidence is available but we suspect there to be important inter-group, i.e. inter-regional, contrasts. The data are given in Table 7.8.

The H_0 under test is of no difference between the mean ranks of the three samples other than might occur by random variation within a common population. H_1 states that there are differences beyond those explicable by random variation and they are attributable to genuine contrasts in village desertion rates between the regions. All such k-sample

Table 7.8 Density of deserted villages in three English regions – the data are prepared for the Kruskal – Wallis test

South-East			Midlands			North		
County	Density[1]	Rank	County	Density	Rank	County	Density	Rank
Kent	1.85	3	Northampton	1.46	2	Northumberland	3.16	9
Surrey	1.07	1	Nottingham	3.03	7	W Yorkshire	2.08	5
Hampshire	3.00	6	Oxfordshire	3.94	10	E Yorkshire	6.20	12
			Leicestershire	3.10	8	N Yorkshire	2.06	4
			Warwickshire	5. 08	11			
		10			38			30

1. In deserted villages per 100km^2.

Source: Beresford, M. and Hurst, J.G. (1971) *Deserted Medieval Villages,* Lutterworth, London

tests are, by their nature, non-directional and essentially two-tailed. We will use the 0.05 significance level. Substitution from Table 7.8 gives:

$$H = \left[\frac{12}{12(12 + 1)}\right] \left[\frac{10^2}{3} + \frac{38^2}{5} + \frac{30^2}{4}\right] - 3(12 + 1) = 3.08$$

Appendix VII shows that the critical H value for $\alpha = 0.05$ and group sizes of 3, 4 and 5 is 5.631. Thus the test statistic does not fall within the rejection, H_0 is accepted and our working hypothesis that village desertion had a regional pattern is thrown into doubt.

The question of tied observations has already been mentioned. It may be added that where they occur in large numbers a multiplication correction (C) must imposed on H which is given by:

$$C = 1 - \frac{(T^3 - T)}{n_s^3 - n_s} \tag{7.23}$$

Equation 7.23

T = number of ties
n_s = number of observations

In most cases C is close to 1.0 but becomes smaller as the proportion of ties increases. The correction is applicable if more than 25 per cent of the observations are tied.

The SPSS **NPAR TESTS** subprogram has the option **KRUSKAL-WALLIS** for which the data should be set out in two columns, one containing the measured variable's data, the other indicating, by number

```
DATA LIST FREE / DESERTED, REGION.
BEGIN DATA.
1.85 1
3.16 3
2.08 3
1.46 2
3.03 2
3.94 2
1.07 1
3.10 2
6.20 3
3.00 1
2.06 3
5.08 2
END DATA.

NPAR TEST /KRUSKAL-WALLIS DESERTED BY REGION (1, 3).
```

```
- - - - - Kruskal-Wallis 1-way ANOVA

    DESERTED
 by REGION

  Mean Rank        Cases

       3.33           3   REGION =    1
       7.60           5   REGION =    2
       7.50           4   REGION =    3
                     --
                     12   Total

                                               Corrected for Ties
   CASES     Chi-Square  Significance  Chi-Square    Significance
      12         3.0872         .2136      3.0872           .2136
```

Figure 7.13 SPSS screen display for data input, running and results of a Kruskal – Wallis test.

code, the group to which that observation belongs. Using the same data as above, the figures can be entered using **DATA LIST**. The two variables are identified as **DESERTED** and **REGION**. The command lines to run the test and the output which is produced are shown in Figure 7.13. The keywords **NPAR TESTS /KRUSKAL-WALLIS** are followed by the measurement variable's name (**DESERTED** in this case) then **BY** and the name of the variable which holds the group allocation (**REGION**). The lowermost and topmost group numbers are then put in parentheses at the close of the line.

The raw data need not be ranked as this task is carried out automatically within the program. The output provides a summary of the ranked and grouped data. It should be noted that the SPSS program assumes the test statistic to be always distributed as χ^2 and indicates the number of cases on which the statistic is based. The random probability of the test statistic is also given (0.2136) and the program automatically corrects for ties and indicates this in the output. As the probability of the test statistic exceeds the significance level (0.05) it fails to fall within rejection region and H_0 is accepted.

```
MTB  >    READ C1 C2
DATA >    1.85 1
DATA >    3.16 3
DATA >    2.08 3
DATA >    1.46 2
DATA >    3.03 2
DATA >    3.94 2
DATA >    1.07 1
DATA >    3.10 2
DATA >    6.20 3
DATA >    3.00 1
DATA >    2.06 3
DATA >    5.08 2
DATA >    END
             12 ROWS READ

MTB  >KRUSKALL-WALLIS C1 C2

 LEVEL          NOBS       MEDIAN   AVE. RANK     Z VALUE
     1             3        1.850         3.3       -1.76
     2             5        3.100         7.6        0.89
     3             4        2.620         7.5        0.68
OVERALL           12

H = 3.09  d.f. = 2 p = 0.214
```

Figure 7.14 MINITAB screen display for data input and running a Kruskal – Wallis test.

The MINITAB **KRUSKAL-WALLIS** option performs the same task (Figure 7.14) but produces an *H*-statistic which can be compared with the tables of critical values. The output also includes the probability of the test statistic (0.214). Once again the specification of columns requires only that the raw data are held in the first of the two (and it need not be ranked) and the group allocations in the second of the two columns which are then specified following the keyword **KRUSKAL-WALLIS**. The output lists also the median value of each group, their mean ranks and the latter's deviations from the overall mean expressed as *z*-values.

7.13 One-way analysis of variance: a *k*-sample test for interval/ratio scale data

The final test in this chapter is one used to discriminate between *k* samples measured using interval or ratio scale data. As a parametric test, the one-way analysis of variance (so called because group discrimination is based on one variable only) makes more demands on both the data and researcher than does its non-parametric counterparts. This test is, however, one of the most powerful available.

The raw data need to be normally distributed and the sample variances should not be grossly dissimilar; the test, nevertheless, is tolerant on both counts. Analysis of variance assesses the likelihood of the *k* samples having been drawn from the same population. It does so by decomposing the total variance of the data into within- and between-groups components, i.e. the variance within each group about their respective group means, and the variance between the groups expressed by the scatter of group means about the mean for the whole body of data. The ratio of these two variances gives an *F*-ratio, the random probability of which can be determined from Appendix VI in the manner outlined in section 7.11. In this instance however, the *F*-ratio is obtained by dividing the between- by the within-groups variances. The degrees of freedom remain v_1 for the greater and v_2 for the lesser variances and depend upon sample size and *k* respectively. If the test *F* exceeds the critical tabled value then H_0 of no difference between the means is rejected. Naturally for any *k* samples drawn from a common population some random variation will exist. The *F* test helps to decide whether the observed differences could have arisen by chance or because the samples come from different populations. But whatever the conclusion it applies equally to all the groups and, by this method, there is no question that one group is 'more different' than the others.

Interest focuses on the three quantities of total, within-groups and between-groups variances. If the between-groups component far exceeds the within-groups, then high *F*-ratios result and suggest that the groups do differ significantly (Figure 7.15). Less strong contrasts produce lower *F*-ratios with more overlap and less distinction between the groups.

In the following example the effects on wheat yields (in tonnes/hectare) of different farming methods were studied. The three land use groups were: (1) crops grown on plots which were neither irrigated nor fertilised; (2) plots which were irrigated but not fertilised; (3) plots which were irrigated and fertilised. Each sample consists of twenty randomly selected observations from a much larger data set, though the test does not require the sub-samples to be of equal size. H_0 states that there is no difference between the three group means. On the other hand, H_1 states that the groups' yields differ beyond that which might be anticipated by sampling from a common population. In this example we will use the 0.01 significance level. Our working hypothesis

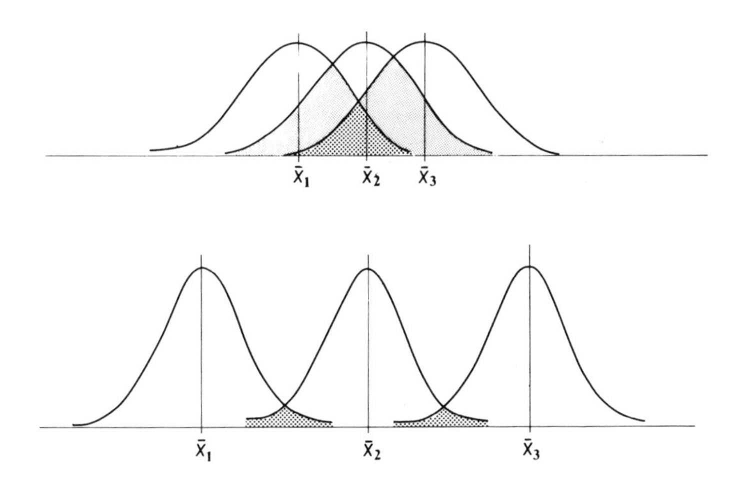

Figure 7.15 Generalised sample normal distributions showing (a) greater and (b) lesser degrees of intergroup overlap.

is that areas employing irrigation farming methods will enjoy higher crop yields than those that do not, and that the use of artificial fertilisers will further enhance the yields.

Having now defined the null and alternative hypotheses and defined the significance level, we can consider how to execute the test. In its most general sense variance is found from:

$$\text{Variance} = \frac{\text{sum of squares}}{\text{degrees of freedom}} \tag{7.24}$$

This qualitative expression can be adapted to the needs of this test in the following manner. The within-groups variance (s_w^2) is found by dividing, for each group, the sum of squared deviations of all observations about their respective group means by the appropriate degrees of freedom. The latter in this case is $n-k$ where n is the total number of observations over all groups and k is the number of groups. As a result:

$$s_w^2 = \frac{\sum_{j=1}^{k} \sum_{i=1}^{j} (X_{ij} - \bar{X}_j)^2}{n - k} = \frac{\text{within-group sum of squares}}{\text{within-group degrees of freedom}} \tag{7.25}$$

where $\bar{X}_j$ are the group means and X_{ij} the observations within each group.

The between groups variance (s_b^2) considers the squared deviations of the group means about the mean for the whole data set ($\bar{X}_t$). These squares, however, are also weighted in terms of the number of observations (n_j) within each group to give:

$$s_b^2 = \frac{\sum_{j=1}^{k} n_j(\bar{X}_j - \bar{X}_t)^2}{k - 1} = \frac{\text{between-group sum of squares}}{\text{between-group degrees of freedom}} \quad (7.26)$$

The degrees of freedom for the between groups variance are given by $k-1$. The total sum of squares is, conveniently, the sum of the within- and between-groups squares and requires no separate calculation (this equality cannot extend to the respective variances which are derived from division by different degrees of freedom). The required F-ratio is always found from:

$$F = \frac{s_b^2}{s_w^2} \quad (7.27)$$

Calculations proceed by first estimating the group means and the mean for the whole data set. The respective sums of squared deviations from the means can then be evaluated which, after division by the appropriate degrees of freedom, will give the variances from which F is determined. Clearly, even for small data sets, the calculations are lengthy and we can consider how MINITAB and SPSS can be used to provide the answers.

With MINITAB the group data are set out in columns (C1 to C3) with a column for each sample. The data were saved in the file **CROPS** which was retrieved from a disc in drive B: of the computer. The instruction **AOVONEWAY** is then followed by the columns holding the samples whose means are to be compared, in this case **C1** – **C3**. The output consists of a conventional *anova* table which lists the source of the variances (FACTOR is 'between groups' and ERROR is 'within groups') with their associated degrees of freedom, sums of squares and variances (listed in Figure 7.16 as MS – mean square). The final F-ratio can then be compared with the tabled value at the selected significance level. The researcher should, however, always take care when using the tables of critical F-values that the degrees of freedom are correctly attributed to the greater and the lesser variance estimates. If preferred, reference can be made to the probability (**P**) of the test statistic which, if less than the significance level, causes H_0 to be rejected.

Supplementary information for each group gives its sample size, mean and standard deviation. A plot showing the 95 per cent confidence intervals for the estimation of each group mean indicates the degree of inter-group overlap.

The subprogram **ONEWAY** within SPSS will perform the same task though its specification differs. In this example the same raw data set is used but set out in a different fashion. Only two columns are now named, **YIELD** (for the measurement variable) and **GROUP** (the assigned group number, here of either **1**, **2** or **3**, is attached to each observation). As a

```
MTB >   RETRIEVE 'B:CROPS'

Worksheet retrieved from file: B:CROPS.MTW

MTB >   AOVONEWAY C1-C3

ANALYSIS OF VARIANCE
SOURCE      DF         SS         MS         F         P
FACTOR       2      25.92      12.96     11.90     0.000
ERROR       57      62.10       1.09
TOTAL       59      88.03
                                INDIVIDUAL 95 PCT CI'S FOR MEAN
                                BASED ON POOLED STDEV
LEVEL     N    MEAN   ST DEV  -----+---------+---------+---------+-
C1       20   5.019    0.764  (------*-----)
C2       20   5.874    0.760             (------*------)
C3       20   6.628    1.451                        (------*-----)
                              -----+---------+---------+---------+-
POOLED STDEV = 1.044             4.90      5.60      6.30      7.00
```

Figure 7.16 MINITAB screen display for the analysis of variance (AOVONEWAY) option.

reminder of this form of data layout readers should consult Figure 7.13. The information was stored beforehand in the system file **CROPS2.DAT** and retrieved using **GET /FILE**. The keyword **ONEWAY** is followed by **VARIABLES** (indicating which of the variables in the data set is to be used to estimate the group means, variances, etc.), followed by the keyword **BY** then the column (indicated by a variable name) in which each individual's group allocation is shown. The upper and lower group membership numbers are then indicated in parentheses. The data layout and consequent instructions are similar to those used in the SPSS Kruskal-Wallis test.

The results (Figure 7.17) are presented in the form of a simple *anova* table. Group descriptive statistics can be requested using the sub-option **/STATISTICS = 1**. If this option is adopted the introductory instruction line must not be terminated with a stop. The random probability of test statistic F is given which, if less than the significance level, requires H_0 to be rejected or, if greater, to be accepted.

In the case of the present examples we see that the group means do indeed differ, between 5.019 tonnes/hectare for untreated plots, rising to 5.874 for irrigated plots and 6.628 for plots that are irrigated and fertilised. But might such groups differences arise by chance within the sampling scheme? To answer this question we should turn to the *anova* table where we see that the ratio of the between (FACTOR in MINITAB) to the within (ERROR) groups variance is 11.9. The critical F-value,

```
GET /FILE 'CROPS2.DAT'.

ONEWAY /VARIABLES YIELD BY GROUP (1, 3)
 /STATISTICS=1.
```

```
          - - - - - - - - - - - O N E W A Y - - - - - - - - - - -

     Variable  YIELD

 By  Variable  GROUP

                               Analysis of Variance
                                 Sum of         Mean           F         F
        Source        D.F.       Square        Square        Ratio     Prob.

Between Groups          2       25.9291       12.9646      11.8853     .0000

 Within Groups         57       62.1759        1.0908

         Total         59       88.1051

          - - - - - - - - - - - O N E W A Y - - - - - - - - - - -

                                Standard     Standard
Group     Count      Mean      Deviation       Error     95 Pct Conf Int for Mean
Grp1         20    5.0195         .7645        .1709       4.6617  To      5.3773
Grp2         20    5.8790         .7627        .1706       5.5220  To      6.2360
Grp3         20    6.6285        1.4513        .3245       5.9493  To      7.3077

Total        60    5.8423        1.2220        .1578       5.5267  To      6.1580

Group       Minimum       Maximum

Grp1         3.4700        6.4300
Grp2         4.5700        7.1000
Grp3         3.3300        8.5700

Total        3.3300        8.5700
```

Figure 7.17 SPSS screen display for running and results of the analysis of variance option (ONEWAY).

taking $\alpha=0.05$, is interpolated from Appendix VIb as 3.14. The test statistic therefore falls well within the rejection region. The *anova* tables also indicate that the random probability of test F is extremely remote, and is less than 0.0005 (hence appearing as 0.000 on the output).

The decision to reject H_0 confirms that different farming practices do indeed lead to different levels of wheat productivity (other crops would require further analyses). In this respect the partitioning of the total data set into groups becomes an explanatory exercise, and the between-groups variances is the 'explained' component in wheat yield variation. The within-groups variance, in contrast, remains that variation about the sample means that is random and 'unexplained'. With this in mind we could rewrite the F-ratio in the following terms:

$$F = \frac{\text{between-groups variance}}{\text{with-groups variance}} = \frac{\text{explained variance}}{\text{unexplained variance}}$$

Naturally, numerical methods and computer applications merely provide us with a measure of the degrees of explanation that have been achieved in numerical processing, the real task of explanation remains with the researcher.

7.14 Conclusions

Earlier chapters have discussed the preliminary treatment of statistical data but, useful though such procedures are, it is largely by hypothesis testing that geographical problems can be analysed. It must be stressed, however, that it is the task of the geographer and not the test to make a sensible interpretation, that is to draw inferences, from the results. In this respect Gould (1971) should be consulted.

Of the tests available, the non-parametric forms are particularly useful in being distribution-free. The more efficient use of data implicit in parametric tests is a quality bought at the expense of more stringent data requirements and lengthier arithmetic procedures. In most cases where large data sets are used computers become all but indispensable. But no matter how alluring computers might be with their ability to process vast amounts of data very quickly we should never overlook the fundamental requirement for non-biased and representative data sets, the individual members of which are independent of one another. Our results and the conclusions we draw from them are only as reliable as the data used to analyse the problem.

While this chapter has introduced a range of tests many have been left unexplored (see, for example, Siegel, 1956). However by this stage the reader should be aware that while individual tests differ in their mathematical procedures they share a common strategy which, if understood, allows one to move easily from one to another, even to those that may initially be unfamiliar. This strategy may be summarised as follows:

1. Formulation of the general research hypothesis;
2. Choice of test appropriate to the data and the aims of the study;
3. Formulation of null and alternative hypotheses;
4. Selection of significance level and definition of rejection region;
5. Calculation of the test statistic;
6. Acceptance or rejection of H_0.

References

Gould, P. (1971) 'Is *statistix inferens* a geographical name for a wild goose?', *Econ. Geog.*, 46, 439–448.

Siegel, S. (1956) *Nonparametric Statistics for the Behavioral Sciences*, McGraw-Hill, New York.

Vlassoff, M. and Vlassoff, C. (1980) 'Old age security and the utility of children in rural India', *Pop. Studies*, 34, 487–500.

Chapter 8

Methods of Correlation Analysis

8.1 Introduction

As we would expect, a great deal of geographical analysis involves studying the relationships between two or more variables, either through time or in different places. This chapter focuses on those statistical techniques that enable us to measure and determine the strength of a relationship between two variables. To measure this we can employ the concept and methods of correlation analysis which, as Table 8.1 demonstrates, can take a number of forms. Once again the decision as to which technique to employ will depend on the type of data with which we are working and the scale of measurement in which they are expressed. Obviously, the strength and efficiency of each of these different types of correlation analysis varies, and it is generally recognised that the product-moment or Pearson's correlation coefficient (r), which uses interval or ratio scale data, is the most powerful. For example, it has been calculated that Spearman's rank correlation, which uses ordinal data, is 91 per cent as efficient as Pearson's r. This means that if, in a sample of 100 cases from a bivariate normal population, the product-moment coefficient is significant, it will require a sample of 110 cases of the same information to achieve the same degree of reliability using the Spearman coefficient. But of course each correlation coefficient has its merits depending on the type of data we wish to examine.

Regardless of the form of correlation analysis we select, the outcome is always expressed as a numerical coefficient that describes the direction

Table 8.1 Types of correlation coefficients and their relation to data characteristics

Type	Measurement scale	Data characteristics
Pearson's (r)	Interval/ratio	Use with both scales
Spearman's rank (r_s)	Ordinal	Both variables must be expressed as ranked data for these two tests
Kendall's tau (τ)	Ordinal	
Biserial (r_b)	Nominal	One dichotomous variable and one variable that has more than two values
Phi coefficient (ϕ or r_ϕ)	Nominal	Both variables must be dichotomous

and character of the relationship between two variables. Importantly, the values of correlation coefficients, derived by any of the methods, can vary only between -1.0 and $+1.0$. These extremes represent respectively the perfect negative or positive relationship between the two variables. In the former case the value of one variable increases as the other decreases, in the latter the two increase in concert. A value of 0.0 indicates the complete absence of any statistical relationship. Correlations can often be qualitatively assessed using *scatter diagrams* on which the two sets of data are plotted in graph form. Figures 8.1a and c show how perfect positive ($r = +1.0$) and negative correlations ($r = -1.0$) might look, while Figure 8.1b indicates the scatter of points that might arise when there is no correlation ($r = 0.0$).

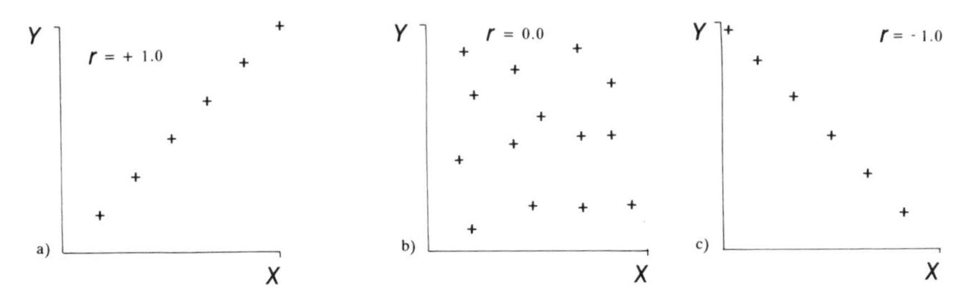

Figure 8.1 Scattergrams showing the differences between perfect positive (a) and negative (c) correlations. Zero correlation might plot as in diagram (b).

Between zero and perfect negative or positive associations there exists a range of less distinct correlations. The following sections will demonstrate how numerical coefficients can be derived to measure objectively the degree of correlation between pairs of variables.

8.2 The product-moment correlation

As with all parametric tests the product-moment correlation coefficient makes assumptions concerning the data to which it can be applied. Firstly, data need to be measured on the interval or the ratio scales; secondly, the two variables should approximate to a normal distribution. Small samples bring their own problems and need to be treated more cautiously, especially with regard to the assumption of data normality.

The product-moment correlation coefficient is based on the idea of *covariance*, a statistical term closely related to the variance we discussed in section 4.7. Variance is used to measure the variability of a sample of data about its mean. Covariance develops this theme and measures the correspondence, or covariation, of two variables together. It will be recalled that variance is estimated on the basis of the sum of squared deviations of individual observations about their mean. Covariance, on the other hand, depends on the sum of the products of each pair of observations about their respective means, thus:

$$\text{Covariance} = \frac{\Sigma(X - \bar{X})(Y - \bar{Y})}{N} \tag{8.1}$$

Consequently, as with the variance and the standard deviation, covariance is an absolute measure that depends on the magnitude of the measurement units and if the same objects were measured on a different scale, say degrees Fahrenheit instead of Celsius, a different estimate of covariance would be achieved. Such, seemingly arbitrary, variations in magnitude can be avoided by making covariance a dimensionless parameter. This is achieved by the correlation coefficient which is found by dividing the covariance of the two variables (X and Y) by the product of their standard deviations. As both covariance (equation 8.1) and standard deviations (equation 4.3) require the division by N, this latter term disappears leaving:

$$r = \frac{\Sigma xy}{\sqrt{(\Sigma x^2 \, \Sigma y^2)}} \tag{8.2}$$

Where, for ease of expression, the terms $(X - \bar{X})$ and $(Y - \bar{Y})$ are replaced by x and y respectively.

As an example we can consider the suggestion that rainfall and altitude are positively correlated and increases in one of the variables are linked with increases in the other. By taking a random sample from the publication *Rainfall 1990* (UK Meteorological Office, 1991) it is possible to examine this proposition. A sample of 30 pairs of observations from the Welsh region was drawn, each pair consisting of a location's altitude (in metres above sea level) and the mean annual rainfall (in millimetres). The individual observations are not listed in their entirety and Table 8.2 provides only the sub-totals needed to calculate the correlation coefficient.

When we substitute the sub-totals from Table 8.2 into equation 8.2 we get:

$$r = \frac{762603}{\sqrt{(334942 \times 4172633)}} = \frac{762603}{1182197}$$

$$r = +\,0.645$$

If we recall that correlation coefficients may vary only within the range $+1.0$ to -1.0, then this result does indeed suggest that some form of association, though not perfect, exists between the two variables. Useful though such a measure might be it does not mark the conclusion of the exercise. We might also calculate the square of the correlation coefficient

Table 8.2 Calculations for product-moment correlation of rainfall and altitude. (The table includes only a sub-sample of the data used in the analysis.)

X Altitude (m ASL)	Y Mean annual rainfall (mm)	$(X-\bar{X})^2 = x^2$	$(Y-\bar{Y})^2 = y^2$	$(X-\bar{X})(Y-\bar{Y}) = xy$
265	1422	13751.5	66409	30220
279	1184	17230.9	388	2586
73	1020	5585.1	241376	36716
156	738	68.3	181732	−3524
...	...	...	...	...
...	...	...	...	...
52	990	9164.9	30381	16686
300	1900	23185.1	541254	112023
4431	34929	334942	4172633	762603

$N = 30 \quad \bar{X} = 147.7 \quad \bar{Y} = 1164.3$

to give the *coefficient of determination* (r^2). This quantity can be thought of as the proportion of the variance of one variable 'explained' by variation of the other. This, however, must be regarded only as a statistical explanation since the scientific principles that actually govern the mechanisms of rainfall increases with altitude must be sought in the science of physics. But such a numerical expression is nevertheless useful as it reflects the degree of order in the relationship. In this case the r^2 of 0.416 shows that 41.6 per cent of the variance of, for example, mean annual rainfall in the region is accounted for by altitude. Conversely, but perhaps equally usefully, this implies that the remaining or residual 58.4 per cent of the variance can be explained only by reference to other influences. As with so many statistical procedures, there is an alternative to equation 8.2 and the correlation coefficient can be estimated also by equation 8.3.

$$r = \frac{N\Sigma XY - (\Sigma X)(\Sigma Y)}{\sqrt{[N\Sigma X^2 - (\Sigma X)^2] \times [N\Sigma Y^2 - (\Sigma Y)^2]}} \tag{8.3}$$

This enables us to divide the calculation into a number of simple steps. But this simplicity notwithstanding, some very large quantities can be accumulated by this method which hampers its use if there is no computational assistance.

As with most of the methods thus far reviewed, there are many occasions when time can be saved and the risk of error minimised by using computer programs. The MINITAB procedures by which we obtain the product-moment correlation coefficient are straightforward. Figure 8.2 shows how data are input using the **READ** command. The present example has only two variables in the data file but many more

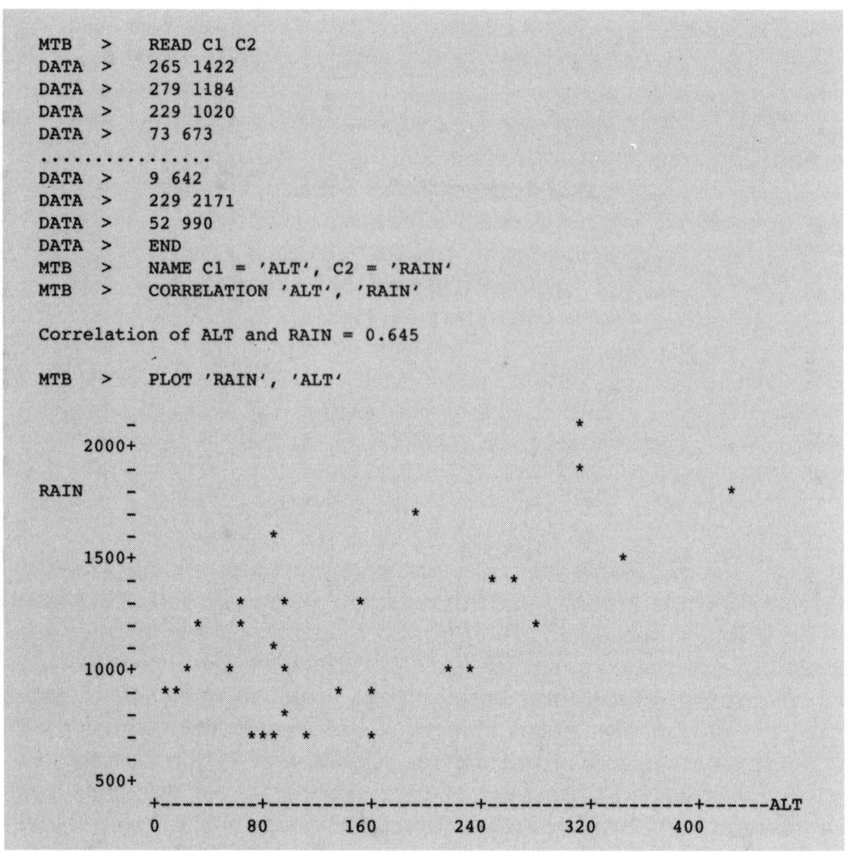

Figure 8.2 MINITAB screen display for CORRELATION and PLOT commands using the rainfall and altitude example. The use of the NAME command attaches terms to the two axes of the graph. Only part of the data sequence is shown.

can be included from which any pair or sequence of pairs can be identified for correlation analysis by reference to their column numbers. The simple instruction **CORRELATION**, followed by the required variable names or column numbers, then produces the correlation coefficient. Figure 8.2 was prepared using the same data as Table 8.2 with, of course, the same result.

Where more than two variables are selected for correlation the results are presented in the form of a *correlation matrix* rather than a single figure. For example if we had a data file of five variables the command **CORRELATION C1–C5** would have produced a matrix or table of all possible paired combinations of the variables.

Figure 8.1 was used to demonstrate how hypothetical pairs of variables with contrasting degrees of correlation might appear if plotted. Scarcely ever will we find perfect correlations of the type shown, but it is often useful to have a visual impression of the degree of association between any two variables. Graphs of this type also allow us to identify points which fall beyond the general trend of the scatter of points. These outliers may be sampling peculiarities but may also identify possible erroneous data points. Such plotting can be tedious if performed by hand, but MINITAB can execute the task using the **PLOT** instruction, followed by the name or columns of the two variables to be plotted as in Figure 8.2. In this application there are advantages to be gained from naming the variables rather than leaving them as C1, C2 etc. This is done with the **NAME** command (see section 2.8). The program will automatically scale, mark and name the axes. The first named variable is plotted on the vertical axis and the second on the horizontal axis. It should be noted that the positive character of the correlation is apparent from the trend of the plotted points. At the same time, as the correlation coefficient has already indicated, the association is not perfect and there is a degree of scatter about that overall trend.

8.3 Significance testing in correlation analysis

It is accepted practice to use correlation coefficients as purely descriptive devices which summarise the numerical association between pairs of variables. However, it is often important to assess the statistical significance of such coefficients. Most coefficients are based on samples drawn from much larger populations and, as we have seen in Chapter 6, there is always the possibility of samples failing to reflect the character of their parent populations. When sampling from *bivariate populations* we may obtain entirely spurious correlation coefficients unconnected with the statistical character of those populations. This is particularly the case with small samples. Fortunately we can employ significance testing methods to determine the probability of any correlation coefficient having arisen by chance alone. If that random probability is remote then the derived coefficient may be held to reflect a genuine and non-random association. Should the probability be high however the correlation should be regarded as doubtful and could have arisen by chance without reflecting the character of the population from which the sample was drawn.

Interest focuses on the null hypothesis (H_0) of no difference between the observed (r) and a zero correlation. We might paraphrase this as an H_0 of no correlation. There are various methods by which this null hypothesis can be tested but the easiest means of doing so is by consulting tables of critical correlations found in Appendix VIII. To establish the significance of the observed correlation we need to know the sample size (n) and the significance level to which we have decided to work. In addition, we must also decide on the tailedness of the test; are we specifying merely that r differs from zero (in either direction) or that it differs in one specified direction. We must also remember that when sampling from bivariate populations the sample size is the number of pairs of observations and not the total number of individual observations. Using the earlier rainfall example, let us take the 0.05 significance level and, furthermore, let us stipulate in the alternative hypothesis (H_1) that we expect observed r to be greater than zero, i.e. that the correlation coefficient is positive and that rainfall and altitude increase together. The tables are arranged by reference to sample size (n), which determines the row, and by significance level and tailedness, which determine the column. From Appendix VIII we obtain a critical correlation coefficient of +0.306. Our observed r must exceed this value in order for us to reject H_0 of no correlation. Section 8.2 demonstrated that as it does exceed this critical value and we can conclude that mean annual rainfall and altitude are, in Wales at least, positively correlated.

The SPSS system shares with MINITAB a simplicity of operation with regard to correlation analysis. Using the same rainfall/altitude example, Figure 8.3 shows the two commands needed to complete the task. The data are compiled as in Figure 8.2 and held in a non-system file **RAINFALL.DAT** in free format. The **DATA LIST FILE** instruction recalls the file, indicates that it is in free format (**FREE**) and attaches the variable names **ALT** and **RAIN** to the two columns. This command must be run with the **F10/RETURN** keys before any further processing can take place. The command **CORRELATIONS** is followed by the sub-command **/VARIABLES** after which the names of the two selected variables are added. The current file contains only those two data sets, but larger data files may contain observations for many more variables; the **VARIABLES** sub-command allows us to select only those pairs in which we have an interest. The results appear on the form of a matrix. Clearly the ALT/ALT and RAIN/RAIN pairings give a correlation of 1.000. Our interest is with the ALT/RAIN correlation coefficient which itself appears on either side of the matrix's principal diagonal. The one-tailed significance of the correlations, with the exception of the diagonal elements, are indicated by asterisks or stars (*). A single star against the correlation indicates significance at the 0.01 level, two stars indicates significance at the 0.001 level.

SPSS also has a plotting facility that allows a scatter diagram of the variables to be quickly produced. Figure 8.4 shows both the instructions

```
DATA LIST FILE 'RAINFALL.DAT' FREE / ALT RAIN.
CORRELATIONS /VARIABLES ALT RAIN.


Correlations: ALT        RAIN
    ALT       1.0000 .     6451 **
    RAIN       .6451 **   1.0000

N of cases:   30      1-tailed Signif: * - .01   ** - .001
```

Figure 8.3 SPSS screen display and results for CORRELATION option.

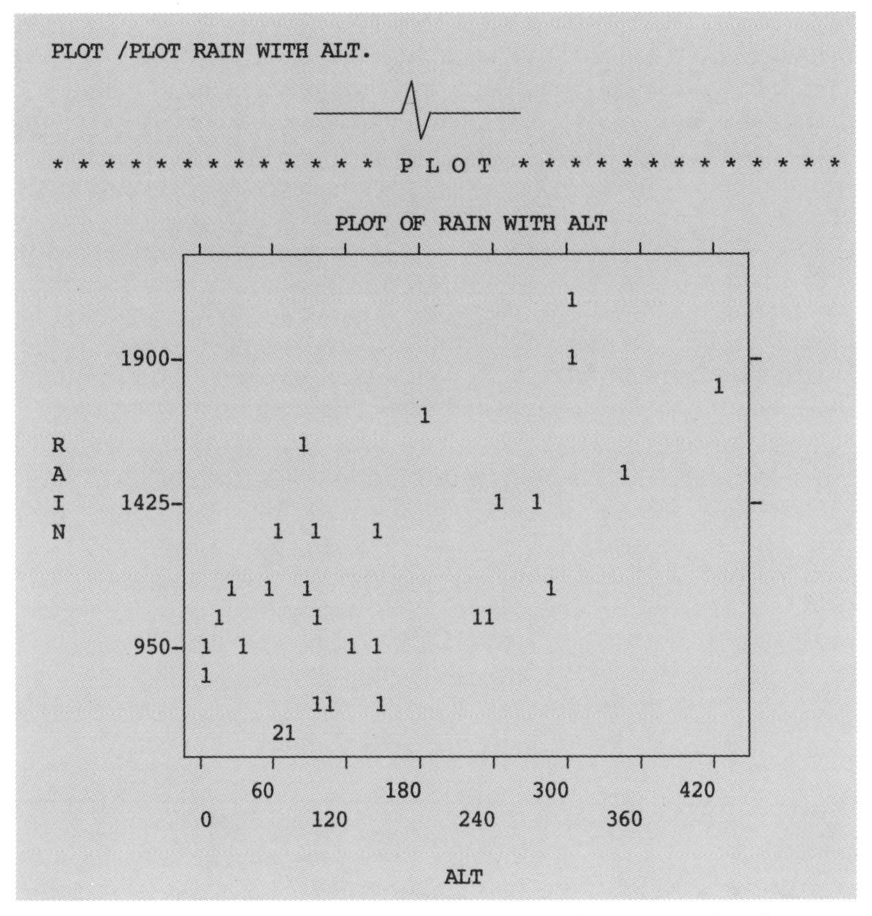

Figure 8.4 SPSS screen display and results for PLOT option using the same non-system data file as in Figure 8.3.

and results of the simple command sequence assuming the data file **RAINFALL.DAT** to be active. The first specified variable after the **PLOT /PLOT** command is used for the vertical axis of the graph, the second for the horizontal. The scaling on both axes is automatic.

The symbol 1 is used to plot individual points. Where two (or more) points are so close as to be indistinguishable the symbol 2 (or 3, etc.) is used instead.

8.4 Correlation analysis of ordinal data

In some circumstances we may wish to assess statistically the relationship between variables measured at the ordinal scale. Alternatively, we might entertain such serious doubts about the normality of our data's distribution that the product-moment coefficient would be an inappropriate measure and need therefore to reduce the data from interval/ratio to ordinal scale measurement. In such circumstances two types of correlation coefficient are available; Kendall's tau and Spearman's rank (Table 8.1). Both measures are equally powerful, although Spearman's rank correlation coefficient (r_s) is the easier to compute and the more widely used. We shall therefore concentrate on the latter which demands only that data be on the ordinal scale and consist of at least five pairs of observations. Each pair of observations consists of two ranks. If the correlation is perfectly positive the sequence of the two paired ranks will be identical. Usually however there will be differences between the paired ranks. The correlation coefficient is derived from those differences in the ranks of the pairs of observations. These differences (D) are squared and then summed for inclusion in equation 8.4.

$$r_s = 1 - \frac{6\Sigma D^2}{N(N^2 - 1)} \qquad (8.4)$$

Equation 8.4

D = differences between ranks of corresponding values of X and Y

N = number of pairs of X and Y values

Where data are converted from the interval/ratio scale ties may occur. In most cases these can be adequately dealt with using the same methods as those described in section 7.10.

In geographical analysis Spearman's rank correlation is used fairly frequently and in a variety of situations. For example, it may be used with interval/ratio data when conditions of normality are not met. It is particularly useful when we are working with small samples from which the character of the population is impossible to determine. Similarly it is used when detailed accuracy of interval type data is in doubt, but the information remains suitable for providing ranked data. This may sometimes be the case when using data from suspect, untested sources, or in some nations with poorly-developed databases and Doornkamp and King (1970) used Spearman's rank correlation coefficient in a geomorphological study in Uganda where data accuracy was questionable. Rank correlation becomes particularly important within research carried out by geographers on cognitive perception and behavioural studies in general and where ordinal data is a common form of measurement (Gould and Ola, 1970).

The use of such information can be demonstrated by the example given in Table 8.3 which compares the environmental dislikes of a sample of people living in the London Borough of Stockwell with those of residents in the rest of Greater London. The various environmental dislikes are ranked within each sample and Spearman's rank correlation can be used to measure whether there is any relationship between the perceptions of the two areas.

Table 8.3 Comparison of environmental dislikes between Stockwell and Greater London

Dislikes	Ranked Stockwell	Ranked Gt. London	Differences in Ranks (D)	D^2
Immigrants	1	2	−1	1
Dirt, litter	2	7	−5	25
Lack of facilities	5	4	1	1
Crime, vandalism	3	5	−2	4
Lack of open space	8	10	−2	4
Children and young people	4	6	−2	4
Council services	6	3	3	9
Traffic problems	7	1	6	36
Schools	10	8	2	4
Noise	9	9	0	0
Shopping facilities	11	11	0	0
				88

Source: Madge, C.H., and Willmott, P., *Inner City Poverty in Paris and London*, Routledge, 1981

Taking the sub-totals from Table 8.3 we can substitute them into equation 8.4 to give:

$$r_s = 1 - \frac{6 \times 88}{11(11^2 - 1)} = +0.60$$

In terms of these dislikes a value of −1.0 would have indicated that the two areas had completely different perceptions of urban environmental problems, whereas identical sets of rankings would have indicated perfect comparability and a coefficient of +1.00. A correlation of 0.60 lies towards the upper end of the scale and suggests that there are degrees of similarity between the perceptions of the two groups. But as our data are based on a sample we must proceed to test the significance of the coefficient. Might it have arisen by chance? The null hypothesis to be tested is again that of no difference between the observed correlation and one of zero.

When we are testing the significance of Spearman's rank correlation we can use one of two methods depending upon sample size. For smaller samples it is better to use prepared tables of the type presented in Appendix IX. These tables are used in the same way as those for the product-moment coefficient. Taking the results from Table 8.3 we can assess the significance of the observed coefficient of +0.60. We will adopt the 0.05 significance level. In addition, our alternative hypothesis asserts that the correlation should be positive, i.e the test is one-tailed. With sample size $n = 11$ we obtain a critical correlation coefficient of $+0.536$. As the observed coefficient is larger, at $+0.60$, we can reject the null hypothesis of zero correlation and conclude that the perceptions of residents of the two areas are indeed correlated.

When using larger samples it is sometimes preferable to convert the rank correlation coefficient to an equivalent t-value where:

$$t = r_s \sqrt{\left(\frac{N - 2}{1 - r_s^2}\right)} \qquad (8.5)$$

The derived t-statistic can then be treated in the same manner as those in Chapter 7. The tailedness of the test must be established, as must the significance level, but in the case of correlation significance testing the t-statistic will have $n - 2$ degrees of freedom. Comparison with the tabled critical value will allow the null hypothesis of zero correlation to be accepted or rejected.

In addition to the **CORRELATION** option, which only allows for the calculation of the product-moment correlation, we can employ SPSS to produce Spearman's rank correlation as an additional statistic within the **CROSSTABS** option. The data can be input either as ranks or as interval/ratio data, the program will automatically rank the latter.

Suppose we had had doubts about the normality of the rainfall and altitude data in section 8.2, we might have preferred to use Spearman's rank correlation which is, in effect, distribution-free. SPSS can use the same data as in Figure 8.3, this information being held in the ASCII file **RAINFALL.DAT**. Following retrieval of that file using the **DATA LIST FILE** instruction, the **CROSSTABS** options is required, together with the names of the two variables to be correlated, as shown in Figure 8.5. In section 7.8 we used **CROSSTABS** to carry out a χ^2 two-sample test but in the present example this is not the object and the optional command **/FORMAT** instructs the machine, using the keyword **NOTABLE**, not to display any tables of frequencies. Of the several possibilities under the **STATISTICS** option we can request **CORR** which will give both the product-moment (Pearson's r) and Spearman's rank correlation coefficients, their respective standard errors of estimate **(ASE1)**, equivalent t-values and random probabilities. The two correlations inevitably differ as they refer to the same phenomena but on different measurement scales. Both, however, are significant with random probabilities well below the significance level of 0.05.

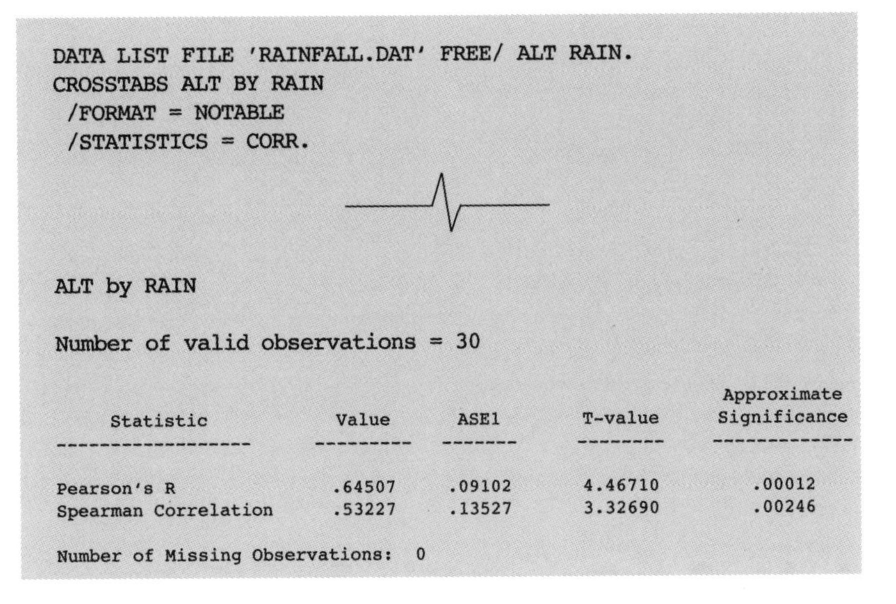

```
DATA LIST FILE 'RAINFALL.DAT' FREE/ ALT RAIN.
CROSSTABS ALT BY RAIN
 /FORMAT = NOTABLE
 /STATISTICS = CORR.

ALT by RAIN

Number of valid observations = 30

                                                              Approximate
     Statistic              Value      ASE1        T-value    Significance
------------------       ---------  -------      --------    -------------

Pearson's R                .64507     .09102      4.46710         .00012
Spearman Correlation       .53227     .13527      3.32690         .00246

Number of Missing Observations:  0
```

Figure 8.5 SPSS screen display and results using CROSSTABS/STATISTICS options to obtain correlation coefficients.

8.5 Measures of correlation for nominal data

We may want to examine the relationship between two variables measured on the nominal scale, in which case we need to consider other types of correlation statistics (Table 8.1). The two most useful measures are the *point biserial coefficient* and the *phi coefficient*, each of which deals with slightly different data sets, as can be illustrated with some simple examples.

Suppose we want to examine the relationship between two variables, one measured on a continuous scale, while the other is a dichotomous variable which can only take the values of, say, 1 or 0. In our example, variable Y represents population totals in a sample of ten villages in South-West England, and variable X indicates the presence or absence of a village post office (Table 8.4). The continuous variable Y can therefore be divided into two sub-groups depending on the value of X, i.e. villages with a post office ($X = 1$) and those without ($X = 0$). We can term these sub-groups Y_0 and Y_1. The point biserial coefficient (r_b) is then given by equation 8.6, in which the means of the sub-groups (Y_0 and Y_1) need to be calculated, together with the standard deviation of all of variable Y (s_Y):

Table 8.4 Raw data of point biserial coefficient between village population and existence of post offices in Wiltshire

Villages	(Y) Population	(X) Post Offices
Bowden Hill	200	1
Corsley	200	1
Bishops Cannings	250	0
Poulshot	250	0
Compton Bassett	250	1
Seend Cleeve	300	0
Hilmarton	300	1
Hormingsham	320	1
Worton	310	1
Broughton Gifford	750	0

Source: W. Wiltshire Structure Plan, 1979

$$r_b = \frac{\bar{Y}_1 - \bar{Y}_0}{s_Y} \sqrt{\left(\frac{N_1 N_0}{N(N-1)}\right)} \qquad (8.6)$$

Equation 8.6

N = total number of observations
N_0 = number of observations with an X value of 0
N_1 = number of observations with an X value of 1

$$\bar{Y}_0 = \frac{\Sigma Y_0}{N_0} \qquad \bar{Y}_1 = \frac{\Sigma Y_1}{N_1}$$

$$s_Y = \sqrt{\left(\frac{N\Sigma Y^2 - (\Sigma Y)^2}{N(N-1)}\right)}$$

From the computational procedures outlined in Table 8.5 we get all the quantities needed for direct substitution into equation 8.6. Hence:

$$\begin{aligned} r_b &= \frac{263.3 - 387.5}{159.4} \sqrt{\left(\frac{6 \times 4}{90}\right)} \\ &= -0.40 \end{aligned}$$

As with other measures of correlation, the point biserial coefficient can vary only between +1.0 and −1.0. We can assess the significance of our results by employing a t-test. The null hypothesis in this case is that of no correlation, i.e the presence of a post office is not determined by village population. In common with all such tests we must select a significance level, say 0.05, and decide on the tailedness of the test;

because the variable used to discriminate the groups is at the nominal scale and might have no natural order it is usually preferable to use the test in its two-tailed form. The degrees of freedom are given by $n-2$, giving us 8 in this case. From Appendix II we obtain a critical t-statistic of ±2.31 while the coefficient's t-statistic can be found from equation 8.7.

$$t = r_b \sqrt{\left(\frac{N - 2}{1 - r_b^2}\right)} \tag{8.7}$$

Using the present data this gives a t-statistic of -1.23. As it therefore fails to fall within either rejection region on the t-distribution we must accept H_0 of zero correlation and conclude that the presence or absence of a post office does not, in this case, depend on population.

The nature of the point biserial coefficient requires that certain assumptions underlie its application. Thus, the values of the continuous

Table 8.5 Calculation of $\bar{Y}_0$ and $\bar{Y}_1$ from Table 8.4

Y_0	Y_0^2	Y_1	Y_1^2
250	62500	200	40000
250	62500	200	40000
300	90000	250	62500
750	562500	300	90000
–	–	320	102400
–	–	310	96100
1550	777500	1580	431000

$\bar{Y}_0 = \frac{1550}{4} = 387.5$ $\bar{Y}_1 = \frac{1580}{6} = 263.3$

variable should be normally distributed. Furthermore, the two sub-samples of the dichotomous variable should not be vastly different in terms of the number of observations. It has been shown that the more equal these sub-groups are, the more accurate the test becomes.

Obviously, the point biserial coefficient is only applicable if one variable is measured on the interval or ratio scale. When we want to assess the correlation between two dichotomous variables other statistics need to be considered. Under these conditions we can use the *phi coefficient*, which represents an extension of the χ^2 test but uses only 2×2 contingency tables (see section 7.9). However, unlike χ^2 values, the phi coefficient can only vary between -1.0 and $+1.0$. The two tests also differ in another fundamental way in that the phi coefficient, like other correlation statistics, tells us about the numerical strength and direction of a relationship between two variables, whereas the χ^2 test is more concerned with assessing the distinctiveness of the two samples.

There are two methods of calculating phi (ϕ) according to whether the χ^2 value has already been derived. If it has not we need to apply equation 8.8.

$$\phi = \frac{AD - BC}{\sqrt{[(A + B)(C + D)(A + C)(B + D)]}} \tag{8.8}$$

Table 8.6 gives the definitions of the terms of the equation which are the same as those used in the 2×2 contingency tables described in section 7.9. We want here to determine whether there is any correlation between the retail centres used for food shopping by car and by non-car owning households. The null hypothesis is that there is zero correlation between the two variables of car ownership and shop location. From equation 8.8 we get:

$$\phi = \frac{(295 \times 83) - (95 \times 27)}{(390 \times 110 \times 322 \times 178)}$$

$$\phi = +\ 0.44$$

This phi coefficient indicates that some degree of association exists between the variables though, as will be shown, it is not easy to determine the significance of this figure.

Table 8.6 Calculation of the ϕ coefficient between car ownership and shopping behaviour

	City shops		Local shops		Total
Households with a car	(*A*)	295	(*B*)	95	390
Households without a car	(*C*)	27	(*D*)	83	110
		322		178	500

If we had already established the χ^2 statistic for this data using equation 7.10 we could have determined the phi coefficient from equation 8.9 in which the χ^2 statistic is divided by the total frequency count (N).

$$\phi = \sqrt{\left(\frac{\chi^2}{N}\right)} \tag{8.9}$$

Unfortunately there is no method of finding confidence limits for ϕ, and in order to test its significance it must be treated as a χ^2 statistic.

We may already have that figure, if not we need only transpose equation 8.9 to give:

$$\chi^2 = \phi^2 N \tag{8.10}$$

If we do this, we obtain a χ^2 statistic of 96.8 which far exceeds the critical value of 6.64 derived from Appendix III (with $\alpha = 0.01$ and remembering that there is only ever 1 degree of freedom in such cases). We must therefore reject H_0 of no correlation between the two variables and conclude that shopping patterns are, in the study area, determined at least partly by car ownership.

As indicated above, the value of ϕ can be calculated from a χ^2 value but this is not generally recommended since negative χ^2 statistics are impossible and the method always gives a positive value. It should also be pointed out that the layout of the contingency table can influence whether the coefficient derived from equation 8.8 is positive or negative (though not its absolute quantity). In order to obtain the correct value the table should be constructed so that A and D in Table 8.6 represent the frequencies of individuals who possess both traits or possess neither trait; while B and C represent the frequencies of individuals who possess one trait but not the other. Therefore, to make practical use of the phi coefficient it is essential that we inspect the data carefully in order that we derive a sensible correlation. In addition to these difficulties care also needs to be taken over the size of frequencies in the contingency table since vastly unequal marginal totals can restrict the final phi coefficient to ranges which are less than $+1.0$ to -1.0. The current example suffers from this difficulty and gives only a modest measure of correlation yet, as we saw above, is highly significant in terms of its χ^2 statistic. These problems notwithstanding, the phi coefficient is often a useful, perhaps the only, means of assessing correlation between variables at the nominal scale.

Thus far we have assumed that our data are based on 2×2 contingency tables, but adaptations to the phi coefficient can be made should we wish to study variables at the nominal scale but measured with more than two attributes. Where this is the case the χ^2 statistic can be greater than sample size with the result that the phi coefficient may exceed 1.0. An alternative statistic not subject to this difficulty is the so-called *contingency coefficient* (C), given by:

$$C = \sqrt{\frac{\chi^2}{\chi^2 + N}} \tag{8.11}$$

In which the χ^2 statistic is derived by the methods described in section 7.8 and N is the total frequency count across all cells of the table.

Once again there are some restrictions on this method and the maximum possible value is partly constrained by the size of the table. A 4×4 matrix, for example, would have a maximum value of +0.87. This problem can in turn be overcome using Cramer's *V* coefficient given by:

$$V = \sqrt{\frac{\chi^2}{N(k - 1)}} \ . \tag{8.12}$$

in which k is the smaller of either the number of rows or columns.

Sections 7.8 and 7.9 have already considered how we can use the SPSS package to derive χ^2 two-sample statistics, we can now develop that theme for measures of correlation. To demonstrate how the phi coefficient, together with the coefficient of contingency and Cramer's *V* statistic are estimated we can again consider the voting habits example used in Figure 7.7 where we found that male and female voters did not differ significantly in the pattern of votes they cast for the different parties. In that context our null hypothesis was one of no difference between the two samples, male and female. Now, however, we are taking a different view of the problem; what, if any, is the degree of correlation, or similarity, between the two patterns of voting?

The data file (**VOTING.DAT** held, in this example, on disc in drive B: of the machine) already exists and we need only recall it with the **DATA LIST FILE** instruction which should, as usual, include the data format (**FREE**), variable names and their code specifications, which here uses A format to indicate that variables are stored as one- and two-alphabetic character codes respectively (see also section 7.8). This line is followed by **CROSSTABS** and the names of the two variables used to provide the frequencies (**VOTE** and **SEX**). We do not need the **CELLS** options used in figure 7.7 as our interest is no longer with the χ^2 statistic itself and the output (Figure 8.6) gives only the absolute cell frequencies. It is the **STATISTICS** option that is now used to provide the correlation coefficients. Specifying **PHI** will give us the phi coefficient together with Cramer's *V* and **CC**, the contingency coefficient. In the current example no difference was found between the phi coefficient and Cramer's *V* statistic as k (the smaller of the number of rows or columns) has a value of 2. The contingency coefficient does, however, differ by a small degree. Nevertheless all three reveal only very small degrees of correlation and their random probabilities are so high (0.40433) that such coefficients are likely to have arisen by chance and that H_0 of zero correlation must be accepted. The terms **ASE1** (the coefficient's standard error) and the T-value are not appropriate for this choice of statistics and hence those entries are left blank on the SPSS listing.

Before concluding this section attention can be drawn to the contrast between the search for correlations and the χ^2 two-sample test as it was applied in section 7.8. In the latter instance we were searching two

```
DATA LIST FILE 'B:VOTING.DAT' FREE / SEX (A1) VOTE (A2).
CROSSTABS VOTE BY SEX
 /STATISTICS PHI CC.
```

VOTE by SEX

		SEX		Page 1 of 1
	Count	F	M	Row Total
VOTE				
	CO	41	34	75 44.9
	LA	33	40	73 43.7
	LD	9	6	15 9.0
	OT	1	3	4 2.4
	Column Total	84 50.3	83 49.7	167 100.0

Statistic	Value	ASE1	T-value	Approximate Significance
Phi	.13220			.40433 *1
Cramer's V	.13220			.40433 *1
Contingency Coefficient	.13106			.40433 *1

*1 Pearson chi-square probability

Number of Missing Observations: 0

Figure 8.6 SPSS screen display and results using CROSSTABS/STATISTICS options to obtain phi, *V* and contingency coefficients of nominal data.

samples for differences with respect to their measured attributes and found that no significant overall differences could be found. The correlation analysis takes another view of the same data and searches instead for significant similarities in the pattern of cell frequencies of the two samples; but none were found. Hence, though male and female voters appear to be members of the same statistical population, that does not simultaneously require that their voting habits are identical or indeed significantly correlated – as this test has shown.

8.6 Problems of correlation techniques in geographical analysis

Since the late 1950s the use of correlation techniques in geographical analysis has been fairly widespread, both as a descriptive tool and as an inferential statistic assessing the significance of a relationship. As well as the statistical assumptions that underlie the use of the various correlation tests, their application to geographical problems is further conditioned by the size of the spatial units on which much of our data are based. In this respect we should recognise that correlation coefficients measure the relationship between variables relative to the scale of the spatial units from which the observations are drawn. The problem is that most spatial units are modifiable, and that in many areas of study the geographer may have a choice of such units. Thus, if we use census data the choice of aerial units in the United Kingdom ranges from small enumeration districts through to a county level of analysis. We know from work carried out over many years that the correlation coefficient between two variables is partially dependent on the size of aerial units used in the study. Yule and Kendal (1950) found that the relationship between potato and wheat yields in England became stronger as the size of spatial units increased. But such variations are not entirely related to the question of spatial scale; they are also affected by the question of spatial contiguity. This refers to the problem resulting from the tendency of adjacent units to resemble each other more closely than those at a greater distance (Duncan *et al.*, 1961). A similar argument can be applied to observations gathered through time intervals.

The solution to such problems is difficult. Robinson (1956) suggested weighting the calculations for the correlation coefficients according to the size of aerial units. But later work by Thomas and Anderson (1965) demonstrated that such weightings were applicable in only a very few circumstances. A study by Curry (1966) disagrees with both the above approaches. Most attempts to resolve these difficulties have failed. Indeed it was the failure of these somewhat rigid mathematical weightings that prompted geographers to take another view of the problem which has taken the form of inquiring into the role of spatial scale and attempting to incorporate the findings into the overall research design. This change in emphasis was partly stimulated by the work of social geographers and their debates over scale problems which became known as the 'ecological fallacy'. This concept describes the situation where patterns of individual behaviour are inferred from larger scale aggregate patterns. For example, suppose in an analysis of crime patterns in central London we found a high statistical correlation between proportions of coloured immigrants and levels of crime within individual districts. This correlation does not necessarily mean that a corresponding proportion of coloured immigrants are criminals. All we have shown is that immigrants tend to live in those parts of London with high crime rates. Some of the crimes could have been committed by people resident beyond the districts in question. It was such problems that prompted geographers to take a more behavioural approach and study the individual as a decision-maker. However, while this may be a satisfactory solution in some circumstances it is not a panacea for all statistical ills.

Similar problems exist in physical geography but have been less vigorously debated. Schumm and Lichty (1965) stressed the importance of both time and spatial scales on statistical interrelationships between stream channel variables. More important in this context is the work of Penning-Rowsell and Townshend (1978) on the role of spatial scale in affecting stream channels. Their study demonstrates that the variables affecting the slope of stream channels vary markedly between the broad regional scale and the local scale as measured by changes in the correlation coefficients.

Two important points emerge from our brief discussion of scale problems and correlation. First, spatial scale is an inherent property in most geographical analysis and we therefore need to understand its effect on statistical tests such as correlation. Leading on from this is the fact that statistical inferences between variables should be made within the context of the aerial units used. Therefore, provided we interpret the results of our correlation analysis within the correct spatial framework, then scale problems can be accommodated in a positive fashion, rather than being negated as some of the early solutions sought to do.

References

Curry, L. (1966) 'A note on spatial association', *Prof. Geogr.*, 18, 97–99.

Doornkamp, J.C. and King, C.A.M. (1971) *Numerical Analysis in Geomorphology*, Arnold, London.

Duncan, O.D., Cuzzort, R.P. and Duncan, B. (1961) *Statistical Geography: problems in analyzing areal data*, Free Press, New York.

Gould, P.R. and Ola, D. (1970) 'The perception of residential desirability in the western region of Nigeria', *Environ. Plan.*, 2, 73–87.

Penning-Rowsell, E.C. and Townshend, J.R.G. (1978) 'The influence of scale on the factors affecting stream channel slope', *Trans. Inst. Br. Geogrs.*, (NS), 3, 395–415.

Meteorological Office (1991) *Rainfall 1990*, HMSO, London.

Robinson, A.H. (1956) 'The necessity of weighting values in correlation of aerial data', *Ann. Assoc. Am. Geogrs.*, 47, 379–391.

Schumm, S.A. and Lichty, R.W. (1965) 'Time, space and causality in geomorphology', *Am. Jour. Sci.*, 263, 110–119.

Thomas, E.N. and Anderson, D.L. (1965) 'Additional comments on weighting values in correlation analysis of areal data', *Ann. Ass. Am. Geogrs.*, 55, 492–505.

Yule, G.U. and Kendal, M.G. (1950) *An Introduction to the Theory of Statistics*, 14th edn, Griffin, London.

Chapter 9
Simple Linear Regression

9.1 Introduction

In Chapter 8 we examined methods of assessing the statistical relationship between two variables. The correlation coefficient measures the strength of such associations and, useful though such an exercise is, it does not allow us to predict the numerical value of one of the variables based on our knowledge of the other. Neither should we make any assumptions or necessarily draw any conclusions about the causal connection between the two variables. We do not, for example, stipulate that one variable determines the behaviour of the other. The importance of regression analysis is that it goes further than correlation and provides us with a method of numerical prediction. In order to do so geographers often consider the question of *direction of causation* and which of the two variables is the *dependent* partner and which is the *independent*, i.e. which of the two variables depends upon, and is influenced by, the other. Statistical convention dictates that the variable whose quantity is to be estimated is termed Y and the variable on which that prediction is made is termed X. For example land values may be seen as the dependent variable (Y) and distance from city centre as the independent or controlling variable (X). In this case the direction of causation is easy to determine with distance controlling the land price but the simplicity of this particular causal connection does not exist in all instances and the choice of dependent variable may simply be a matter of what we want to predict.

Simple linear regression is a valuable predictive and modelling tool allowing the geographer to recreate in numerical terms the way in which one variable controls another. It is however a further example of a parametric method and as such requires the data to be normally distributed. Simple linear regression, as the name suggests, also assumes that the nature of the relationship is indeed linear and equal increments in the predictor (independent) variable bring about consistent responses in the dependent variable. Expressed more simply, a graphical plot of the two variables would produce a scatter of points that tends towards a straight line. It would be expected nonetheless that points would be dispersed to some degree about that imaginary line as in, for example, Figure 8.2.

In essence linear regression methods 'fit' a straight line through a scatter of points. In section 2.5 we saw that any straight line can be described in algebraic terms that allow it to be plotted accurately on a sheet of graph paper. It was also demonstrated that term Y (which in the context of regression is also our dependent variable) is related to term X (the independent variable) through two constants a and b. If we know the numerical values of the latter we can chose any value of X and calculate the corresponding value of Y (see, for example, Table 2.7). Several such pairs of co-ordinates plot as a perfect straight line. Raw geographical data will, on the other hand, always plot as a scatter of points through which the straight regression line must pass. Many different results are possible if we rely on visual judgements to locate the line which best approximates the scatter of points. Each of those subjectively-selected lines will have its own a and b coefficients. Clearly we require an objective means of determining the *best fit* line and the criterion we use is that of *least squares* whereby the line passes through the points in such a way that the sum of squared deviations of Y about that line are at a minimum. This definition also requires that the sum of the individual (unsquared) deviations of Y about the line should be zero. There is no ambiguity about such a definition and there is only one line, defined by only one pair of a and b values, for each data set that conforms to these requirements.

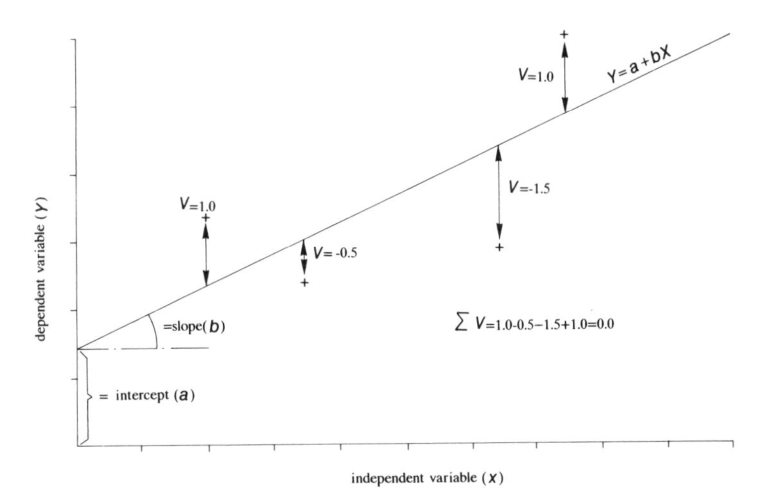

Figure 9.1 Graphical representation of a regression model showing the slope of the regression line (b), the intercept term (a) and the manner in which individual observations will differ by varying amounts (V) from the best-fit line. In the perfect case the sum of all such differences will be zero.

Figure 9.1 is a simplified illustration of these principles in which we have termed the individual deviations as V. We should also notice that coefficient b, known in the current context as the *regression coefficient*, determines the slope of the line, and coefficient a is known as the *intercept term*. The latter is the point at which the line cuts through, or intercepts, the Y axis. In algebraic terms it is the value of Y when X is zero.

9.2 Estimating the linear regression model

Equation 9.1 is the full expression for a linear regression model. We can see that it is almost identical to the equation of the straight line. The only difference is that the regression model includes an error term (e) that recognises the degree of scatter about the line.

$$Y = a + bX + e \tag{9.1}$$

Equation 9.1

Y = dependent variable
X = independent variable
a = intercept term
b = regression coefficient
e = error term

In the following example we will examine the relationship between average daily dietary intake of protein and average male life expectancy. The data are a random sample for 32 nations around the World, the individual observations being the national averages for each of the countries. Reason suggests that life expectancy is determined, at least partly, by the dietary intake of protein, hence the former is the dependent variable (Y) and the latter the independent variable (X). Figure 9.2 shows how we have used the MINITAB options of **PLOT** and **CORRELATION** to provide a preliminary scatter diagram and product-moment correlation of the bivariate sample. The graph has been made clearer by use of the **XLABEL** and **YLABEL** subcommands which attach labels to be printed against the axes. The results show that there is indeed a linear tendency in the scatter of points while the correlation of +0.829 is significant at the 0.01 level. From these preliminary findings we can proceed to estimate the best-fit regression line and, later, to test its statistical significance.

Our first task is to determine the value of the a and b coefficients of the least-squares line which passes through the scatter of points. There is no need at this stage to evaluate the error term (e) in equation 9.1. As with the correlation analysis (section 8.2) there are two principal means by which this can be achieved. Equations 9.2 and 9.3 can be used where N is the number of pairs of observations and X and Y are the individual

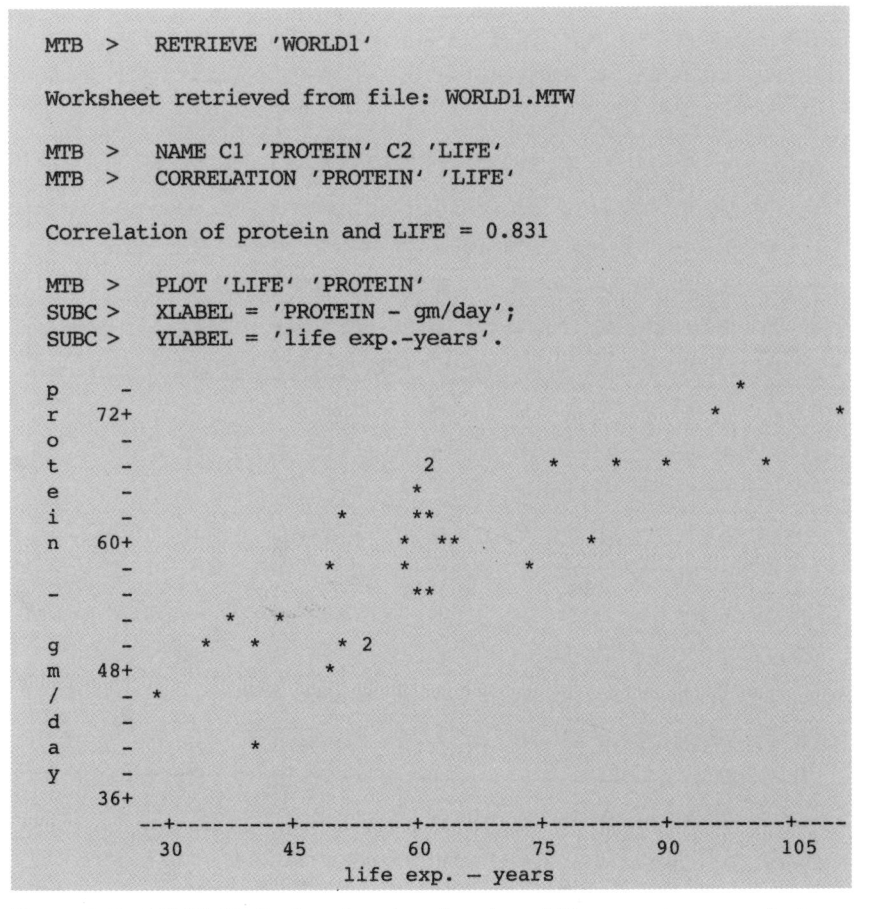

Figure 9.2 MINITAB display showing the plot of life expectancy against protein intake for a sample of 32 nations.

observations. However, this method can produce some undesirably large sub-totals in the squared terms.

$$a = \frac{(\Sigma Y)(\Sigma X^2) - (\Sigma X)(\Sigma XY)}{N\Sigma X^2 - (\Sigma X)^2} \tag{9.2}$$

$$b = \frac{N\Sigma XY - (\Sigma X)(\Sigma Y)}{N\Sigma X^2 - (\Sigma X)^2} \tag{9.3}$$

If estimating the coefficients manually a preferable method is that based on the variations of individual observations about their respective means. Thus, if $x = X - \bar{X}$ and $y = Y - \bar{Y}$ the regression coefficient b is given by:

$$b = \frac{\Sigma xy}{\Sigma x^2} \tag{9.4}$$

Our criterion for the best-fit line dictates also that it must pass through the *data centroid* – the point on the graph where the $\bar{X}$ and $\bar{Y}$ co-ordinates intersect. It follows therefore from equation 9.1 and its transposition that the intercept term (a) is found from:

$$a = \bar{Y} - b\bar{X} \tag{9.5}$$

Table 9.1 shows how the data might be arranged and processed to give the sub-totals needed by equations 9.4 and 9.5.

Table 9.1 Abridged table showing how sub-totals Σx^2 and Σxy can be calculated from which the least-squares regression equation is estimated. Variable X is national average protein intake in grams per person per day. Variable Y is national male life expectancy in years. (The table includes only a sub-sample of the data used in the analysis.)

	Protein intake	Life expectancy	$(X - \bar{X}) = x$	$(Y - \bar{Y}) = y$	x^2	xy
1	59.7	61.57	−3.5656	2.6616	12.71	−9.490
2	51.0	47.00	−12.2656	−11.9084	150.45	146.064
3	53.0	61.90	−10.2656	2.9916	105.38	−30.710
4	56.2	50.85	−7.0656	−8.0584	49.92	56.938
5	39.0	50.28	−24.2656	−8.6284	588.82	209.375
...	...	...	...	...	...	...
...	...	...	...	...	...	...
...	...	...	...	...	...	...
29	110.0	71.80	46.7344	12.8916	2184.10	602.479
30	101.5	68.13	38.2344	9.2216	1461.87	352.580
31	34.3	50.83	−28.9656	−8.0784	839.01	233.997
32	50.9	56.52	−12.3656	−2.3884	152.91	29.535
Means	63.266	58.908		Total	12761.0	4378.2

Regression coefficient (b) = 4378.2/12761.0 = 0.3431
Intercept term (a) = 58.908 − 0.3431 × 63.266 = 37.20
Source: Statistical Yearbook 1987, United Nations, New York (1990).

In this example a positive regression coefficient (b) was established, but negative coefficients will be encountered in just the same manner in which negative correlations exist.

Having evaluated the least-squares regression line we can now plot it on the scatter diagram. To do so we need only two points on the line. These can be determined by substitution of any two values of X into the regression equation. At the upper end of the range we might take a value of 105 grams per day of protein, and at the lower end 45 grams per day. Notice that for these and any other predictive purpose we are not limited in our choice to the observed values only. We can now determine the life expectancy (Y) that result from such protein consumptions.

$$Y_1 = 37.20 + 0.3431 \times 105 = 73.22 \text{ years}$$

and

$$Y_2 = 37.20 + 0.3431 \times 45 = 52.64 \text{ years}$$

These two pairs of co-ordinates are sufficient to plot the regression line (Figure 9.3).

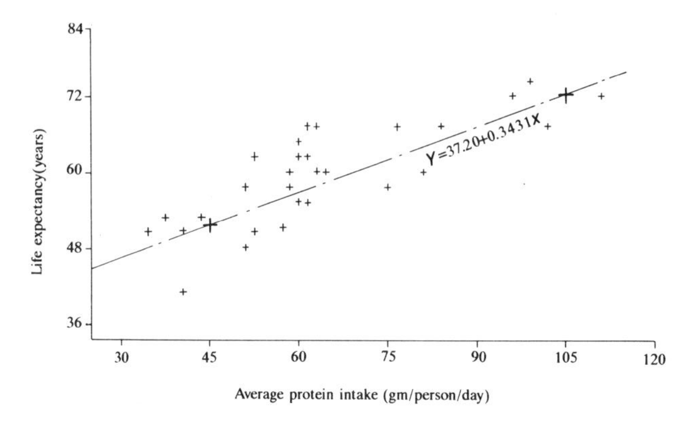

Figure 9.3 Least-squares regression line plotted on the scatter diagram of life expectancy/protein data. The two calculated points on the line are shown by bold crosses.

By fitting a regression line to the data points we can see more clearly that there is a close positive association between diet and life expectancy. But we can proceed further; if a Y value fell on the line it would be wholly explained, or perfectly determined, by its accompanying X value. However, even with such a high correlation as exists here the points do not fall on the line and the scatter reflects the degree of *unexplained variance*. To understand this concept we must remember that the dependent variable has a variance, or scatter, about its mean. Regression analysis seeks to account for that variance through the behaviour of a predictor variable. This concept can be illustrated by reference to any one point on the graph. For all observed X terms there is an accompanying Y value. The latter will differ from its mean by a certain amount given by $Y - \bar{Y}$. At the same time we can estimate the predicted Y value from the regression model thus:

$$\hat{Y} = a + bX \tag{9.6}$$

Where $\hat{Y}$ is the *best estimate* of Y. But this figure will also differ from the mean of Y by the quantity $\hat{Y} - \bar{Y}$. Importantly this latter variation, that of $\hat{Y}$ about its mean, can be regarded as being 'explained' by the behaviour of X. The quantity $\hat{Y} - Y$ (the difference between the observed and predicted Y values) is known as the *residual* or error term. It is also a measure of the variability of Y 'unexplained' by X. These points are

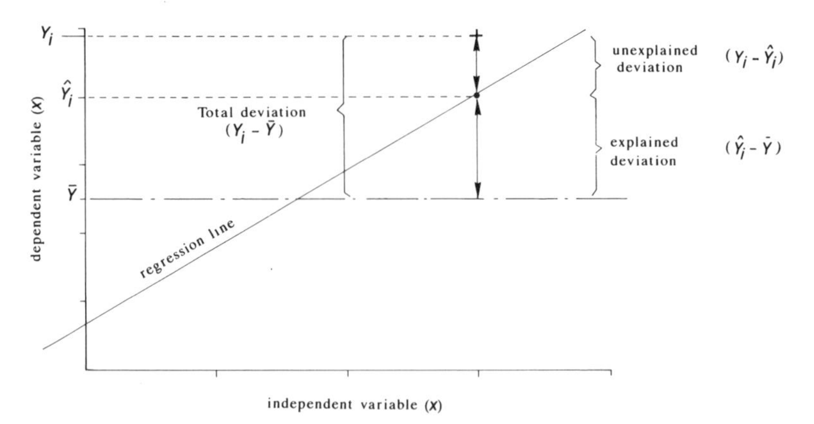

Figure 9.4 Explained and unexplained components of linear regression.

illustrated in Figure 9.4. In a simple bivariate population such as we have here, the unexplained component might be regarded as random variation about the regression line though in reality at least part of the latter might be explicable in terms of a second or other predictor variables. It is, for instance, self-evident that life expectancy does not depend exclusively upon diet, but also upon other factors such as income, standards of public health and living conditions. Fortunately, there are methods by which we can explore more than one predictor variable but we will defer discussion of these multivariate methods until Chapter 11.

We can now see that the regression equation provides the researcher with a means of predicting the value of a dependent variable that corresponds to any value of the independent variable that we chose. Regression analysis is, in this fashion, quite different from correlation analysis. Furthermore, correlation coefficients are scale-independent and the same degree of correlation will be found in a bivariate sample no matter what units we chose for the measurement of the two variables. In the worked example from section 8.2 exactly the same result would have been obtained had the units been inches rather than millimetres of rainfall. This is not the case in regression where the regression coefficient (*b*) is scale-dependent and if the units of measurement are changed the regression equation parameter also changes. Neither is the regression coefficient confined, as correlation coefficients are, to the range -1 to $+1$ and can assume any value. Hence in the above example the correlation coefficient was $+0.830$ but the regression coefficient was $+0.3431$. We can understand this important distinction more clearly if we remember what the regression indicates – that is the measured response of the dependent variable to one unit of change in the single independent variable. In the current example it demonstrates that within the range of observed data life expectancy increments by 0.3431 of a year for every additional gram of protein consumed per day. Hence, if the units of either variables change, so must the resulting regression coefficient and intercept term.

Regression analysis is one of the most powerful methods we have introduced in this book but its usefulness is limited if its preconditions are not met. Thus, both variables should be normally distributed. The dependent terms Y should also have a constant degree of scatter about the regression line. It is also important that the pairs of observations are independent of other pairs; for example, that one large, or one small, pair of observations is not in some way linked to others in the sample making them also large, or small.

9.3 Significance testing in simple regression

Regression equations are estimated from samples drawn from unknown populations. As such they are subject to sampling variations and it is important to identify the reliability of sample estimates a and b of the population parameters α and β. We need also to determine the reliability of the estimates of Y made from them. A number of alternatives are available to do this, but the method here selected has the advantage of being adaptable and will be used again in later chapters on multiple and non-linear regression. We have already touched on its principles in the previous section, where it was shown that the variance of the dependent variable can be decomposed into that 'explained' by the behaviour of the independent term and into that which remained 'unexplained'. We used a similar argument in section 7.13 where we introduced the one-way analysis of variance test. In all such cases we know that:

$$\text{Variance} = \frac{\text{sum of squares}}{\text{degrees of freedom}}$$

We can then use the F-test (see section 7.13) to determine the significance of the ratio of explained to unexplained variance the result of which tells us how effective the regression equation is in accounting for the variability of the dependent term.

In the case of regression modelling, the explained variance is known as the regression variance and the unexplained is known as the residual or error variance. For every observed Y value there is a corresponding 'best estimate' ($\hat{Y}$) that can be determined by substitution of the appropriate X term into the regression equation, there is also a corresponding residual which is the amount by which the former two differ ($\hat{Y} - Y = e$). All three sets of data, $\hat{Y}$, Y and e will have their respective variances. The total variance can be determined from the original data using equation 4.3, the regression ($s_{\hat{Y}}^2$) and residual (s_e^2) variances are found from the following:

$$s_{\hat{y}}^2 = \frac{\Sigma(\hat{Y} - \bar{Y})^2}{k} \qquad (9.7)$$

$$s_e^2 = \frac{\Sigma(\hat{Y} - Y)^2}{n - k - 1} \qquad (9.8)$$

Equations 9.7 and 9.8

$\hat{Y}$ = estimated Y values
$\bar{Y}$ = mean of observed Y values
Y = individual Y values
n = number of observations
k = number of predictors
(always 1 in simple regression)

which combine to give the F-ratio:

$$F = \frac{s_{\hat{y}}^2}{s_e^2} \qquad (9.9)$$

As usual in such circumstances we must set up a null hypothesis, which is here one of 'no explanation' of the variability of Y in terms of X. When two variables are zero-correlated the best-fit regression line through the scatter of points will be horizontal ($b = 0.0$) but pass through the data centroid ($\bar{X}$, $\bar{Y}$). Such a line is one about which there is a minimum sum of residual squares, but it leaves the total variance of Y equal to the residual variance, i.e. there is no explanation. This equality vanishes as soon as b departs from zero and any degree of linearity appears in the scatter of data points. Our problem is to decide at what point the degree of explained variance is significantly greater than that of the residual variance. With a decrease in the scatter of points about the regression line the explained variance grows at the expense of the residual or unexplained variance. The consequent F-ratio can be compared with the tabled figures in Appendix VI to determine its significance and whether H_0 of no explanation can be rejected.

Our interest lies in determining the total, regression and residual sums of squares. But because the sum of the latter two items is also the total sum of squares we need only calculate any two of the three. Table 9.2 shows how we might set out the calculations if performing them by hand. In this example we are calculating only the regression and residual sums of squares. Before doing so, we must take every pair of X and Y observations and, by substituting the former into the regression equation,

Table 9.2 Abridged worksheet showing how the significance of the regression equation predicting life expectancy from protein intake can be tested. Only those sub-totals needed in the final calculations have been listed.

Protein (X)	Life (Y)	Predicted life $(\hat{Y})$	Residual $(Y - \hat{Y})$	$(Y - \hat{Y})^2$	$(\hat{Y} - \bar{Y})$	$(\hat{Y} - \bar{Y})^2$
59.7	61.57	57.6831	2.0169	4.07	−1.2254	1.497
51.0	47.00	54.6981	−3.6981	13.68	−4.2103	17.710
53.0	61.90	55.3843	−2.3843	5.68	−3.5241	12.405
56.2	50.85	56.4822	−0.2822	0.08	−2.4262	5.877
39.0	50.28	50.5809	−11.5809	134.12	−8.3275	69.315
62.0	55.00	58.4722	3.5278	12.45	−0.4362	0.189
...	...	...	...	...	...	...
...	...	...	...	...	...	...
...	...	...	...	...	...	...
110.0	71.80	74.9410	35.0590	1229.13	16.0326	257.107
101.5	68.13	72.0247	29.4753	868.80	13.1162	172.087
34.3	50.83	48.9683	−14.6683	215.16	−9.9401	98.766
50.9	56.52	54.6638	−3.7638	14.17	−4.2447	18.000
Mean	58.91					
Totals				674.3		1502.1

Analysis of variance table

Source	Sum of squares	Degrees of freedom	Variance	F-ratio
REGRESSION	1502.1	1	1502.1	66.83
RESIDUAL	674.3	32 − 1 − 1	22.477	
TOTAL	2176.4	32 − 1		

produce a corresponding $\hat{Y}$ value. The regression sum of squares is based on the individual differences between the best estimates of $\hat{Y}$ and the mean of the observed Ys. The residual sum of squares is based on the differences between the estimated and observed Ys. From the sub-totals an *anova* (analysis of variance) table can be prepared from which the total sum of squares and the test F-statistic can be abstracted. In this example we will use the 0.05 significance level. The regression and the residual degrees of freedom are given by k (the number of predictor terms which, in simple regression, is always 1) and $n - k - 1$ respectively (where n is sample size). Remembering from section 7.13 to allocate the degrees of freedom correctly to the greater and lesser variances, we find from Appendix VI that the critical F-value (with 1 and 30 degrees of freedom) is 4.17. As the test F-value exceeds this we can reject the null hypothesis of no explanation and conclude that protein intake may indeed be used to make a reliable estimate life expectancy.

A useful additional feature of this method is that the correlation coefficient and the coefficient of determination (section 8.2) can be abstracted from the *anova* table. The coefficient of determination (r^2) is given by the ratio of regression to total sum of squares, the correlation coefficient (r) being its square root. Thus:

$$r = \sqrt{\left(\frac{\text{regression sum of squares}}{\text{total sum of squares}}\right)} = \sqrt{\left[\frac{\Sigma(\hat{Y} - \bar{Y})^2}{\Sigma(Y - \bar{Y})^2}\right]} \qquad (9.10)$$

In this example we obtain:

$$r^2 = 1502.1/2176.4 = 0.690$$

and, hence:

$$r = \sqrt{0.690} = +0.831$$

As a result we see that 69.0 per cent of the variation in the dependent variable can be 'explained' by variation in the independent variable. Readers will have noted that the square root of r^2 could have been either $+0.830$ or -0.830. The correct quantity is found by checking the sign of the regression coefficient, which in this instance is positive.

9.4 Confidence limits in simple regression

Whenever we use a regression equation to produce an estimate or prediction of Y it is useful to establish confidence limits about that estimate. If those limits are narrow we can conclude that the estimate is reliable, otherwise we must be more circumspect. We have already introduced the general idea of confidence limits in Chapter 6 where we examined the reliability of sample means. Those same principles can be extended, with modification, to regression analysis.

One of the important features that govern the reliability of any estimate of Y is the scatter of observed points about the regression line, i.e. the residual variance. In section 9.3 we saw how to estimate this quantity, the square root of which is known as *the standard error of the residuals* (s_e). Referring to the life expectancy example in Table 9.2, where the residual variance is listed as MS (mean square) of the 'error', we have:

$$s_e = \sqrt{22.477} = 4.74 \text{ years}$$

This is of course an estimate, based on our sample, of the corresponding population parameter. We must not forget that that line too is only an estimate, again based on our sample, and is itself subject to sampling errors. If we were to draw another sample of 32 nations we would not expect to obtain precisely the same estimates for a and b. Any such sample will only produce an estimate of the population parameters α and β. The consequences of sampling errors in both terms are illustrated in Figures 9.5a and b. Errors in the intercept term (a) cause the line to be moved vertically though remaining at a fixed slope, while errors in the regression coefficient (b) cause it to be 'rotated' about the data centroid. The combined effects of these sampling errors is to create curved, to be precise – hyperbolic, confidence limits either side of the least-squares line. The most immediate consequence of this is that the confidence intervals become wider (and our estimates less reliable) as we move away from the data centroid (Figure 9.5c).

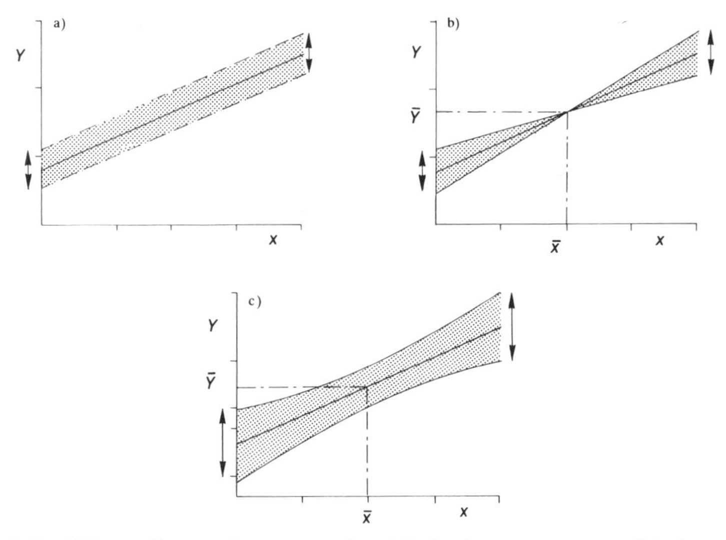

Figure 9.5 Effect of sampling errors for (a) the intercept term, (b) the regression coefficient and (c) the combined effect of the two sources of uncertainty.

Because the confidence limits are curved a different calculation is required for each new estimate we make of Y. In addition, if we want to estimate the confidence limits for any individual predicted value of Y we need to take into account both the inherent variability of the dependent term about the least-squares line and also the variability in the regression line's parameters. If we denote the standard error (or prediction interval) of any individual estimate by $SE_{\hat{Y}}$ then:

$$SE_{\hat{Y}_i} = s_e\sqrt{\left[1 + \frac{1}{n} + \frac{(X_k - \bar{X})^2}{\Sigma(X_i - \bar{X})^2}\right]} \qquad (9.11)$$

Equation 9.11

s_e = standard error of residuals
n = number of observations
X_k = selected X value
$\bar{Y}$ = mean of variable X
X_i = individual observations of X

We can examine the use of this method of determining confidence limits by taking two values of X, one close to the mean (60.00 grams/day) and one towards the extreme upper end of the observed range (105.00 grams/day). For each we can calculate the corresponding life expectancies, which are 57.79 and 73.23 years respectively. We need now to determine the confidence intervals about these estimates using equation 9.11. Although the equation has an undoubtedly intimidating appearance the quantities required for its evaluation are not difficult to obtain. We already have the standard error of the residuals (s_e), the sample size n is 32, while $\bar{X}$ and $\Sigma(X-\bar{X})^2$ were evaluated in Table 9.1. The specific terms for X_k are chosen by the researcher. In the case of X_k of 60.00 (from which $Y = 57.79$) we obtain by substitution:

$$SE_Y = 4.74\sqrt{\left[1 + \frac{1}{32} + \frac{(60.00 - 63.266)^2}{12761.0}\right]}$$
$$= 4.74 \quad \sqrt{(1 + 0.0313 + 0.00084)} = 4.82 \text{ years}$$

Similarly for X_k of 105.0, from which $Y = 73.23$ years, we get:

$$SE_Y = 4.74\sqrt{\left[1 + \frac{1}{32} + \frac{(105.00 - 63.266)^2}{12761.0}\right]}$$
$$= 4.74 \quad \sqrt{(1 + 0.0313 + 0.1365)} = 5.12 \text{ years}$$

These two examples demonstrate clearly how the standard error for the estimate of individual values of Y increases with distance from the data centroid where they are at a minimum.

The next step in our analysis is to determine the factor by which the standard error must now be multiplied in order to establish, say, the 95 per cent confidence limits. In earlier applications of the standard error principle (for example, section 6.3) we multiplied by the appropriate z-value which, for the 95 per cent (0.95) confidence interval would have been ±2.58. Now, however, we must use the t-distribution with $n-1$ degrees of freedom. In general the confidence limits may be written as:

$$\hat{Y}_k \pm SE_{\hat{Y}} \times t \qquad (9.12)$$

Where t is the critical statistic for the degrees of freedom and selected significance level. From Appendix II we find, in this example, critical t to be 2.04. As a result we can be 95 per cent certain that the population male life expectancy corresponding to a diet of 60.00 grams of protein per day is within the limits:

$$57.79 \pm 4.82 \times 2.04 = 47.96 \text{ to } 67.62 \text{ years}$$

and 95 per cent certain that for an intake of 105.00 grams of protein per day the limits are:

$$73.23 \pm 5.12 \times 2.04 = 62.79 \text{ to } 84.67 \text{ years}$$

The width of these limits is a salutary reminder of the sampling errors that can accumulate in such studies, especially where sample sizes are small. Larger samples will certainly narrow the confidence intervals and reduce their degree of curvature by giving more reliable regression parameters. In addition we must not forget that we are trying to explain the behaviour of one variable (life expectancy) on the basis of only one predictor variable (protein intake). The uncertainty inherent in using only one predictor is evident in the scatter of points about the regression line. This can only be reduced by acknowledging the possible contribution of additional predictor variables, as we shall do in Chapter 11.

In conclusion it should be added that if we wanted to estimate the confidence limits for the mean of all possible $\hat{Y}$s rather than one individual estimate the procedure would be the same, but the limits would be narrower depending only on the sampling errors of the equation parameters. The equation used in such circumstances is closely related to 9.11 and is given by:

$$SE_{\hat{Y}} = s_e\sqrt{\left[\frac{1}{n} + \frac{(X_k - \bar{X})^2}{\Sigma(X_i - \bar{X})^2}\right]} \qquad (9.13)$$

9.5 Analysis of residuals

In the previous sections reference has been made to residuals – the differences between the observed and estimated values of the dependent variable. They represent that component of the variation in the dependent variable that cannot be accounted for by variation in the independent variable. The study of residuals may direct us towards other variables that can be used in multiple regression to improve the predictive capacity of the regression equation; it also provides information on the reliability of the regression model.

It is often useful in work of this type to express the residuals not as absolute values but in standardised form. By this means the residual quantity ($Y_i - \hat{Y}$) is expressed as a proportion of the standard error in much the same way as raw data were standardised in section 5.8. Draper and Smith (1966) refer to the standardised residuals as *unit normal deviates*. They are given by:

$$\text{Standardised residual} = \frac{Y_i - \hat{Y}}{s_e} \qquad (9.14)$$

Geographers can take take particular advantage of this analysis by mapping out the residuals and examining their spatial distributions. The use of residuals in this form is widespread but their application is especially well illustrated by Clark (1967) in his work on farming patterns in New Zealand. A paper by Thomas (1968) reviews the full extent to which residuals can be used in map form. His work highlights three important areas of use: the formation and modification of hypotheses concerning the spatial association of variables and the search for new variables, the establishment of regional boundaries and units and, finally, the identification of specific areas for intensive field study and further investigation. Such areas will be generally those with high residual values where the regression model fails to predict accurately variations in the dependent variable. There are, however, further preconditions concerning residual behaviour that need to be fulfilled before the regression equation can be regarded as both unbiased and the best possible estimate of the population parameters. These points are discussed by Draper and Smith (1966) and Poole and O'Farrell (1971). It is, firstly, important that the residuals should be normally distributed about the regression line. Secondly, the degree of scatter of points about the regression line should not vary. As a result the standardised residuals should have zero mean and unit variance over the whole range of observed independent terms. This is the requirement of *homoscedasticity*. If the residuals do not conform to this scheme the data are said to be *heteroscedastic* and are unreliable for regression purposes. Figures 9.6a and b illustrate how these two conditions might appear when the data are plotted in graph form. One further requirement connected with residuals is that they should not be serially correlated, a condition more correctly known as *autocorrelation*. Residual autocorrelation is revealed by the arrangement of negative and positive items along the regression line.

If these conditions of homoscedasticity and zero autocorrelation are met we can be confident that our sample estimates of a and b are not subject to undue error or bias. A visual check of the graphs of the data is often sufficient to identify severe departures from these conditions. Should we be in any doubt concerning these points the residuals can be tested using a null hypothesis of no autocorrelation. We can review one of these methods later, but before doing so we need to know a little more about the nature of residual autocorrelation which can be either positive

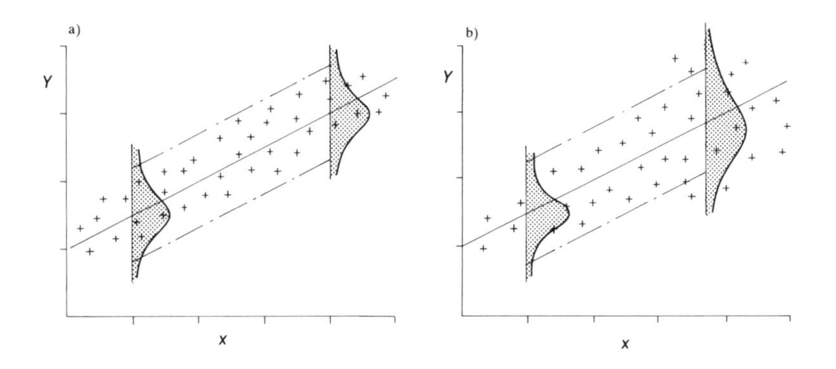

Figure 9.6 Representation of conditions of (a) homoscedasticity (equal spread of residuals) and (b) heteroscedasticity (varying spread of residuals).

or negative in form. If we were to plot a graph of residuals against the predicted *Y* items from which they are derived then positive autocorrelation occurs when positive and negative residuals occur in long runs (Figure 9.7a). Negative autocorrelation, no less of a problem, occurs when positive and negative residuals alternate in regular succession along the *Y* axis (Figure 9.7b). Zero autocorrelation is a state midway between these two conditions with residuals randomly scattered along the *Y* scale.

One of the more widely used measures of autocorrelation is the Durbin–Watson *d*-statistic, the derivation of which is relatively simple and based on the ordered sequence of residuals:

$$d = \frac{\text{sum of successive squared differences}}{\text{sum of squared residuals}}$$

In algebraic form this is represented by:

$$d = \frac{\Sigma(e_i - e_{i-1})^2}{\Sigma e_i^2} \qquad (9.15)$$

In the case of positive autocorrelation the adjacent residuals do not differ widely and *d* tends towards zero. If on the other hand negative autocorrelation is present then the differences between the adjacent ordered residuals are large and *d* tends towards its maximum possible value of 4.0. The value of *d* tends towards 2.0 as a condition of zero autocorrelation is approached.

The decision to accept, or reject, the null hypothesis of zero autocorrelation is not easily arrived at. Appendix X lists the bounds for

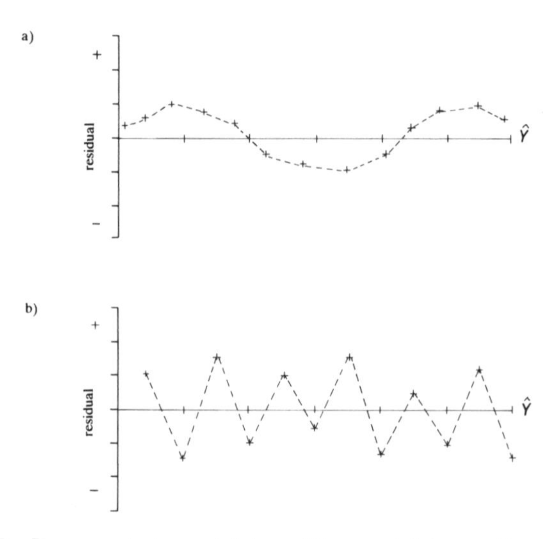

Figure 9.7 Representation of (a) positive and (b) negative serial autocorrelation in regression residuals.

the critical regions on the *d*-distribution. The bounds are obtained by reference to sample size *n* (the number of residuals), the number of independent terms *k* (always 1 in simple regression) and the significance level. Upper (d_u) and lower (d_l) values are given in the tables and are used to define two rejection regions for H_0. One is the rejection region for significant positive autocorrelation, the other for significant negative autocorrelation. In addition, a zone of acceptance of H_0 is defined about $d = 2.0$, leaving two areas along the distribution where the test is indeterminate. This system is depicted and summarised graphically in Figure 9.8.

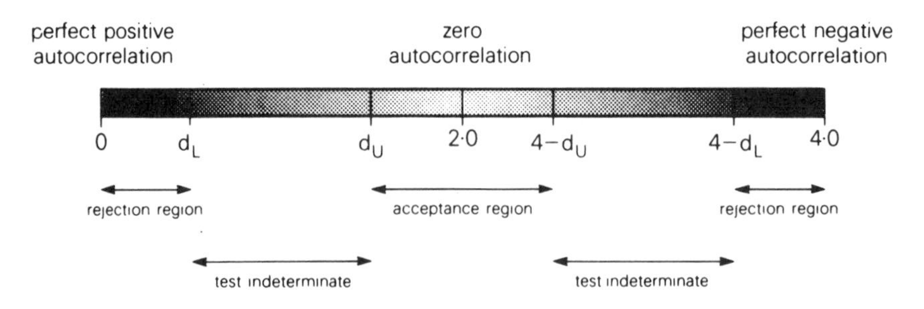

Figure 9.8 Rejection and acceptance regions for the Durbin – Watson statistic.

We will now test for autocorrelation in the residuals of predicted life expectancy. Either absolute or standardised residuals can be used, we will employ the former. A sample of the workings are shown in Table 9.3. It should be noted that for the purposes of subtracting successive items the

Table 9.3 Derivation of the Durbin – Watson *d*-statistic. Only some of the data are listed, but the sub-totals apply to the whole data set.

Predicted (Y)	Observed Y (Y)	Residual (Y − Y) = e	Squared residual e^2	Successive difference $(e_i - e_{i-1})$	Squared difference $(e_i - e_{i-1})^2$
46.9440	44.90	−2.0440	4.178		
				3.906	15.257
48.9683	50.83	1.8617	3.466		
				0.141	0.020
50.2378	52.24	2.0022	4.009		
				2.303	5.303
50.5809	50.28	−0.3009	0.091		
				9.993	99.860
50.9240	40.63	−10.2940	105.966		
...	...	...	...	...	...
...	...	...	...	...	...
...	...	...	...	...	...
70.9954	74.16	3.1646	10.015		
				7.059	49.829
71.0247	68.13	−3.8947	15.168		
				0.754	0.569
74.9410	71.80	−3.1410	9.866		
Totals			674.30		1591.35

Durbin – Watson *d* statistic = 1591.35/674.30 = 2.36

residuals are generally ordered according to the predicted and not the observed values of *Y* to which they correspond, though ordering by some other factor, such as time, might be possible.

From Table 9.3 we can use the sub-totals and equation 9.15 to give:

$$d = \frac{1591.35}{674.30} = 2.36$$

If we adopt the 0.01 significance level for the test we can consult Appendix Xb to find the lower (d_l) and upper (d_u) bounds which define the rejection and acceptance regions for *d*. As $k = 1$ and $n = 32$ we have $d_l = 1.160$ and $d_u = 1.282$. Using the method outlined in Figure 9.8 the zones on the *d*-distribution are listed in Table 9.4.

As a result we accept the null hypothesis of zero autocorrelation and are confident that, in this regard, our data conform to the test requirements. The problem of test indeterminacy can be reduced by using larger samples which narrow those two bands where *d* can offer no guidance.

Table 9.4 Rejection and acceptance regions for the Durbin – Watson d statistic for $n = 32$ and $k = 1$.

Range of d	Definition
0.0 – 1.160	Reject H_0 (significant positive autocorrelation)
1.161 – 1.282	Test indeterminate
1.283 – 2.718	Accept H_0 (no autocorrelation)
2.719 – 2.840	Test indeterminate
2.841 – 4.000	Reject H_0 (significant negative autocorrelation)

9.6 Spatial autocorrelation

Finally we must attend to the specifically geographical question of spatial autocorrelation. When data are drawn from spatially contiguous units, such as English counties or American states, we are confronted with a further problem – that of spatial, as opposed to serial, correlation. Expressed otherwise do residuals tend to group spatially, perhaps with positive residuals in one area and negative residuals in another? If such a pattern does emerge then the assumption of residual independence has been infringed irrespective of the conclusions of any Durbin–Watson test that may have been carried out.

Once again we may detect severe difficulties of spatial autocorrelation merely by mapping the residuals, but a more objective measure is always preferable. A relatively simple means of achieving this is through an adaptation of the χ^2 test (section 7.6).

A simple example is provided by the regression equation that describes the rainfall in the 48 contiguous American states by reference to their mean altitudes. The equation, significant at the 0.01 level, is:

$$Y = 45.33 + 0.062X$$

where Y is annual rainfall in inches and X is mean altitude in feet. If we wish to examine the question of autocorrelation of the residuals they must first be mapped (Figure 9.9) and from that map all the information needed for the autocorrelation test can be abstracted. From Figure 9.9 we can see that there is a tendency for the states with positive rainfall residuals, i.e. those with more rainfall than might be expected from the model, to be grouped in two principal areas. We must decide whether this visual assessment of the degree of spatial autocorrelation stands up to objective statistical analysis. In this example the residuals, which are measured on the ratio scale, are reduced to nominal scale attributes (positive or negative). This represents a lose of 'information' but renders the test far simpler than its parametric counterparts (Dacey, 1968).

Having once mapped the residuals we can proceed to test the null hypothesis of zero spatial autocorrelation. Any one state can be categorised as being one with a positive residual (group *P*) or one with a negative residual (group *N*). Equally all states are contiguous with a number of further states; California, for example, is contiguous with Oregon, Nevada and Arizona. Each contact can be one of three types, either positive with positive (*PP*), negative with negative (*NN*) or positive with negative (*PN*). If we take each state in turn we can make a running count of the number of contacts of all three types. At the conclusion, however, all contacts will have been measured twice and the total number registered in each class and overall must be divided by 2.0 to give the correct figure. If we denote the actual number of *PP*, *PN* and *NN* contacts by *X*, *Y* and *Z* respectively and the overall number of contacts by *L*, we get:

$$X = (\Sigma PP)/2 \tag{9.16}$$

$$Y = (\Sigma PN)/2 \tag{9.17}$$

$$Z = (\Sigma NN)/2 \tag{9.18}$$

$$L = (\Sigma L_k)/2 \tag{9.19}$$

where L_k is the total number of counted contacts made by all units k.

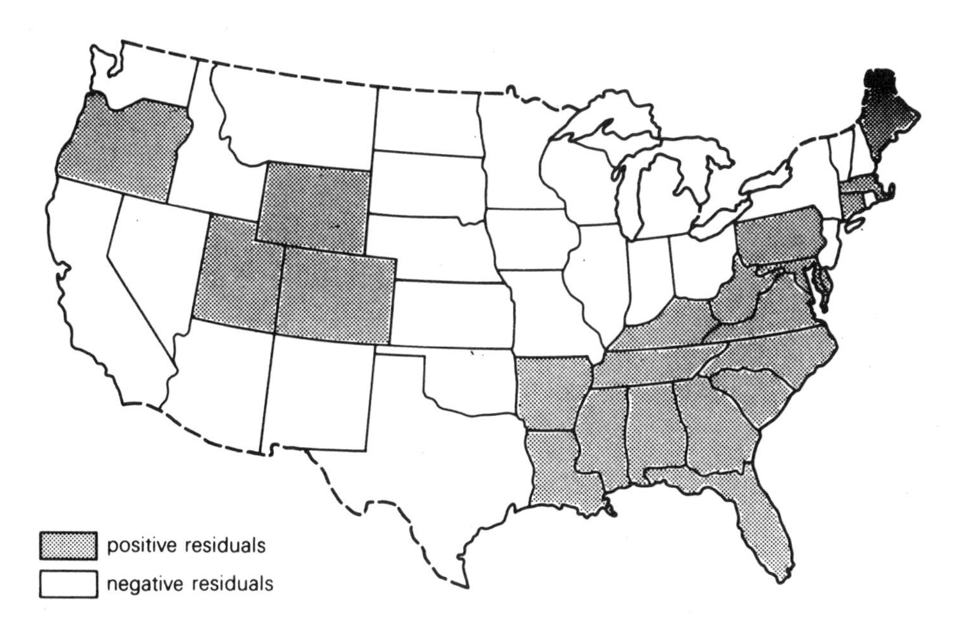

Figure 9.9 Map of residuals from the rainfall-altitude regression equation for US states.

Employing the principles of the χ^2 one-sample test we can determine the expected number of contacts of each type under a null hypothesis of zero autocorrelation. If the spatial distribution of the residuals is indeed random with no autocorrelation the probability of any one of them being positive must be 0.5 and the possibility of it being negative is also 0.5. Hence:

$$p(P) = p(N) = 0.5$$

It follows that for any junction of two spatial units – states in this case – the probabilities of each of the three possibilities are:

$$p(PP) = p(P)^2 = 0.25$$

$$p(PN) = 2 \times p(P) \times p(N) = 0.50$$

$$p(NN) = p(N)^2 = 0.25$$

Notice that the junction of positive and negative residuals can occur either as *PN* or as *NP* with either state taking on either role, hence the need to multiply that probability by 2. As the outcome must be one of the four possible pairings their probabilities sum to 1.0. We can establish the expected frequencies (E) under H_0 for all three categories through multiplying their respective probabilities by the total number of contacts (L). Thus:

$$E(X) = p(PP)L \quad (9.20)$$

$$E(Y) = p(PN)L \quad (9.21)$$

$$E(Z) = p(NN)L \quad (9.22)$$

With the observed and expected frequencies a χ^2 table can be constructed and analysed by the methods discussed in section 7.6.

The counting of observed contacts is a demanding task but one that is unavoidable. Not all states have the same number of contacts, Missouri has eight but Florida only two, and short boundaries can be easily overlooked. From Figure 9.9 the total number *PP*, *PN* and *NN* contacts were found to be 60, 74 and 76 respectively, giving thereby:

$$X = 60/2 = 30$$

$$Y = 74/2 = 37$$

$$Z = 76/2 = 38$$

$$L = 210/2 = 105$$

The corresponding expected frequencies are:

$$E(X) = 0.25 \times 105 = 26.25$$

$$E(Y) = 0.50 \times 105 = 52.50$$

$$E(Z) = 0.25 \times 105 = 26.25$$

From which the χ^2 table can be prepared. Table 9.5 shows that the test statistic exceeds the critical χ^2 value. We may therefore reject our null hypothesis of zero autocorrelation and conclude the distribution of residuals to be spatially non-random.

Table 9.5 χ^2 test for spatial randomness of US states rainfall residuals

	X	*Y*	*Z*
Observed frequency (*O*)	30	37	38
Expected frequency (*E*)	26.25	52.50	26.25
$(O - E)^2/E$	0.54	4.58	5.26

Degrees of freedom = 2; test statistic = 10.38; critical value at 0.05 significance level = 5.99

It must not be concluded that such a result is necessarily unhelpful. It may cause us to question the reliability of our regression equation and of the data but, at the same time, it may suggest means by which our understanding might improve. In this example the wetter than expected south-eastern states suggests that exposure to the humid airstreams from the Gulf of Mexico modifies the altitudinal effect and that subsequent studies should recognise the importance of exposure to humid air masses as a rain-forming factor. On the other hand we must not conclude that the absence of spatial autocorrelation necessarily indicates the absence of further important variables, merely that they do not operate on the purely spatial dimensions.

9.7 Simple regression using MINITAB and SPSS

It will have been apparent from the worked examples that regression analysis requires a notable degree of data manipulation and that computer assistance becomes near-indispensable when students are confronted with large data sets. The MINITAB and SPSS systems both offer means of quickly determining and testing regression equations.

Figure 9.10 shows how we could have processed data from the earlier example in MINITAB. The data are entered with a column for each variable. In this example the two variables (columns) have already been entered in the file **WORLD1** and upon recall named **LIFE** and **PROTEIN**. We have only two variables here but we are free to enter as many as are wished from which the **REGRESS** command line will select the two required for analysis. The dependent variable column or name appears first, followed by '1' (which indicates that only one predictor

```
MTB >   RETRIEVE 'WORLD1'
 WORKSHEET SAVED 11/13/1992

Worksheet retrieved from file: WORLD1.MTW
MTB >   NAME C1 'PROTEIN' C2 'LIFE'
MTB >   REGRESS 'LIFE' 1 'PROTEIN';
SUBC >  DW;
SUBC >  RESIDUALS C3.

The regression equation is
LIFE = 37.2 + 0.343 PROTEIN

Predictor        Coef       Stdev     t-ratio        P
Constant       37.203       2.784       13.36    0.000
PROTEIN       0.34309     0.04197        8.17    0.000

s = 4.741        R-sq = 69.0%      R-sq(adj) = 68.0%

Analysis of Variance

SOURCE        DF          SS          MS         F        P
Regression     1      1502.1      1502.1     66.83    0.000
Error         30       674.3        22.5
Total         31      2176.4

Unusual Observations
Obs.  PROTEIN     LIFE       Fit  Stdev.Fit   Residual    St.Resid
  20       61   67.530    58.131      0.843      9.399       2.01R
  25       40   40.630    50.926      1.287    -10.296      -2.26R
  29      110   71.800    74.943      2.133     -3.133      -0.74 X

R denotes an obs. with a large st. resid.
X denotes an obs. whose X value gives it large influence.

Durbin-Watson statistic = 2.36
```

Figure 9.10 Result of regression analysis using the MINITAB option and subcommands.

variable is being specified) then the column number or allocated name of that predictor variable. No further entries need be included and the results would appear immediately. In this example we do, however, take advantage of the sub-command structure to request that all standardised **RESIDUALS** are entered into column 3 **(C3)** of the data file for later inspection and that the Durbin – Watson statistic **(DW)** is also calculated.

The results begin with the least-squares regression equation, followed immediately by information on each of the coefficients where the intercept term *a* is described as the 'Constant' and the regression coefficient by the name of the independent variable **(PROTEIN)**. The standard errors for each estimate are given (Stdev) and although we haven't discussed the methods by which they may be obtained, each also has a t-value and associated random probability (P). The t-value is obtained from equation 9.23:

$$t = \frac{b - (\text{hypothesised value})}{\text{standard deviation of } b} \tag{9.23}$$

where the hypothesised value is usually 0.0. This allows us to determine the significance of both coefficients (d as well as b).

The term 's' is the standard error of the residuals (see equations 9.8 and 9.11) and 'R-sq' is the percentage of explained variance (equation 9.10). This latter figure can be corrected to give an 'adjusted' and unbiased estimate if the two contributing sums of squares are divided by their respective degrees of freedom. The analysis of variance table then follows. The program also identifies unusual observations. Those marked with an X are unusual with respect to the predictor term, those with R are unusual with respect to the residual term. Observed and fitted dependent terms are given together with the standard deviation (i.e. standard error) of the fit and the absolute and standardised residuals. The standard deviation of fit is derived using equation 9.13 and, hence, it varies according to X as described in section 9.4. It should, furthermore, be noted that the standardised residuals in the final column are based on the latter estimates of the standard error of Y but calculated with that particular extreme value excluded from the data set, thereby giving it greater emphasis.

The three items that are picked out by this procedure are those for Malaysia, where the observed life expectancy (67.53 years) is far longer than that expected from the model (58.13 years), for Guinea, where observed life expectancy (40.63 years) is far less than is expected (50.92 years) and New Zealand where although the residual is not large the protein intake is exceptionally high. The identification of these anomalies is often helpful in furthering the investigation of the factors controlling the dependent variable. The Durbin–Watson statistic (2.36) appears on the final line of the output. If the complete set of residuals stored in column 3 (**C3**) needs to be examined a separate **PRINT** instruction must be used.

The SPSS system can also be used to provide simple regression equations but it is particularly well-suited to multiple regression analysis and is discussed and introduced in Chapter 11. The instructions required for multiple regression analysis are exactly the same as those for simple regression, the latter requiring only the specification of one instead of two or more predictor variables.

9.8 Closed number systems in correlation and regression analysis

Geographers make frequent use of closed number sets such as proportions or percentages, often in an attempt to bring varying sized areas to a common and comparable scale. The term 'closed number' refers to the situation where all data sum to either 1 (if they are proportions) or 100 (in the case of percentages). However, the use of percentages can raise problems in the application of correlation and regression techniques. As long ago as the early 1960s Krumbein (1962) demonstrated how correlation coefficients based on percentages differ from those using open number sets for the same data. Two problems were identified.

The first of these relates to the fact that the range of possible values is fixed between 0 and 100 and predicted values either side of these limits are nonsensical. This difficulty is not absent from all open number sets where there may be vague upper and lower limits on the magnitude of dependent variables, but it is more marked when using closed number sets. We may state that the regression model applies only to the observed range of dependent values and there is much to commend this course of action. But when using closed number sets we can adopt a second solution involving the conversion of the percentage values into an infinite ratio scale using a *logistic* or *logit* transformation (with natural logs (*ln*)) where:

$$L_j = ln \frac{P_{cj}}{(100 - P_{cj})} \tag{9.24}$$

Equation 9.24

P_{cj} = percentage value of observation
L_j = logit value for observation

In this way any closed number set converts to a set which ranges from zero to plus infinity using the ratio $P_{cj}/(100-P_{cj})$. The further step of taking the latter's natural logarithm expands the set to one which can occupy the range minus to plus infinity. The independent and dependent variables may now conform to the requirements of normality and of being measured on the interval/ratio scale. When required this transformation can be carried out on MINITAB and on SPSS by using their respective **LET** and **COMPUTE** instructions as described in section 5.12, the correlation and regression analysis then being carried out on the transformed data. In all such cases however the geographical interpretation of the regression model may be complicated by the transformation and the consequent loss of numerical clarity.

A further problem concerns the use of percentages in the dependent variable (or variables). Where data are in complementary classes, such as male and female, the cumulative total over the two for any given sample must be 100 per cent. The correlation between them is, therefore, predetermined and as one increases the other must decrease. Hence we must always take care never to use both such classes. The same problem, though multivariate in character, persists where three or more such classes exist. For example, using the British census socio-economic classes 1 to 5, once the percentage of people in classes 1 to 4 are known, the percentage in class 5 is predetermined. In all such cases we are, in effect measuring the same variable twice. The problem disappears as the number of complementary classes grows. Unfortunately, there are no easy solutions to such difficulties and number sets of this character must be avoided in correlation and regression analysis.

9.9 Use of linear regression models in geography

Regression analysis is one of the most widely employed statistical techniques in geography. It is used in three ways: to establish a predictive numerical model, to test a model or hypothesis and, lastly, to describe the relationship between variables.

A review of geographical literature shows that all three applications are widespread, but most emphasis has been given to using regression in a descriptive and explanatory fashion. For example, Haggett (1964) used regression to describe the relationship between physical and economic variables and forest cover in Brazil. Similarly, Doornkamp and King (1971) used the technique to explore aspects of drainage basin morphometry. Johnston (1981) has used regression coefficients to examine trends in voting patterns over time, relating the slope coefficients to the r^2 values (the coefficients of determination), to test an ecological model of voting behaviour. Obviously, it is possible to assess statistically the differences in such coefficients, as was demonstrated by Garner (1966) in a study of shopping centre characteristics.

More recently, however, some geographers have taken a more critical look at regression analysis and have suggested a certain amount of misuse of the technique (Mark and Peucker, 1978). This work cast doubt on the use, in certain circumstances, of regression for describing the relationship between variables. The problems seem to arise when the independent variable is subject to measurement error. In most uses of regression we relax this condition and assume that all the unexplained or residual error is ascribed to the dependent variable (Y) and that the independent term (X) is error-free. Mark and Peucker argue that this is not the case in many studies and that the alternative technique of 'functional' or 'structural' analysis should be used. This technique, unlike regression, divides the unexplained error between both the independent and the dependent variables. As yet such a technique has been little used in geography and remains outside the scope of this book. However, it is something that researchers will perhaps be giving more attention to in the future. Fortunately, these problems with regression are not important when the coefficients of determination (r^2) are high, i.e. close to 1.0, because the error terms of X and Y may be relatively small.

A final problem concerns the identification of the dependent and independent variables. In some situations this so-called process–response relationship is not entirely clear, especially if we are merely using regression to explore the generally numerical relationships between variables. Under such circumstances we may end up plotting two regression lines, one of Y on X and another of X on Y. But more crucial is the fact that statistical dependence does not necessarily imply that there is a geographically valid, physical relationship between two variables. For example, a high correlation and close-fitting regression line can either indicate that there is a link between variables or that they are merely responding simultaneously to a third, unknown, factor. If this is the case our analysis needs to take into account this other variable. Many geographical problems are indeed of this *multivariate* character and the following chapters will show how these more complex techniques can be used.

References

Clark, W.A.V. (1967) 'The use of residuals from regression in geographical research', *New Zealand Geog.*, 23, 64–67.

Dacey, M.F. (1968) 'A review on measures of contiguity for two and *k*-coloured maps', in B.J.L. Berry and D.F. Marble (eds) *Spatial Analysis: a reader in statistical geography*, Prentice-Hall, Englewood Cliffs.

Doornkamp, J.C. and King, C.A.M. (1971) *Numerical Analysis in Geomorphology*, Arnold, London.

Draper, N.R. and Smith, H. (1966) *Applied Regression Analysis*, Wiley, New York.

Garner, B.J. (1966) *The Internal Structure of Retail Nucleations*, North Western University Studies in Geog., No. 12.

Haggett, P. (1964) 'Regional and local components in the distribution of forested areas in south east Brazil: a multivariate approach', *Geogrl. J.*, 130, 365–377.

Johnston, R.J. (1978) *Multivariate Statistical Analysis in Geography*, Longman, London.

Johnston, R.J. (1981) 'Regional variations in British voting trends, 1966–1979: tests of an ecological model', *Regional Studies*, 15, 23–32.

Krumbein, W.C. (1962) 'Open and closed number systems in stratigraphic mapping', *Bull. Ann. As. Petrol. Geol.*, 46, 2229–2245.

Mark, D.M. and Peucker, T.K. (1978) 'Regression analysis and geographic models', *Canadian Geogr.*, 22, 51–64.

Poole, M.A. and O'Farrell, P.N. (1971) 'The assumptions of the linear regression model', *Trans. Inst. Br. Geogrs.*, 52, 145–158.

Thomas, E.N. (1968) 'Maps of residuals from regression: their characteristics and uses in geographic research' in B.J.L. Berry and D.F. Marble (eds) *Spatial Analysis: a reader in statistical geography*, Prentice-Hall, Englewood Cliffs.

Chapter 10
Nonlinear regression

10.1 Introduction

Thus far we have assumed in both correlation and regression that the relationship between any two variables is linear and a scatter diagram of the data would plot approximately as a straight line. But such linearity does not always exist and it is important to know how to deal with nonlinearly-related variables. Some approaches to nonlinear analysis are far more demanding than those we have already outlined. Yet much can be done using simple adaptations of the linear model and geographers should not shrink from exploring nonlinear associations as such a condition does not preclude causal connection between the variables in question. Consider for example the case of Figure 10.1 where the data plot suggests a close degree of association between X and Y but one that is decidedly nonlinear and may require special treatment in regression analysis.

The following sections will show how many nonlinear relationships can, through the transformation of one or both of the variables, be 'linearised'. In this transformed condition the two variables can be treated by the methods described in the preceding two chapters. While this strategy precludes the use of those curves and their attendant equations which cannot be linearised it leaves sufficient options to meet the needs of most geographers.

10.2 First steps in curve fitting

The most efficient means of detecting nonlinearity is to plot the scatter of the data using the plotting procedures in MINITAB or SPSS (sections 8.2 and 8.3). In addition, the nonlinearity of some relationships appears only over a wide range of data and to assume otherwise on the basis of a limited sample may be to overlook a potentially fruitful line of analysis. This problem is illustrated if we imagine ourselves to have plotted Figure 10.1 using only data from either end of the range of X values; how different might our perception of the relationship have then been?

However the identification of nonlinearity marks only the start of an analysis that demands careful thought on the part of the researcher. When using linear regression the only equation with which we are concerned is that of a straight line, $Y = a + bX$. But when examining nonlinear relationships there are many curves from which to select our

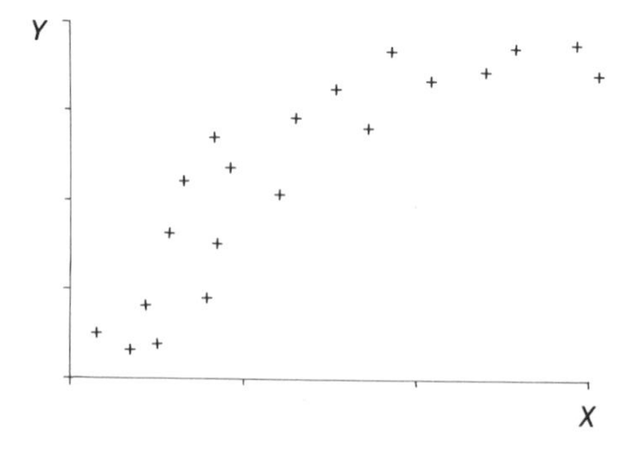

Figure 10.1 Scatter diagram representing a hypothetical nonlinear association.

'best fit' equation, and our choice may have important implications for the interpretation of the results. The need to select the most appropriate curve is a problem that can be overcome by either of two approaches. Firstly we may suspect that two variables behave in a particular way. The linear model assumes, for example, that equal increments in X produce correspondingly equal increments in Y over the whole data range. In nonlinear models this equality might no longer apply. To take one example; we know that population density does not decline linearly with distance from city centre and instead of decreasing by a fixed absolute amount for each unit of distance it tends to decline by a fixed proportion. For example, over the first kilometre density may decline by 10 per cent of its initial (city centre) value, over the second kilometre it then falls by 10 per cent of the density at one kilometre and so on. By this means the decrease at each kilometre step becomes ever smaller and describes what is known as an *exponential curve*. This type of curve is important and is one that we will be considering as it covers a wide range of phenomena.

Such a deductive approach has the advantage of directing the researcher to a specific curve or family of curves. Its disadvantage is that we must have a good initial understanding of the nature of the two variables under study and the manner in which they are causally connected. Very often, however, this degree of understanding is absent leaving the researcher to employ a more empirical approach whereby a range of curves are studied and fitted to the data in order to find the most appropriate of them. Clearly, such an approach can be extremely time-consuming.

In practice a compromise procedure may be adopted with attention confined to a limited range of curves. This is not necessarily a disadvantage as the flexibility of the curves and their consequent ability to summarise different trends is considerable. Three popular curves can be identified and summarised as follows:

$$\text{Simple power curve } Y = aX^b \tag{10.1}$$

$$\text{Simple exponential curve } Y = ae^{bX} \tag{10.2}$$

$$\text{Simple logarithmic curve } Y = a + b\log X \tag{10.3}$$

All three are similar to the linear model in possessing one predictor and one dependent variable whose behaviour are described by two constants (*a* and *b*). Only the exponential equation differs by its inclusion of the base of natural logarithms *e* which is equivalent to 2.7183 (section 2.3) and is not to be confused with the use of the same symbol in describing error terms in regression models. All three curves are also, in most circumstances, linearisable through the transformation of one or both variables. More importantly, once linearised the conventional least-squares methods described in Chapter 9 can be used to determine the best-fit values for *a* and *b*. By this method a straight line is fitted through the now linear scatter of points. As a final step it is possible to 'detransform' the data to plot the curve that corresponds to the derived least-squares line.

The means by which we can evaluate and plot curves from equations have already been introduced in section 2.5. Figures 2.4 and 2.5 also demonstrated how, in the case of logarithmic equations, the curve may be linearised by the purely graphical device of using logarithmic paper. This has the effect of plotting *Y* not against *X* but against its log-transformed equivalent, log*X*. The latter variable could equally be termed *V*, which would give the recognisably linear expression:

$$Y = a + bV \qquad (10.4)$$

Figure 10.2 is a redrafting of Figure 2.5 and shows the nature of the link between *V* and *X*. The latter, original, values are shown in brackets beneath their logarithmic counterparts (*V*) which were treated as a 'new' linear variable and plotted on a uniform scale on the horizontal axis. Any scatter of points that approximated to such a pattern could be treated as linear and the least-squares coefficients estimated and substituted into equations 10.3 and 10.4. Plotting a graph of the latter equation would produce a straight line, and of the former a logarithmic curve.

For the power and exponential curves the linearising procedures differ only slightly from this scheme. Power curves are linearised by log-transforming both the *X* and the *Y* variables, while exponential curves plot as straight lines when *Y* alone is logged. These requirements are summarised in Table 10.1.

It must not be overlooked, however, that the geographer's raw data consist of imperfectly related variables which plot as scatters of points and not as perfect curves or straight lines. The curves fitted to pass through the points merely summarise the general trend of the data and to be successful in that task the appropriate choice of curve must be made. Preliminary checks should be made by plotting the transformed data along the lines suggested in Table 10.1. Hence, for example, if we plot log *Y* against log *X* and obtain a linear scatter of points we might be justified in fitting the power curve equation to that data. But if non-linearity persists we should consider an alternative curve.

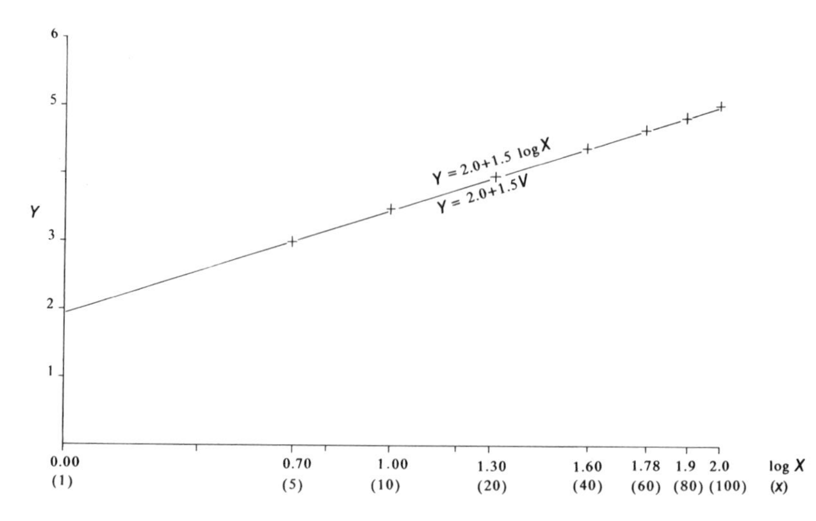

Figure 10.2 Graph based on Figures 2.4 and 2.5 showing how logarithmic relationships can be linearised. The logarithms and the arithmetic values from which they are derived (in brackets) are shown on the horizontal axis.

Table 10.1 Summary of transformations to linearise different curves

Expression	X term	Y term
$Y = a + b \log X$	Convert to logs	No transformation
$Y = ae^{bX}$	No transformation	Convert to logs
$Y = aX^b$	Convert to logs	Convert to logs

The option of summarising non-linearly related variables is a useful one for geographers but problems may be found with the data when the methods here described are employed. Most importantly logarithms cannot be taken of zero or of negative quantities. One way of avoiding this problem might be to add a constant term to X in order to render all values positive. Thus if X ranges between zero and -14.5 we might modify the logarithmic (or any other equation) by re-expressing it in the form:

$$Y = a + b(X + 15.0)$$

In this way X is increased to a positive quantity before it is logged. The standard procedures for establishing the values of a and b can then be used provided that in any subsequent interpretation the inflated character of X is not overlooked. It must also be stressed that nonlinear relationships do not reduce the importance of correctly determining the direction of causation in the model and the terms X and Y must be attached to the appropriate independent and dependent variables respectively.

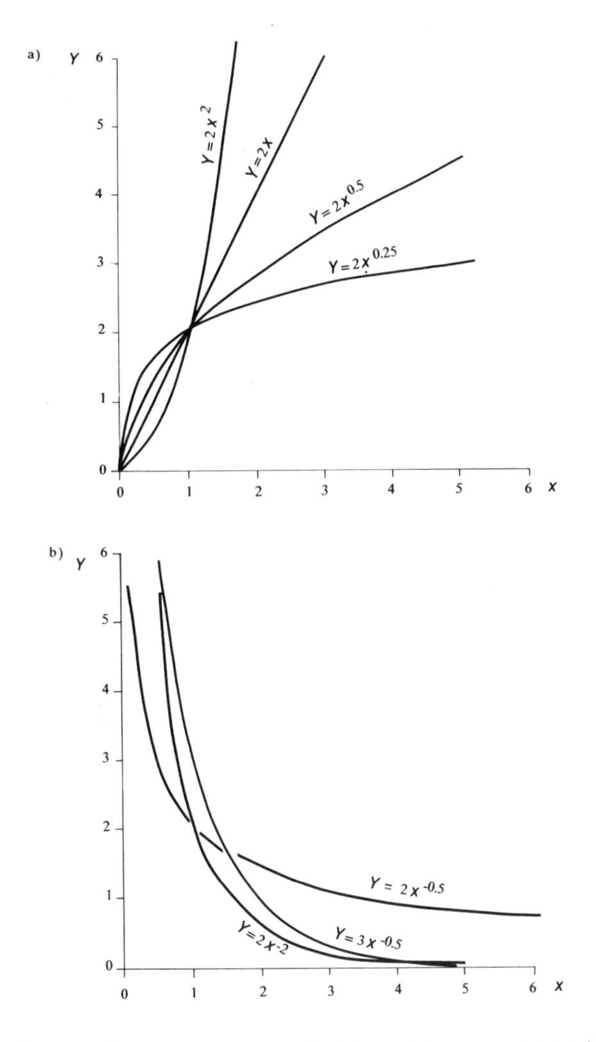

Figure 10.3 Forms of power curves with (a) positive powers and (b) negative powers.

10.3 The simple power curve $Y = aX^b$

In this first example we will consider the simple power curve and, in particular, the way in which the two coefficients determine its form. Figures 10.3a and b show that its curvature is governed by the magnitude and sign of coefficient b, while coefficient a determines its location on the co-ordinate system being the point on the Y scale where $X = 1.0$. In addition the curve is concave-upwards for positive b terms in excess of 1.0, but convex-upwards for b values between 0.0 and 1.0. The line is straight if b is equal to 1.0. If b is negative the curve is asymptotic with respect to both axes, i.e. as the axes extend it approaches ever more closely to them but contact is never made. In the case of neither positive nor negative *b* coefficients does the curve extend across the axes into 'negative' areas of the co-ordinate system.

The simple power curve can be linearised by log-transforming both variables. Thus, if data do approximate to a power curve that tendency will be revealed through the linearity of a graph plot of the logs of both sets of data. Such power expressions are used widely in geography but are especially important in geomorphology where many aspects of hydrology and hydraulic geometry are known to be nonlinear. The following example is based on data from a study of the flow characteristics of a river in Washington State, USA, and examines the relationship between river discharge ('flow') and water velocity ('vel') at one point along the river's course. MINITAB is used to transform and to plot the data. The instructions, which include not only the naming of the variables but also labelling of the *X* and *Y* axes, are shown in Figure 10.4a and b which

```
MTB  >   RETRIEVE 'USRIVERS'

Worksheet retrieved from file: USRIVERS.MTW

MTB  >   NAME C1 = 'flow'
MTB  >   NAME C4 = 'vel'
MTB  >   LET C5 = LOGT('flow')
MTB  >   LET C6 = LOGT('vel')

MTB  >   PLOT 'vel' 'flow';
SUBC >   XLABEL = 'flow - cu.ft/s';
SUBC >   YLABEL = 'vel - ft/s'.

        -
     6.0+                            *    *  2
        -                              *                          *
v       -          *                         *
e       -      *      * *       *
l       -                 *
     4.0+          *  ***
-       -    *** **   *
        -   **  *   * **
f       -   * ** *2
t       -  2*2*
/    2.0+ * 2
s       - 33
        - 5
        -  *
        - *
     0.0+
          +---------+---------+---------+---------+---------+------
          0        60       120       180       240       300
                              flow - cu.ft/s
```

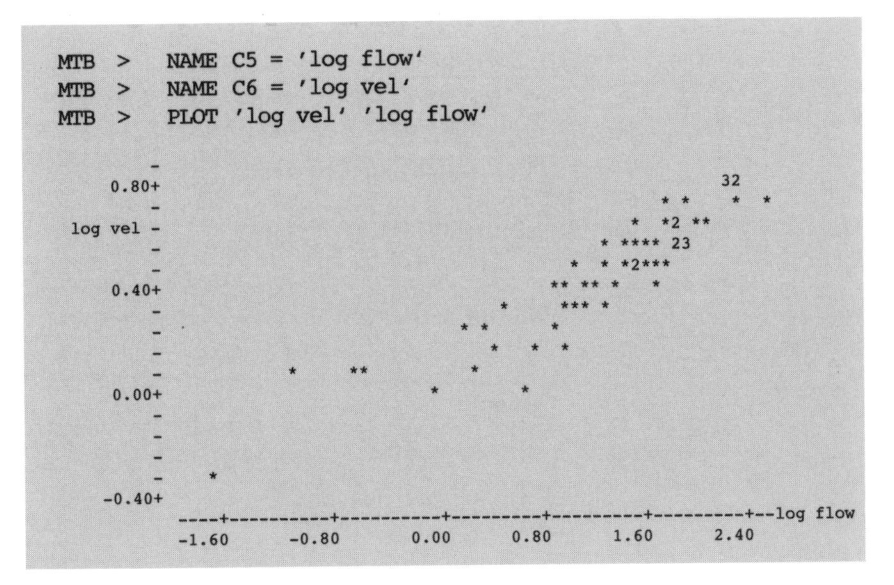

Figure 10.4 MINITAB-derived plot showing (a) instructions for retrieving and transforming the data file with a plot of the untransformed data, and (b) a plot of the data after transformation.

include also the two plots. The **LET** command is used to log-transform the two variables and to store them in columns 5 and 6 of the data matrix. It is readily apparent that the raw data do indeed plot as a curve with increases in velocity becoming less marked as discharge increases. In contrast, the plot of the transformed data is clearly linear and suggests that the relationship is indeed one that can be approximated by a power expression. Readers should note that we have taken stream velocity to be dependent upon river discharge.

We can now proceed to treat the logged data as we would any other sample and estimate the least-squares line applicable to Figure 10.4b. We will use the method exemplified in section 9.2 based on equations 9.4 and 9.5 to which we need make only minor changes to acknowledge the transformed character of the data. In effect we are finding the straight line parameters for the following expression:

$$log Y = \log a + b \log X \tag{10.5}$$

which becomes more recognisable if we designate $\log Y$, $\log X$ and $\log a$ as U, V and a_1 respectively so that the equation now reads:

$$U = a_1 + bV \tag{10.6}$$

As a consequence equations 9.4 and 9.5 must be rewritten as equations 10.7 and 10.8:

$$b = \frac{\Sigma uv}{\Sigma v^2} \tag{10.7}$$

Equation 10.7

b = power coefficient
Σuv = sum of all uv products
Σv^2 = sum of all v^2
$u = U - \bar{U}$; $v = V - \bar{V}$

$$a_1 = \bar{U} - b\bar{V} \tag{10.8}$$

Equation 10.8

a_1 = log of coefficient a
$\bar{U}$ = mean of logs of Y
$\bar{V}$ = mean of logs of X
b = power coefficient

Table 10.2, which can be compared usefully with Table 9.1, shows how the working might be set out if the calculations are performed manually. The appropriate sub-totals can then be included in equations 10.7 and 10.8 to give linearised coefficients. In this form the least-squares straight line can be plotted through the scatter of points in Figure 10.4b by taking any two values of V (log$X$) and substituting them into equation 10.6 to give the corresponding U (log$Y$) values as outlined in section 9.2.

It may be equally informative however to plot the curve that describes the data in Figure 10.4a. But to do so the linear equation must be 'detransformed' by taking the antilogs of all logged terms. Hence U and V go back to Y and X while the intercept term becomes antilog a_1. The regression coefficient, on the other hand undergoes no numerical change but, in accordance with the principles of operations with logarithms introduced in section 2.3, becomes a power term. The expression now appears as

$$\log Y = 0.126 + 0.2511 \times \log X$$

from which we get the power equation:

$$Y = 1.337X^{0.2511}$$

By substituting a range of X values into this equation and deriving thereby the corresponding Y terms we can obtain sufficient pairs of values with which to plot the curve (Figure 10.5). Whenever such

Table 10.2 Calculations and some of the data used in estimating the least squares regression line relating stream velocity (in feet per second) to discharge (in cubic feet per second).

X discharge	Y velocity	V ($\log X$)	U ($\log Y$)	v $(V - \bar{V})$	u $(U - \bar{U})$	v^2	uv
124.0	4.70	2.093	0.672	0.832	0.229	0.692	0.191
96.7	4.45	1.985	0.648	0.724	0.205	0.524	0.148
12.8	1.93	1.107	0.286	−0.154	−0.157	0.024	0.024
190.0	5.80	2.279	0.763	1.018	0.320	1.036	0.326
...	...	...	...	...	...	...	...
...	...	...	...	...	...	...	...
...	...	...	...	...	...	...	...
0.07	1.10	−1.155	0.041	−2.416	−0.401	5.837	0.969
0.29	1.10	−0.538	0.041	−1.799	−0.401	3.236	0.721
9.6	2.66	0.982	0.424	−0.279	−0.019	0.078	0.009
Means		1.261	0.443				
Totals						43.267	10.862

Regression coefficient *(b)* = 10.862/43.267 = 0.2511
Intercent term (a_1) = 0.443 − 0.2511 × 1.261 = 0.126
Source: *Morphology and Hydrology of a Glacial Stream – White River, Mount Rainies, Washington*, US Geol. Survey Prof. Paper 422–A, Washington DC (1963)

calculations are carried out it must be remembered that the power term precedes the multiplication in order of execution. As the power

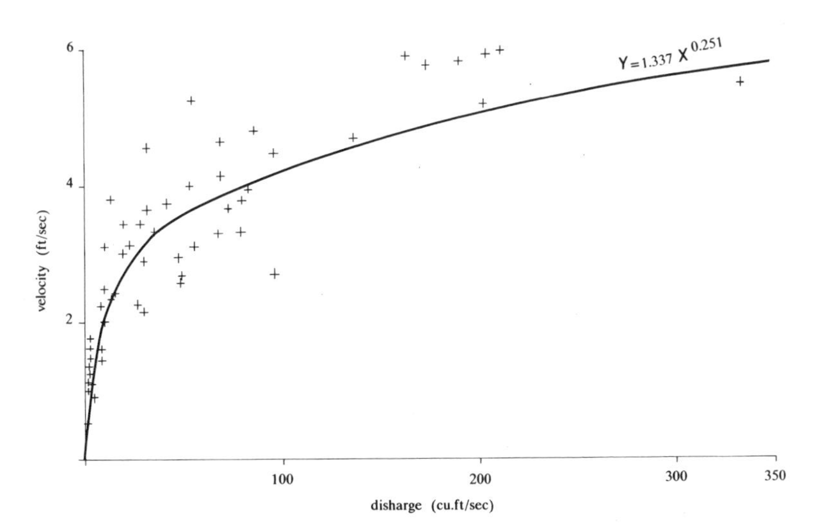

Figure 10.5 Fitted curve describing the relationship between stream velocity and discharge derived from the log-transformed least-squares line.

coefficient is less than 1.0 the resulting curve is convex-upwards. The character of the simple power equation also dictates that the Y value for $X = 1.0$ is the value of coefficient a.

Any analysis of the residuals along the lines introduced in Chapter 9 should, it must be noted, take place on the log-transformed data and results. In this condition the assumptions of linearity are fulfilled and the standard methods can be applied to determine, for example, the Durbin–Watson statistic for autocorrelation.

10.4 The simple exponential curve $Y = ae^{bx}$

Although it includes the base of natural logarithms (e) the simple exponential curve is readily linearisable. Figure 10.6 illustrates the principal characteristics of this curve in which coefficient b governs the degree of either positive or negative slope, with coefficient a now assuming the role of an intercept term and having the value of Y when X is zero. The simple negative exponential curve is asymptotic to the X

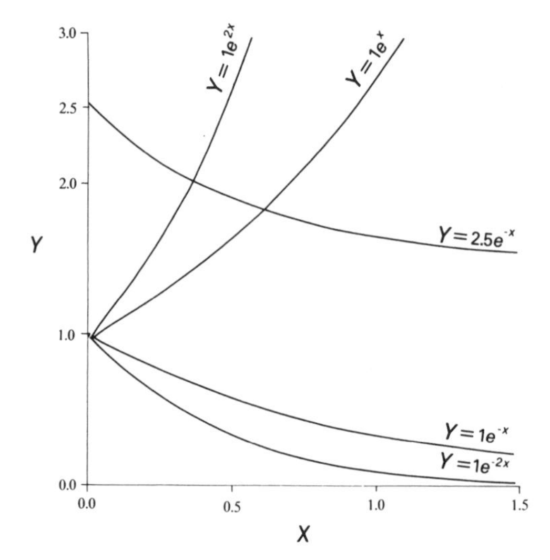

Figure 10.6 Forms of exponential curve.

axis and in Figure 10.6 their downward sweep approaches, but never meets, the X axis.

To identify an exponential relationship the logs of Y should be plotted against the untransformed values of X. If the pattern of points is linear then an exponential expression may summarise the association between the variables. This is the converse of a logarithmic curve which can be linearised by log-transforming the X variable (see Figure 10.2).

Such forms of association are widespread, especially in the physical sciences but are found also in human geography. It is, for example, widely supposed that population density declines in a generally exponential manner with distance from city centres and this suggestion forms the basis of the example that follows. There are many cities where the population densities are at their greatest close to the centre. On the other hand, no matter how far from the city the study might extend negative population densities will be impossible. In respect of both features the negative exponential curve is a prima facie candidate as it can never cross the X-axis and declines from its greatest value at the point where X (distance from city centre) is zero.

The process of linearising the exponential equation requires no change to the independent term X, but the whole of e^{bX} needs to be rearranged. At the same time the intercept term a, which is expressed in the units of Y, and Y itself need to be log-transformed. These changes give:

$$\log Y = \log a + (b \log e)X \qquad (10.9)$$

Equation 10.9

$\log Y$ = common log of the variable Y
$\log a$ = common log of the intercept term
b = exponential coefficient
$\log e$ = common log of the constant e 0.4343
X = untransformed variable X

This superficially intimidating expression can be simplified by allocating new symbols to the transformed components, thus:

$$U = a_1 + b_1 X \qquad (10.10)$$

Equation 10.10

U = $\log Y$
a_1 = $\log a$
b_1 = $b \log e$
X = untransformed variable X

In this form the equation is recognisably linear and differs from the standard linear form only in the symbols used to describe its terms. As with the linearised power expression we can employ the conventional procedures for determining the least-squares coefficients a_1 and b_1. The latter is given by:

$$b_1 = \frac{\Sigma xu}{\Sigma x^2} \qquad (10.11)$$

Equation 10.11

Σxu = sum of all xu products
Σx^2 = sum of all x^2
b_1 = $b\log e$
$x = X - \bar{X}$; $u = U - \bar{U}$

while a_1 is found from:

$$a_1 = \bar{U} - b_1\bar{X} \qquad (10.12)$$

Equation 10.12

a_1 = log of intercept term
$\bar{U}$ = mean of logs of Y
$\bar{X}$ = mean of X
b_1 = $b\log e$

Our example takes data from the UK census and examines the decline in population density away from the city of Kingston-upon-Hull. The dependent term is population density, which is determined by distance from the city centre. A random sample of 27 civil parishes and urban wards was taken. The linearity of the plot of the logs of the population density against distance (Figure 10.7) suggests that the exponential equation might be applicable and also indicates a high degree of correlation between the variables. The MINITAB package was used to obtain these results. Figure 10.7 shows how the dependent term was log-transformed and the three variables named before correlation and plotting using the now-familiar commands.

A sub-sample of the data and the working needed for the manual calculation of the coefficients is shown in Table 10.3. The negative b_1 term shows that population density decreases with distance from city centre. While the a_1 term is the log of the population density at the city centre. The equation can be detransformed to represent the original curve. This is done by taking the antilog of a_1. The conversion needed to identify the value of b is less straightforward. The coefficient b_1 is equivalent to the whole expression $b\log e$, as a result $b = b_1/\log e$, and

$$b = b_1/0.4343 \qquad (10.13)$$

For the current example we obtain

$$a = \text{antilog } 2.0686 = 117.11$$

and

$$b = -0.1396/0.4343 = -0.3214$$

to give the exponential expression

$$Y = 117.11e^{-0.3124X}$$

This equation can be used to evaluate points from which the curve can be plotted (Figure 10.8) to pass through the scatter of points represented by the untransformed sample data. In evaluating Y from any X term it must be remembered that e is always 2.7183 and that only after e has been raised to power bX should it be multiplied by a. It should also be recalled from section 2.2 that any number raised to the power of zero is 1.0.

```
MTB >   RETRIEVE 'HULLPOPS'

Worksheet retrieved from file: HULLPOPS.MTW

MTB >   NAME C1 = 'density'
MTB >   NAME C2 = 'distance'
MTB >   LET C3 = LOGT('density')
MTB >   NAME C3 = 'log dens'
MTB >   CORRELATION C2 C3

Correlation of distance and log dens = -0.911

MTB >   PLOT 'log dens' 'distance'

          -
      2.0+  * *
          -     *  **  *
 log dens-      2
          -            *
          -           **         *
      1.0+                               *
          -                                       *
          -                                        *
          -                 *   *     *
          -                            *              *
      0.0+                                  * *
          -                                              *
          -                                  *                    *
          -                                          *  *
          -
     -1.0+
          --------+---------+---------+---------+---------+--------distance
                 3.5       7.0      10.5      14.0      17.5
```

Figure 10.7 MINITAB instructions and plot of log-transformed population density against distance from city centre.

Table 10.3 Calculations and a sub-sample of the data used in estimating the least-squares regression line relating population density to distance (in people per hectare) from city centre (in kilometres).

Y pop. density	X distance	U (log Y)	x $(X-\bar{X})$	u $(U-\bar{U})$	x^2	ux
49.51	2.35	1.695	−6.770	0.901	45.838	−6.099
51.77	3.50	1.714	−5.620	0.920	31.580	−5.170
93.82	1.75	1.972	−7.370	1.178	54.322	−8.682
25.55	4.75	1.407	−4.370	0.613	19.100	−2.679
...	...	...	...	...	...	...
...	...	...	...	...	...	...
...	...	...	...	...	...	...
0.48	12.30	−0.319	3.180	−1.113	10.112	−3.539
2.20	11.80	0.342	2.680	−0.452	7.182	−1.211
0.28	16.00	−0.553	6.880	−1.347	47.334	−9.267
Means	9.120	0.794				
Totals					800.56	−111.86

Regression coefficient (b_1) = −111.86/800.56 = −0.1397
Intercept term (a_1) = 0.794 − (−0.1397 × 9.120) = 2.068
Source: *Census 1971 England and Wales County Report: Yorkshire East Riding*, Part I, HMSO London 1973.

10.5 Significance testing in nonlinear regression

The curves obtained in the two earlier examples are, in their linear forms, least-squares lines and as such may be legitimately tested to assess their statistical significance. This is as important a task as it was in the intrinsically linear case since it allows the reliability of the regression model and any estimates derived from it to be determined. The various tests proceed exactly along the lines already described in Chapter 9 with the exception that where transformations to either variable have been made it must be the transformed data and their derived statistics that are used. In this respect all the assumptions and requirements of regression modelling specified in Chapter 9 apply with equal force and cannot be overlooked because either or both variables are transformed. As an example we can consider the case of the water velocity/discharge power relationship found in section 10.3 to be:

$$Y = 1.337X^{0.2511}$$

It is, however, the linearised form that will be the focus of attention:

$$\log Y = 0.126 + 0.2511 \log X$$

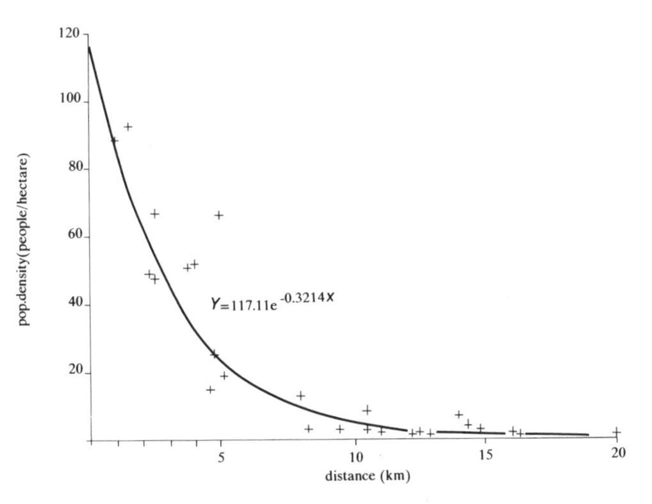

Figure 10.8 Fitted exponential curve describing population density decay with distance around the city of Kingston-upon-Hull, England.

This is more readily recognised as a linear expression if the transformed variables are redesignated as U and V:

$$U = 0.126 + 0.2511V$$

from which the total, regression and residual sums of squares can be derived. If all Y terms in equations 9.7 and 9.8 are replaced by U the regression and error sums of squares are found from:

$$\text{Regression sum of squares (SSR)} = \Sigma(\hat{U} - \bar{U})^2 \qquad (10.14)$$

$$\text{Error sum of squares (SSE)} = \Sigma(\hat{U} - U)^2 \qquad (10.15)$$

The degrees of freedom by which each of these sums of squares must be divided to obtain the corresponding variances are the same as in the linear case; $k = 1$ for the regression variance (because there is only one predictor term) and $n - k - 1$ for the residual variance. Once again it is necessary to calculate the $\hat{U}$ term for every V (log predictor) value in the sample. The data might then be set out as in Table 10.4 from which the analysis of variance table can be prepared. The null hypothesis under test is that there is no explanation of the dependent variable (stream velocity) in terms of the predictor variable (discharge). The critical F-value at the 0.01 significance level (with 1 and 55 degrees of freedom) is 7.13. The test F-statistic exceeds this value and the null hypothesis is, hence, rejected and the regression equation confirmed to be significant. It must be stressed that the model under test is linear only with regard to the logs of the variables and the significance tests apply only to that transformed model.

With this advice in mind we can now consider the question of

Table 10.4 Abridged worksheet showing how the significance of the stream velocity and discharge regression equation can be tested. Only those sub-totals needed in this and later calculations are included.

	log discharge (V)	log velocity (U)	predicted log velocity $(\hat{U})$	residual $(U - \hat{U})$	$(U - \hat{U})^2$	$(\hat{U} - \bar{U})$	$(\hat{U} - \bar{U})^2$
	2.093	0.672	0.652	0.020	0.0004	0.209	0.044
	1.985	0.648	0.625	0.023	0.0005	0.182	0.033
	1.107	0.286	0.404	−0.119	0.0141	−0.039	0.001
	2.279	0.763	0.698	0.065	0.0042	0.256	0.065
	...	...	...	...	...	...	...
	...	...	...	...	...	...	...
	...	...	...	...	...	...	...
	−1.155	0.041	−0.164	0.205	0.0420	−0.607	0.368
	−0.538	0.041	−0.009	0.050	0.0025	−0.452	0.204
	0.982	0.424	0.373	0.051	0.0026	0.070	0.005
Means	1.261	0.443					
Totals					0.4683		2.727

Analysis of variance table

Source	Sum of squares	Degrees of freedom	Variance	F-ratio
REGRESSION	2.7270	1	2.7270	320.31
RESIDUAL	0.4683	57 − 2	0.0085	
TOTAL	3.1963			

determining the confidence limits about any individual predicted velocity estimate. The principles are those already described in section 9.4 but, once again, all the work is done with data in the logarithmic form, only when the tasks are complete are the quantities antilogged to give the arithmetic (untransformed) confidence limits. Interest is focused on the logged predictor terms (V). Let us consider the confidence limits of the velocity predicted from a discharge value of 17.58 cubic feet per second. In logarithmic form this quantity becomes 1.245 which, when used in the linearised model obtained above gives a log velocity of 0.439. The standard error of the estimate of U can be found using a simple adaptation of equation 9.11 in which all X terms are replaced by V equivalents we get:

$$SE_{Ui} = s_U \sqrt{\left[1 + \frac{1}{n} + \frac{(V_k - \bar{V})^2}{\Sigma(V_i - \bar{V})^2}\right]} \qquad (10.16)$$

Equation 10.16

SE_{Ui} = standard error of the line at point V_k
s_U = standard error of the residuals
n = sample size
V_k = selected value of V
$\bar{V}$ = mean of transformed variable V
$\Sigma(V - \bar{V})^2$ = sum of squared deviations of V

The standard error of the residuals is found from the square root of the residual variance in Table 10.4 and is based on $n-k-1$ degrees of freedom and gives the figure of 0.0922. The sum of squared deviations of log discharge about the mean has already been obtained when deriving the quantities used to estimate the regression parameters in Table 10.2 where $\Sigma v^2 = (V - \bar{V})^2$. Hence we have:

$$SE_{Ui} = 0.0922 \sqrt{\left[1 + \frac{1}{57} + \frac{(1.245 - 1.261)^2}{43.267}\right]}$$
$$= 0.0922 \sqrt{1.0175} = 0.093$$

The 95 per cent confidence limits about the predicted value are estimated by multiplying the standard error by appropriate t-statistic (with $n-2$ degrees of freedom) as explained in section 9.4. From Appendix II we find t is 2.00. The required 95 per cent confidence limits are therefore:

$$0.439 \pm 2.00 \times 0.093 = 0.253 \text{ to } 0.625$$

As a result we can be 95 per cent certain that the log of the velocity at the selected V value will be between 0.253 and 0.625. Only now can these quantities be antilogged to give the limits of 1.791 and 4.217 feet per second about the predicted value of 2.748 feet per second. This confidence interval is relatively narrow and reflects in part the high degree of correlation between the variables but will, as discussed in Chapter 9, become wider as the predictor term moves away from the mean and assumes very high or very low values. It will also be noted that because of the nature of antilog transformations the predicted U term does not lie midway between the confidence limits as it does in the transformed state.

In conclusion the data from the analysis of variance table can be used to determine the correlation coefficient of the two variables. The coefficient of explanation is given by:

$$r^2 = \left(\frac{\Sigma(\hat{U} - \bar{U})^2}{\Sigma(U - \bar{U})^2}\right) \qquad (10.17)$$

the square root of which is the correlation coefficient of the linearly related logs of the two variables. The result is:

$$\sqrt{(2.7270/3.1963)} = \sqrt{0.853} = +0.924$$

Had the raw untransformed data been used the correlation coefficient would have been different, in this case +0.823. This is an important point and demonstrates that nonlinear relationships can be properly examined only if they are reduced to a suitable linearised form. Failure to acknowledge nonlinearity on the other hand may merely obscure the underlying character of the association between the variables.

10.6 Using MINITAB and SPSS to evaluate nonlinear coefficients

In exactly the same manner in which MINITAB was used to determine the least-squares coefficients for intrinsically linear relationships, so too can we use it in the nonlinear case – provided that the relationship can be successfully transformed to a linear condition. The transformation of variables is an easy matter in both MINITAB and in SPSS and has been introduced in section 5.12 where skewed data were normalised using options within both systems. Once having been appropriately transformed, the 'new' variables can be treated by standard methods. Let us examine how the stream discharge and velocity relationship might have been studied.

Figure 10.9 shows the sequence of commands required to fit a least-squares line through the linearised scatter of points represented by the logs of stream velocity when plotted against the logs of stream discharge. The raw data are held in the MINITAB file **USRIVERS**. Having retrieved this file the two variables can be transformed using the **LOGT** option as was done earlier when plotting the data (Figure 10.4). The procedures from this point are identical to those introduced in section 9.7. In this example we have also given names to the variables. The results list the least-squares coefficients and their statistics, the analysis of variance for regression and residual variances and identifies also the extreme cases for observed or residual values. The layout and content of the results are exactly as described in section 9.7. It must not be forgotten however that these findings refer only to the transformed data. The programme does not detransform the information, that task is left to the investigator should it be necessary.

MINITAB and SPSS, as was demonstrated by Table 5.4, allow for a variety of standard and user-specified transformations. As a result we are not restricted to log-transformations alone when attempting to linearise relationships and the conventional regression procedures can used to study variables transformed by any method including natural logs,

```
MTB >   RETRIEVE 'USRIVERS'

Worksheet retrieved from file: USRIVERS.MTW

MTB >   NAME C1 = 'flow'
MTB >   NAME C2 = 'vel'
MTB >   LET C3 = LOGT('flow')
MTB >   LET C4 = LOGT('vel')
MTB >   NAME C3 = 'log flow'
MTB >   NAME C4 = 'log vel'
MTB >   REGRESS 'log vel' 1 'log flow'

The regression equation is
log vel = 0.126 + 0.251 LOG FLOW

Predictor        Coef        Stdev     t-ratio        p
Constant      0.12633      0.02150        5.88    0.000
LOG FLOW      0.25105      0.01403       17.90    0.000

s = 0.09227      R-sq = 85.3%      R-sq(adj) = 85.1%

Analysis of Variance

SOURCE         DF          SS           MS          F        p
Regression      1      2.7270       2.7270     320.31    0.000
Error          55      0.4683       0.0085
Total          56      3.1953

Unusual Observations
Obs. LOG FLOW   log vel      Fit Stdev.Fit  Residual   St.Resid
  26    -1.15    0.0414  -0.1636    0.0360    0.2050     2.41RX
  28    -1.70   -0.3010  -0.3002    0.0433   -0.0008    -0.01 X
  43     0.63   -0.0177   0.2849    0.0151   -0.3026    -3.32R
  44     0.94    0.1790   0.3631    0.0130   -0.1841    -2.02R

R denotes an obs. with a large st. resid
X denotes an obs. whose X value gives it large influence.
```

Figure 10.9 MINITAB instructions for deriving the least-squares line for the log-transformed variables of stream velocity and discharge (see also Table 10.2).

squares, square roots and so on. The advantages of this adaptability will be apparent in the following section.

10.7 Other forms of linearisable curves

The variety offered by the simple logarithmic, power and exponential equations may well be sufficient to meet the needs of most geographers. Nevertheless they are by no means the only such equations to which recourse might be made. This section briefly examines some of the alternatives. The principles which govern their application to geographic data are identical to those that we have already studied and we have not therefore included worked examples. But the reader is urged not to forget that regression methods and significance testing can only be carried out on the linearised and transformed data. In common with the earlier examples, transformation back to the original data form can take place only at the very last stage of the analysis. Finally, we will again restrict our attention to models with one predictor variable, though that item may appear more than once in the equation should greater variation in the form of the fitted curve be required.

A nonlinear expression such as:

$$Y = a + \frac{b}{X} \tag{10.18}$$

is not one that we have hitherto examined but can be linearised by plotting Y against the reciprocals of X (given by $1/X$), which are then used to determine the least-squares values for coefficients a and b. In other words the reciprocals of X become the new variable. In exactly the same way, the expressions:

$$Y = a + bX^2 \tag{10.19}$$

and

$$y = a + b\sqrt{X} \tag{10.20}$$

can be linearised by taking respectively the squares and the square roots of the X terms and treating them as 'new' variables. The list of curves that can be treated in this fashion can be greatly extended. However by this stage it should be apparent that they are all subject to the same general treatment which requires only that there is a single independent term X which is linked to the single dependent term Y by two coefficients.

Simple curves of this form have great value in describing the plots of geographic data. The most frequent problem is posed by the wide choice of equations that can be used to accomplish this task. These expressions do, nevertheless, share one important characteristic in that their derived curves are constrained to 'bend' or 'flex' in one direction only. Valuable though such curves often are there may be occasions when data plots will not reveal this simple arrangement. Greater degrees of flexibility can be introduced into the fitted curve by including additional terms in X. These expressions are said to be *polynomials*, the best example of which is the *quadratic* curve

$$Y = \mathrm{a} + bX + cX^2 \tag{10.21}$$

Notice that the X term now appears twice, once in its linear form and again in its squared form. There is no reason why we should not use this

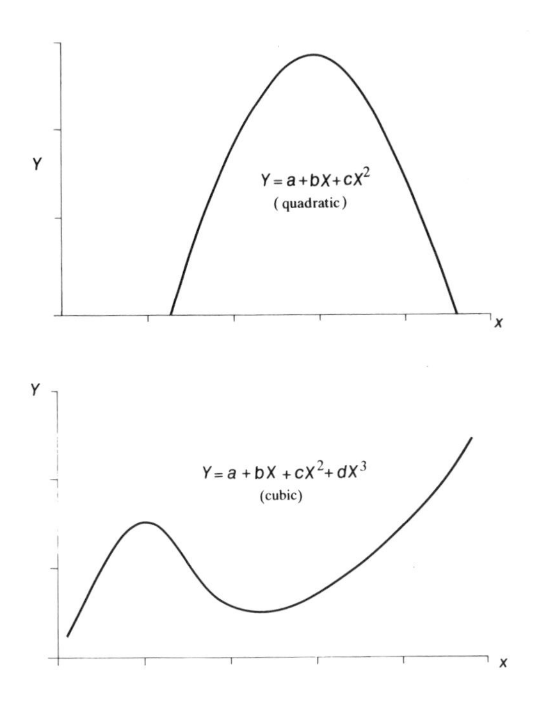

Figure 10.10 Generalised quadratic and cubic curves.

predictor variable twice in this manner though it does present more challenging mathematical problems when the coefficients need to be evaluated. The typical curve derived from the quadratic equation is shown in Figure 10.10 and is described as a *parabola*. Should we want to introduce yet more flexibility into the curve this can be done by including further terms in X. Each term will however be accompanied by its own constant. In the case of the cubic curve (equation 10.22) they are given by b, c and d. The cubic curve (Figure 10.10) is given by

$$Y = a + bX + cX^2 + dX^3 \tag{10.22}$$

in which the X cubed item allows for the additional degree of curvature.

It might be tempting to introduce a large number of terms into the model to account for ever-greater degrees of variation in the plotted data. This temptation should be resisted as their interpretation becomes ever more difficult and there is much to be said for economy of description in these cases. In addition the computational effort required to evaluate high-order curves is rarely rewarded by significant improvements in the levels of explanation that they provide.

The importance of polynomial expressions of the type under discussion is that they are linearisable in each of the X terms. If the raw observations are transformed by raising them to the appropriate power all terms become linear with respect to their association with Y. For example in equation 10.21 the X^2 is replaced by a transformed variable, say Z, which represents the squares of the original X values. The same procedure would be used in equation 10.22 but in addition a second transformed variable, we might term it W, represents the cubed values of X. The first of the three X terms, being already linearly related to Y, requires no transformation and the adapted version of equation 10.22 would be:

$$Y = a + bX + cZ + dW$$

Each of the transformed variables can, from the point of view of the calculations needed to estimate their least-squares coefficients, be regarded as a separate variable. But in recognising this characteristic we have moved prematurely into the area of multiple regression, the discussion of which is properly delayed until the following chapter. Although we include no worked example of this equation, Table 10.5 shows how we can specify the transformations in MINITAB and in SPSS and thereby set up 'new' variables which can be used in multiple regression analysis.

Table 10.5 Examples of MINITAB and SPSS instructions used to linearise polynomial functions.

Polynomial	MINITAB	SPSS
$Y = a + bX + cX^2$	LET 'Z' = 'X' * 'X'	COMPUTE Z = X * X.
$Y = a + bX + cX^2$	LET 'Z' = 'X' * 'X'	COMPUTE Z = X * X.
$+ dX^3$	LET 'W' = 'Z' * 'X'	COMPUTE W = Z * X.

Note: For the purposes of this table the variables are assumed to have been given single character names. Thus 'new' variable Z is set to X^2, while W is set to X^3. All subsequent calculations include these 'new' variables in a multivariate model.

Finally, though they will not be discussed in this book, there are a very large number of expressions and their dependent curves that cannot be linearised. While neither MINITAB nor SPSS can properly evaluate the coefficients of these curves there are a number of more specialised programs that can be employed for this purpose. Such programs use iterative techniques to converge upon a least-squares solution. Significance testing of such solutions is less straightforward however, but this activity takes us far beyond the scope of this volume and readers are referred to texts dealing specifically with these issues such as Daniel and Wood (1980). However, nonlinear relationships are not uncommon in geographical studies and it is important not to side-step some of the difficulties in this area. Indeed, these obstacles may work to our advantage for, as Mather and Openshaw (1974) have observed, nonlinear analysis means that 'one has to think before computing (which is often the reverse of normal practice!)'.

References

Daniel, C. and Wood, F.S. (1980) *Fitting Equations to Data*, Wiley, New York.
Mather, F. and Openshaw, S. (1974) 'Multivariate methods and geographical data', *The Statistician*, 23, 283–308.

Chapter 11
Multiple Regression and Correlation

11.1 Measuring multivariate relationships

Many of the problems studied by geographers are of a complex nature, often involving a consideration of a number of interacting variables. In such circumstances bivariate statistics are rather inadequate tools of analysis and multivariate methods need to be used. As with other statistical techniques the selection of a particular method depends on the type of data, the nature of the problem and the objectives of the research. Thus, some multivariate techniques are concerned with structural simplification and the summarising of a large number of variables or observations by a smaller number of synthesised parameters; others can often prove to be the key to exploring relationships, leading to the subsequent generation and testing of hypotheses. But, from any point of view, the importance of multivariate techniques in general cannot be overstressed. We have already observed in Chapter 9 that we would be fortunate indeed if the behaviour of any one geographical variable could be fully explained by only one other variable, and useful though simple regression is, it may serve only to focus our attention on the need for additional variables that might increase the degree of 'explanation' that the simple model can offer.

As Figure 11.1 shows there is a wide range of multivariate techniques available, and in geography these are used in a variety of ways. First, we can distinguish those statistics concerned with the analysis of dependence, in which one or more variables are singled out and examined in terms of their dependence on others. For example, in multiple regression an attempt is made to explain the variation in one dependent variable through variations in the explanatory or independent variables. Similarly, *canonical correlation* techniques, though not discussed here, extend such analysis to examine interdependence between two main groups of variables – predictor and criterion variables. As such it provides something of a link to a second group of techniques which are concerned with the analysis of interdependence that exists between

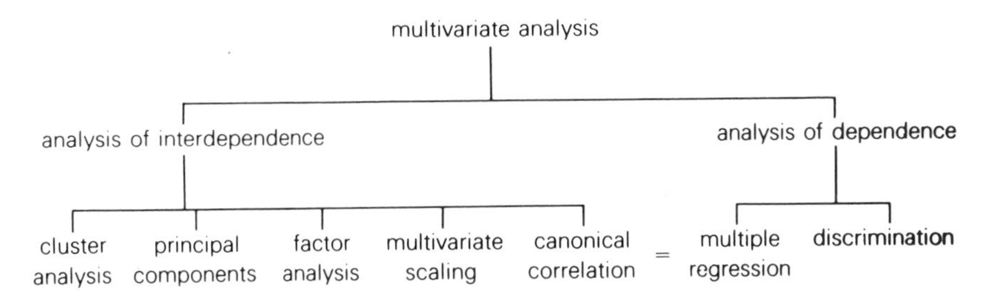

Figure 11.1 The typology of multivariate methods.

variables; this may range from being completely independent to close interdependence with each variable being either a linear or nonlinear function of the others – a condition often described as *multicollinearity*. Thus, we may use *principal component analysis* or *factor analysis* to transform our original data to a new and smaller number of uncorrelated synthesised 'variables' each of which represents the collective behaviour of groups of original variables within which strong correlations exist. These methods are explored more fully in Chapter 12. Finally, though this approach is not pursued in this text, we may use multivariate techniques as classification tools, where concern is to identify similarity between objects or variables. Cluster and discriminant analysis fall into this category.

A great number of the multivariate techniques are based on the assumption that the data approximate to a multinormal distribution. It is, in this respect, fortunate that most multivariate statistical methods are fairly reliable under conditions of departure from normality (Chatfield and Collins, 1980). But it must be pointed out that very few multivariate methods can deal satisfactorily with non-parametric data. The ability to use any of these multivariate methods is also conditional on having access to computer facilities. Hitherto, albeit with growing demands on the researcher's time and arithmetic skill, analysis has been possible 'by hand'. This is no longer a practical proposition and we will place greater emphasis on the use of computer packages than has been the case in earlier chapters.

11.2 Multiple linear regression

The multiple regression model is a direct extension of the simple linear form. It attempts to predict and to explain the variation of a single dependent variable (Y) from a number of predictor terms. The latter may, or may not, be correlated between themselves; it is generally better, however, if they are not. The multiple regression equation takes the form:

$$Y = a + b_1X_1 + b_2X_2 + \ldots b_jX_i \pm e \qquad (11.1)$$

Equation 11.1

a = intercept value
b_1 to b_j = partial regression coefficients
e = error term

where each of the X terms and their associated regression coefficients represent the individual predictor variables.

With bivariate regression and correlation it was always possible to represent the data in graphical form. This cannot be done with multiple regression, with the possible exception of the two-predictor model. Nevertheless, the concept of least-squares is still mathematically valid and used to obtain the best-fit 'surfaces' through multi-dimensional space rather than best-fit lines in two-dimensional space. This can be illustrated in Figure 11.2 which represents a best-fit plane in three-dimensional space in which each of the dimensions is defined by one of the variables. No matter how many dimensions (variables) we include within the model, the error term remains, as it was in the simple model, the difference between the observed and expected values of Y. Least-squares estimates for the regression coefficients linking the two predictors with Y and defining thereby the geometry of the plane, can be found using the equations listed in Table 11.1.

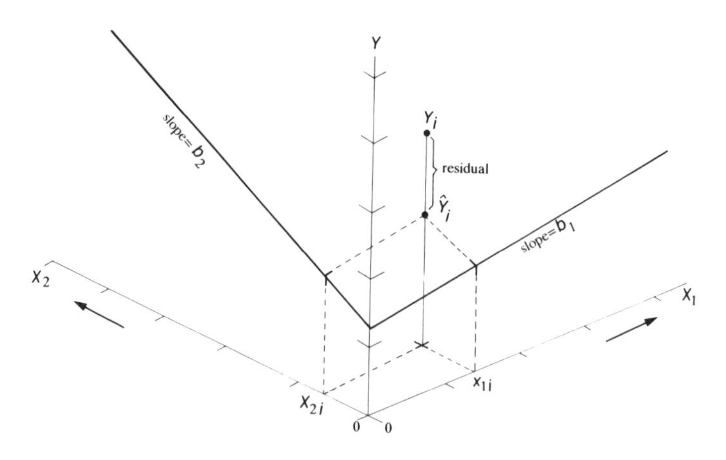

Figure 11.2 Representation of a regression plane in three-dimensional space. Any point (Y_i) can be located by reference to its value by the two variables (represented by the axes), for example X_{li} and X_{2i}. The residual value is the degree to which the latter point lies above or below the regression 'surface' at that location.

Table 11.1 General equations for estimating the least-squares partial regression coefficients ($b_{01.2}$ and ($b_{02.1}$) and the intercept term (a) in a two-predictor regression model. This method requires the simple regression coefficients of Y on $X_1(b_{01})$, Y on X_2 (b_{02}), X_1X_2 (b_{12}) and X_2 on X_1 (b_{21}). In the following equations subscripts i, j, and k can be substituted by 0, 1 and 2 as necessary for calculation of the two partial coefficients. In most cases 0 denotes the dependent variable and 1 and 2 the two predictors. This notation will be used again in section 11.3 for partial correlation coefficients, where it is explained more fully, though a simpler form is employed through most of the text.

$$b_{ij.k} = \frac{b_{ij} - (b_{ik})(b_{kj})}{1 - (b_{jk})(b_{kj})}$$

and

$$a = \bar{Y} - b_{ij.k}\bar{X}_j - b_{ik.j}\bar{X}_k$$

Clearly, graphical representation in more than three-dimensional space is impossible though the geometry of the best-fit surfaces can still be found by developments of the equations in Table 11.1. It requires little effort to appreciate that computational assistance is all but indispensable in calculations of this type and later sections will show how they can be performed using MINITAB and SPSS.

In multiple regression the slope coefficients (b_1, b_2, etc.) are known as *partial regression coefficients* since they measure values that are obtained by controlling for the other independent variables in the model. This concept is more clearly understood if we first consider the character of a two-predictor regression equation which, when written in full, takes the form:

$$Y = a + b_{01.2}X_1 + b_{02.1}X_2 \pm e \qquad (11.2)$$

The subscript 0 represents the dependent variable, and $b_{01.2}$ is the partial regression coefficient representing the slope of the regression line of Y on X_1. The second predictor (b_2) is not excluded from the model but the effect is to hold its value constant. Similarly, the coefficient $b_{02.1}$ is the slope value of the regression of Y on X_2 but now holding X_1 constant. In graphical terms (see Figure 11.2) the two partial coefficients are the slopes of the regression plane where they intersect the two vertical planes that represent each predictor item. It helps to note that the two regression coefficients linking Y with X_1 the various Xs (predictor variables) in their respective simple models would assume different values to their partial counterparts.

It should be observed, however, that these partial regression coefficients are expressed in units the magnitudes of which depend upon the units of the Y and X terms. They are said to *scale-dependent* (see section 9.2) and cannot be directly compared, i.e. simply because one partial regression coefficient has a larger value than any other does not imply that it is correspondingly more important or more significant. Direct comparisons of the importance of the independent variables are more appropriately made through the medium of the *beta weights*. For each partial term these are determined by equation 11.3, where s_{xi} and s_y are respectively the standard deviations of the independent variable under consideration and the dependent variable. The process can be regarded as the 'standardisation' of the partial regression coefficient b_i. In this case we may now conclude that the larger the numerical value of the beta weight (B_i either positive or negative), the greater is its importance in accounting for the behaviour of the dependent term.

$$B_i = b_i(s_{xi}/s_y) \qquad (11.3)$$

The use of these measures can be illustrated by a simple example. An attempt was made to explain the variation in violent crimes in 20 US cities by reference to population size and to the proportion of black residents. The results of the regression analysis give the equation $Y = 357 + 0.266X_1 + 40.2X_2$, where X_1 is the total population, X_2 is the percentage of black residents and Y is the number of detected crimes. The coefficient 0.266 shows the increase in crime rates for changing population but with the proportion of black residents being held at a constant level. The coefficient 40.2 on the other hand shows the change in crime rate with increasing proportions of black residents but with absolute population now being held constant. The numerical contrasts between these two partial coefficients might mislead us into concluding that the proportion of black residents has an overwhelming influence on the crime rate. But these are unstandardised coefficients between which direct statistical comparisons cannot be made because of the difference in the units in which the variables are measured. When the two coefficients are standardised the corresponding beta weights are for b_1 0.388 and for b_2 0.573. We can now conclude with greater certainty that crime rate increases more rapidly with growing proportions of black population than it does with with growing absolute population, though the difference between the two is by no means as marked as might have been assumed from the unstandardised values.

11.3 Multiple and partial correlation

In the previous section attention was focused on the structure of the multiple regression equation and the importance of each of the independent variables. However, in many instances our initial interest may be in the explanatory power of the whole model rather than its individual components. This can be found from the *multiple coefficient of explanation* (R^2) which is derived in the manner described in section 9.3. Attention is concentrated on the total sum of squares of the dependent variable, given by $\Sigma(Y - \bar{Y})^2$, and how it can be decomposed into two components; that accounted for by the now multiple regression equation, the 'explained' component, given by $\Sigma(\hat{Y} - \bar{Y})^2$ and the residual or error term, found from $\Sigma(\hat{Y} - Y)^2$. The *multiple correlation coefficient* (R) is found from the square root of the multiple coefficient of explanation, thus:

$$R^2 = \frac{\Sigma(\hat{Y} - \bar{Y})^2}{\Sigma(Y - \bar{Y})^2} = \qquad (11.4)$$

$$\frac{\text{variation explained by all independent variables}}{\text{total variation of observed } Y \text{ values}}$$

The equation represents the proportion of variation in the dependent term explained by collective variation in the predictor terms. The only difference between equation 11.4 and equation 9.10 is that in the latter case the predicted values (Y) are based on the simple model, and in the former case on a model with two or more independent terms. The derived multiple correlation coefficient is usually denoted as $R_{0.1\ldots n}$ – the subscript 0 again indicating the dependent variable and the others the individual n predictor terms.

In the earlier example of crime rates in US cities the R^2 value was 0.589, indicating a 58.9 per cent explanation of crime in terms of the two selected predictors. On subjective grounds we might suggest this to be a reasonable level of explanation. Fortunately we can test the significance of the regression model to provide us with a more objective assessment of its reliability. In section 9.3 the F-ratio was determined by reference to the explained (regression) and unexplained (residual) variances, each of which were obtained by dividing their corresponding sums of squares by their respective degrees of freedom. The same principles apply in multiple regression with the exception that the regression and residual sums of squares have their degrees of freedom adjusted by the number of predictor terms (k) which will now always be greater than 1. The equations and expressions needed to evaluate the sums of squares and variances are summarised in Table 11.2.

In the same way as we examined partial regression coefficients so too can we study *partial correlation coefficients*. These measure the correlation between the dependent variable and each of the individual

Table 11.2 Sums of squares formulae for testing the significance of multiple regression equations. The degrees of freedom are determined by reference to sample size (n) and the number of predictor variables (k)

Source of variation	Sum of squares	Degrees of freedom
Regression	$\Sigma(\hat{Y} - \bar{Y})^2$	k
Residual	$\Sigma(Y - \hat{Y})^2$	$n - k - 1$
Total	$\Sigma(Y - \bar{Y})^2$	$n - 1$

independent variables while holding all others constant. They share with simple, or *zero-order* correlation coefficients the characteristic of varying only between +1 and −1 depending upon the strength and nature of the relationship. The partial correlation coefficient between the dependent term and the first of, say, three independent variables is written as $r_{01.23}$. This notation can be extended simply by adding further subscripts to the right of the dot, each indicating the inclusion of a further variable which is being controlled. The total number of controlled variables designates the *order* of the correlation. Hence, as noted above, a simple correlation is one of zero-order. A first-order correlation will have one control ($r_{01.2}$), a second-order correlation will have two controls ($r_{01.23}$) and so on. First-order partial correlation coefficients can be found from equation 11.5 using the zero-order correlations between each of the possible pairings of the three predictor variables.

$$r_{01.2} = \frac{r_{01} - (r_{02})(r_{12})}{\sqrt{(1 - r_{02}^2)}\sqrt{(1 - r_{12}^2)}} \qquad (11.5)$$

Equation 11.5

r_{01} = correlation between Y and X_1
r_{02} = correlation between Y and X_2
r_{12} = correlation between X_1 and X_2

We can return to our earlier crime example to see how these principles might be applied. The zero-order correlations are listed in Table 11.3 which, by substitution into equation 11.5, give $r_{01.2} = +0.442$ and $r_{02.1} = +0.640$.

With just two independent variables these are the only possible combinations. The number of such partial correlations grows rapidly however with the number of independent variables and their methods of estimation are correspondingly more demanding. We will see later how computers come to the aid of the researcher at such times.

Table 11.3 Zero-order correlations for variables in crime studies of US cities

	Y	X_1	X_2
Reported crimes (Y)	–		
Total population (X_1)	0.552		
Percentage black population (X_2)	0.700	0.371	

In this example both partial coefficients are positive and measure the degree of association between each pair of dependent and independent variables, the remaining member of the latter being held constant in both cases. Correlation coefficients, irrespective of their order, are scale-independent and provide a second direct method of measuring the separate influences of independent variables. Partial correlations are not, however, identical with beta weights as they represent different facets of the multivariate associations. Beta weights indicate how much change in the dependent variable is produced by a standardised change in an independent variable. In contrast, partial correlations measure the proportion of variation in the dependent variable accounted for by each of the independent variables. By squaring the partial correlation coefficients we can estimate the proportion of variance explained by the independent variable when all others are controlled.

The relationships between zero-order and partial correlation coefficients are not, however, always obvious. In many cases the move from zero-order to a partial correlation may result not only in notable changes in the magnitude of statistical associations between variables, but changes in sign are even possible. As we shall see, explaining partial correlations is no easy task and considerable care needs to be taken over their use. One point that needs to be clarified at this stage is that the multiple correlation coefficients obtained from any given data set does not equate to the sum of the partial or the zero-order terms. We found above, for example, that the two partial coefficients add to 1.104 and not to 1.0. While in Table 11.3 the zero-order coefficients between Y and the two independent variables add to 1.252. Why should this be the case? To answer this question takes us into an important area that occupies the following section.

Let us consider first the zero-order correlations. Table 11.3 shows not only that notable correlations exist between the dependent and the independent terms but also that the two latter items are themselves correlated. Thus there is a degree of 'double-counting' in the zero-order coefficients. In the extreme case of perfect correlation between the two predictors this double-counting would be at a maximum. Partial correlations, on the other hand, endeavour to overcome this difficulty through the control they place on the third (or further) variables. In this they are however, arguably, too successful since they examine only the variation attributable exclusively to one predictor but there remains a proportion of the variation, attributable to the combined effects of the independent terms, which the partial correlations overlook. In respect of all these difficulties researchers are advised, though it cannot often be realised, that predictor variables should not be correlated between themselves. Such a strategy eliminates this thorny problem of *multicollinearity*.

11.4 Problems of multicollinearity

The multiple regression model makes the same demands and is based on the same assumptions as the simple linear model. However, meeting such requirements can be more difficult in the case of multiple regression since a greater number of variables need to be considered. Take, for example, the assumption that the variables' frequency distributions are approximately normal; for multiple regression studies all variables, but particularly that of the dependent term, need to be examined. In some instances data transformations may be required and the whole regression equation can begin to take on an unfamiliar appearance with, possibly log values being introduced with consequences of the type discussed in the previous chapter. While such transformations may be statistically legitimate they may complicate the process of interpretation. In a similar manner all the requirements stipulated in Chapter 9 with regard to the residuals of the regression model remain equally valid in the setting of multivariate analyses.

These cautionary observations notwithstanding, it is the problem of multicollinearity that raises the most common difficulties. The term refers to the situation in which high correlations exist between the independent variables. Thus when two predictors are highly correlated their zero-order correlations with the dependent variable will suggest similar degrees of explanation. The difficulty is that the two independent variables can be considered as explaining the 'same' variability and either might be used alone in a simple regression expression without serious prejudice to the model. This problem can be identified either by an examination of the correlation matrix of independent variables, or by comparing the zero-order with the multiple correlation coefficients. If the multiple correlation coefficient obtained by adding one predictor to a simple model reveals only a slight increase on the bivariate correlation it suggests one of two things. On the one hand the new variable may have little or no correlation with the dependent term and is unlikely therefore to added significantly to the multiple correlation. Conversely, it may have a high degree of zero-order correlation but is also correlated with the other predictor and for that, quite different, reason adds little to the predictive capacity of the regression equation. It can therefore be seen that improvements in the utility of the regression model are most readily achieved by including apposite but uncorrelated predictors.

Multicollinearity also causes a considerable amount of ambiguity in the interpretation of the effects of individual variables. Under these circumstances very small differences in the multiple correlation coefficient obtained by adding new variables may, at the same time, bring about large differences in the partial terms and the relative importances of the independent variables becomes difficult to interpret. Generally, the higher the degree of multicollinearity, the less reliable are the partial coefficients (Kendall, 1975; Johnston, 1978). Only by careful inspection of the data and choice of variables can such problems be minimised.

Depending on the degree of multicollinearity, three possible statistical techniques are available to overcome the problem. One is to use an objective method, such as stepwise regression, where variables are added, or subtracted, progressively from the data set until an optimum solution, determined by reference to the explained sum of squares or its direct derivatives, is found. This important line of study is developed in a later section. Alternatively, if all variables are included then factor analysis or principal components analysis may be used to group together sets of two or more independent variables in orthogonal (uncorrelated) synthesised variables. Multiple regression may then proceed using the 'new' variables. Such procedures are, however, fraught with manipulative and interpretational difficulties. For this reason the following section not only introduces the means by which we can use the computer packages to process the data, it also retains a focus on traditional regression methods and permits us to examine the question of significance testing in multivariate models. In doing so, as we will demonstrate, it becomes possible to disentangle at least some of the problems presented by multicollinearity.

11.5 SPSS and multiple regression analysis

Chapters 9 and 10 demonstrated how MINITAB could be used to generate the least-squares regression line with only one independent variable. The same system can produce least-squares multiple models but we will take the opportunity to demonstrate how the SPSS system deals with the demands of regression analysis.

In Chapter 9 it was suggested that while we might use average protein intake to predict life expectancy there would, self-evidently, be other variables which might also help to determine how long people are expected to live. The following example in multiple regression takes that simple case and examines the role of additional variables; in doing so, the example highlights a number of difficulties that all researchers encounter. The first of these is availability of data. Our example takes average data for a random sample of 32 nations. For many nations a huge amount of data exists to cover any working hypothesis that we might wish to explore. On the other hand there are equally as many for which the data are either absent or of dubious accuracy. We might, for example, want to establish the contribution that immunisation programmes make to life expectancy. It is an obvious line to follow, but not an easy one. In many cases the proportion of the population that have been immunised is simply not known. In addition the term covers a range of diseases. With which of them are we concerned; typhus, cholera, influenza, etc? The matter is by no means as simple as we might imagine when embarking on even a straightforward study such as this one. In the absence of data we may have to abandon any consideration of that variable or, more helpfully, try to find a surrogate. In this example we do have data for the number of qualified doctors per head of population and we can use this as at least a first approximation to the level of health provision in each nation, though clearly it cannot convey the whole picture. The other variables for which data are available and which on a primae facie basis might influence life expectancy are: proportion of illiterates (which may determine ease of communicating new ideas on farming, public health and other similar matters), rate of natural population increase (which indicates the degree of demographic development and the effectiveness of important policies such as birth control), tractors per head of population (indicating the degree of mechanisation and efficiency of national farming practices) and population density. The chosen variables, their units and the keywords used to describe them in the SPSS system are given in Table 11.4.

Table 11.4 Definitions and key to variables used in the multiple regression examples.

Keyword	Definition and units
LIFE	Average male life expectancy in years
PROTEIN	Average daily protein intake in grams per person
POPINCR	Natural population increase per thousand
DENSITY	Population density in people per square kilometre
POP	Total population
ILLITS	Illiterates as a percentage of the total population
TRACTORS	Number of tractors in each country
DOCTORS	Number of qualified doctors per 10,000 people
TPERHEAD	Number of tractors per million people

Source: Statistical Yearbook 1987, United Nations, New York (1990)

As a first step we might want to establish the degrees of correlation which exist not only between the dependent variable and the independent terms but that which exists between the independent terms themselves. SPSS can produce such a correlation matrix, the key instructions for

```
DATA LIST FILE 'WORLD2.DAT' FREE/ protein, live, popincr,
 density, pop, illits, tractors, doctors.
COMPUTE tperhead = tractors/pop.
SAVE /OUTFILE 'WORLD3.DAT'.
CORRELATIONS /VARIABLES ALL WITH ALL
 /OPTIONS 3.
```

Correlations:	PROTEIN	LIFE	POPINCR	DENSITY	POP	ILLITS
PROTEIN	1.0000**	.8308**	-.7189**	-.0340	-.0267	-.6063**
LIFE	.8308**	1.0000**	-.7884**	.2262	.0426	-.6741**
POPINCR	-.7189**	-.7884**	1.0000**	-.1075	.1249	.7350**
DENSITY	-.0340	.2262	-.1075	1.0000**	-.0794	-.1026
POP	-.0267	.0426	.1249	-.0794	1.0000**	.0221
ILLITS	-.6063**	-.6741**	.7350**	-.1026	.0221	1.0000**
TRACTORS	.2044	.2336	-.1892	-.1128	.7619**	-.2728
DOCTORS	.8041**	.7191**	-.7358**	-.0098	-.0210	-.6780**
TPERHEAD	.5524*	.4563*	-.5576**	-.0817	-.0238	-.4601

Correlations:	TRACTORS	DOCTORS	TPERHEAD
PROTEIN	.2044	.8041**	.5524*
LIFE	.2336	.7191**	.4563*
POPINCR	-.1892	-.7358**	-.5576**
DENSITY	-.1128	-.0098	-.0817
POP	.7619**	-.0210	-.0238
ILLITS	-.2728	-.6780**	-.4601*
TRACTORS	1.0000**	.3433	.4239
DOCTORS	.3433	1.0000**	.6280**
TPERHEAD	.4239	.6280**	1.0000**

```
N of cases:   32        2-tailed Signif:    * - .01    ** - .001

" . " is printed if a coefficient cannot be computed
```

Figure 11.3 SPSS instructions and screen display for the CORRELATION option. This sequence of instructions sets up the data file, computes a new variable, saves the new data file and then produces the correlation matrix of specified variables.

which are given in Figure 11.3. The general manner in which instructions are introduced should by now be familiar. The data file (**WORLD2.DAT**, written free-field format) is recalled and the variables then named. In addition to the variables described in Table 11.4, a new item had to be defined using the **COMPUTE** option as the original data source provides the number of tractors per nation and the total population, but not the number of tractors per head and it is necessary to divide the former by the latter to derive a properly weighted figure. This new variable must be estimated before the **CORRELATION** command is introduced. In this example, so as to avoid having to set up the data set on every occasion that we use it, the **SAVE** command holds all the preceding lines in an SPSS system file (see section 2.9). The subcommand **OPTIONS 3** which follows the request for the correlation subprogram requires the two-tailed significances to be indicated in the matrix – by default the one-tailed significances would have been printed. In this example we have requested, using the **ALL WITH ALL** instruction, that the correlation coefficients of all possible pairs are produced. Should it have been necessary we could have listed the specific pairings for which correlations were required. Suppose, for example, we had wanted only the correlation coefficients between **LIFE** and **PROTEIN**, **POP** and **DOCTORS**, the instruction line would have read:

CORRELATIONS /VARIABLES LIFE WITH PROTEIN, POP, DOCTORS

Because, with our specification, the correlation matrix is a large one it cannot be printed in a single unit and is accommodated within two sections. Readers should make due note of the manner in which the significance of the coefficients is indicated.

In addition to average protein intake, natural population increase, proportion of illiterates, doctors per head of population and tractors per head of population are also significantly correlated with average male life expectancy. A moment's reflection will indicate why total population and total number of tractors do not correlate with life expectancy, though we might be a little surprised that population density seems to have no clear degree of association. But, as we explained in the previous section, this is only part of the picture, and one which is obscured by the significant correlations that exist between the independent terms. We cannot hope to unravel the complex picture of collinearity that we see here without resorting to multiple regression and its derivative partial coefficients.

```
GET /FILE 'WORLD3.DAT'.
The SPSS/PC+ sytem file is read from
    file WORLD3.DAT

REGRESSION / VARIABLES (COLLECT)
  /STATISTICS R, ANOVA, COEFF, ZPP
  /DEPENDENT LIFE
  /METHOD ENTER PROTEIN, POPINCR
  /DEPENDENT LIFE
  /METHOD ENTER PROTEIN, ILLITS
  /DEPENDENT LIFE
  /METHOD ENTER PROTEIN, DOCTORS
  /DEPENDENT LIFE
  /METHOD ENTER PROTEIN, TPERHEAD
  /DEPENDENT LIFE
  /METHOD ENTER PROTEIN, POPINCR, ILLITS.
```

```
 * * * *  M U L T I P L E   R E G R E S S I O N  * * * *

Block Number  1.  Method:  Enter    PROTEIN  POPINCR

Equation Number  1   Dependent Variable..  LIFE

Variable(s) Entered on Step Number
   1..     POPINCR
   2..     PROTEIN

Multiple R               .87510
R Square                 .76580
Adjusted R Square        .74965
Standard Error          4.19244

Analysis of Variance
                        DF    Sum of Squares     Mean Square
Regression               2        1666.71265       833.35633
Residual                29         509.71897        17.57652

F =    47.410305              Signif F = .0000

---------------------- Variables in the Equation ----------------------

Variable              B        SE B        Beta    Correl  Part Cor   Partial

POPINCR       -3.572096    1.167292   -.395636  -.788409  -.275003  -.494059
PROTEIN         .225626     .053393    .546335   .830767   .379753   .617332
(Constant)    52.514998    5.576759

--------- in ---------

Variable           T  Sig.T

POPINCR       -3.060  .0047
PROTEIN        4.226  .0002
(Constant)     9.417  .0000
```

Figure 11.4 SPSS instructions for system file retrieval and subsequent REGRESSION option work specifying a number of different models.

One means of exploring the correlation structure with a view to defining the most efficient model, i.e. that which provides the best explanation with the smallest number of predictor terms, is to add variables successively to the simple model that we already have. Of all the predictor variables at our disposal, protein intake has the best correlation, followed by natural population. We can use SPSS to produce the least-squares model using these latter two items as predictor variables.

Having saved the data instructions in the system file **WORLD3.DAT** we need only use the **GET FILE** command henceforth with no need either to redefine the variable names or to recompute the new variable (Figure 11.4). The regression subprogram is indicated by the command

REGRESSION ,followed by the variables list. In this case we have specified **(COLLECT)**, which instructs the computer to expect the variables for each regression model to be individually identified as each model is set up (though it can use only those variable names already defined in the system file unless further **COMPUTE** instructions are used to set up additional new variables). By default the output would contain all the results listed in Figure 11.4 with the exception of the zero-order and partial correlation coefficients. The subcommand **STATISTICS** is used to have these included in the output (by specifying **ZPP**) but we must, having used this command to override the default, make sure to include also the key requests **R**, **ANOVA** and **COEFF** in order that the multiple correlations, analysis of variance tables and the regression coefficients and their associated *t*-statistics are also included. Having included **STATISTICS** at this early stage the instruction will apply to all regression models specified in that set.

We have specified five such regression models, in each of which the dependent variable must be listed first and is always **LIFE**, with the independent terms being introduced on the next line. There are a choice of procedures under the heading **METHOD** and for the moment we will use only **ENTER**, an instruction that requires the program to include only those independent variables which follow on the same line. The final regression instruction concludes with the stop (**.**) and, though not shown, the whole session would conclude with **FINISH**. though before reaching that stage we could, if we wished, redefine our **REGRESSION** variables or make use of any other option.

Figure 11.4 lists the detailed results for only the first of the models, that predicting **LIFE** from **PROTEIN** and **POPINCR**. A summary of the remaining models is given in Table 11.5. Although not the most efficient way of achieving the objective of defining the optimum model – we will see later how that can be done – this procedure does emphasise important points concerned with multiple regression modelling.

Table 11.5 Summary of regression analyses using different combinations of predictor variables

Predictor variables	R^2	Partial correlation	Regression coefficient	Beta weight	*t*-statistic
PROTEIN	0.736	0.718	0.275	0.667	5.56
ILLITS		−0.385	−0.091	−0.269	−2.25
PROTEIN	0.700	0.611	0.295	0.715	4.16
DOCTORS		0.154	0.163	0.144	0.84
PROTEIN	0.690	0.780	0.832	0.833	6.72
TPERHEAD		−0.005	−3.7E−06	−0.004	−0.03
PROTEIN	0.772	0.604	0.218	0.528	4.01
POPINCR		−0.368	−2.921	−0.326	−2.10
ILLITS		−0.161	−0.039	−0.116	−0.86

It has been shown (section 9.3) that the simple regression equation describing the link between life expectancy and protein intake is a significant one. We can also see from Figure 11.3 that this pairing is the strongest of the zero-order correlations. We might therefore use it as a basis on which to construct multiple models of two, perhaps even three or more variables. But which of the variables should be included in the model? There appears, at this stage at least, little point in examining those variables such as total population which have non-significant correlations with life expectancy. It is far better to look at those others with significant zero-order correlations. The best of those variables is POPINCR for which $r = -0.7884$, which suggests that nations with rapid population increases are likely to be those with low life expectancies. A summary of the screen display showing the least-squares parameters of this model are shown in Figure 11.4. All of those statistics already discussed in this chapter are listed. The multiple correlation coefficient (Multiple R) and the multiple coefficient of explanation (R Square) are derived by the methods described above. The program also adjusts the latter to take into account the possibility of sampling errors. The standard error of estimate of Y is also listed and is followed by the analysis of variance table from which the F-value and its random probability (Signif F =) are derived. The regression and residual variances appear under the heading Mean Square. In our example, the F-value is large (47.410305) with a vanishingly small random probability. It is, hence, unlikely to have arisen by chance and the null hypothesis of no explanation can be rejected. This conclusion is confirmed by reference to Appendix VIb where the critical F-ratio with 2 and 29 degrees of freedom is 3.33 (at the 0.05 level).

Comparing these results with those in Figure 9.10 we can see that the degree of explanation, in terms of R^2, and of the regression variance has increased following inclusion of POPINCR. The latter, though itself correlated with protein intake ($R = -0.7189$) might be expected to replicate the effect of PROTEIN, but can be seen to make an individual contribution to the equation. This is confirmed in the section of the screen display headed 'Variables in the Equation' under which the individual regression coefficients are listed but which also includes the beta weights (Beta) which indicate that while PROTEIN (beta = 0.546) is the more important of the two, the magnitude of beta weight for POPINCR (beta = −0.395) is not substantially smaller. In addition the t-statistics for both partial regression coefficients have random probabilities (Sig T) less than the critical 0.05 from which the conclusion is drawn that both contributions are statistically significant. In addition, having specified a listing of the partial correlations we can see the degree to which each of the two predictors influences life expectancy when the other is included in the model but controlled. The partial correlation of −0.494 between POPINCR and LIFE measures the degree to which the two variables are correlated in an imaginary setting in which a large number of nations have the same protein intake. Life expectancy shortens in nations with greater natural population increases, though the strength of that correlation is now much reduced from its zero-order counterpart (listed under Correl) in the output. A similar picture emerges for the partial correlation between PROTEIN and LIFE where the zero-order correlation of 0.830 falls to 0.617 when POPINCR is included but controlled. The item Part Cor (sometimes known as the *part correlation coefficient*) is the square root of the change in the coefficient of explanation that results from the inclusion of the new variable, the others in the set (only one in this example) having already been taken into account.

Although Table 11.5 only summarises the same output for other models in the regression set we see that no other combination of two predictors gives such a high degree of explanation. In only one of these additional two-variable cases does the second variable, when added to PROTEIN, make a significant contribution to the model despite their high zero-order correlations. The partial regression coefficients and associated *t*-statistics show that life expectancy decreases significantly as illiteracy rates (ILLITS) rise. On the other hand the control on life expectancy exerted by the number of doctors or the number of tractors per head of population, while individually important, vanishes after protein intake is taken into account. These results may surprise us, and this fact alone reveals the importance of multiple regression modelling when seeking to unravel the complex relationships between variables. Ideally we should make the same analysis for all possible additional pairings of PROTEIN with other variables, though on the basis of the conclusions thus far reached we might not expect to uncover any other important models.

At this stage, however, we might turn to consider the possible roles of three-predictor models. It is, in the nature of things, impossible for the R^2 values to fall with the inclusion of new variables at any stage in the modelling process. At worst there will be no change. The question to be decided is whether the increase is sufficiently great for us to conclude the partial contribution of the new term to be significant. For most purposes we can use the significance of each regression coefficient's *t*-statistic to help us determine that point. But even in this relatively simple example the number of possible three-predictor models is quite large and to test each of them would take much time – a problem which grows quickly with the number of variables in the sample. Taking the three variables shown thus far to have had the most important contribution to determining LIFE, i.e. PROTEIN, POPINCR and ILLITS, Table 11.5 reveals important changes when all of them are included in the model. The variable ILLITS, already shown to be important when used with PROTEIN, now finds itself to have no contribution to make. Changes of this character illustrate the need to explore many possible combinations of variables. Where several variables are included in the data set there is a clear requirement for a more efficient method of establishing the optimum model than that described above. This brings us to what is known as *stepwise regression*. Such a procedure can be discussed in general terms but can be carried out only with computational assistance.

11.6 Multiple regression using stepwise and other methods

The strategy adopted in the previous section was a simple one in which variables were added selectively to the equation model. The selection criterion was the zero-order correlation between the independent and dependent terms. The variable with the highest correlation was included first, the next high second and so on. There is, however, nothing to be gained by successively adding items to the model until all the variables have been included. Some of them may have no connection with the dependent term and others, though appearing to have a significant zero-order correlation, may be so strongly correlated with other significant independent terms that their partial correlations become very small. Remembering that at each step the inclusion of a further variable can never reduce the coefficient of explanation, the important question is to determine the significance of the increase in the degree of explanation offered by any one additional variable. The method we adopted to answer this question was to examine the output listing of the t-statistics associated with each predictor variable within the model. If the statistic was not significant at the selected level (we took 0.05) then it was excluded from the model. By this selective and additive method we established the optimum and most efficient equation in terms of balancing the degree of statistical explanation against the number of predictor terms. But other methods exist to accomplish this task.

In addition to the above *forward inclusion* method there is also a *backwards elimination* procedure. The latter scheme starts with an equation that includes all the measured independent terms. These are taken in turn and treated as if they were the last to be included. Most computer programs make the exclusion, or inclusion, decision on the basis of changes in the regression (explained) sum of squares, the significance of which can be determined using an adaptation of the analysis of variance procedure. The variable making the lowest contribution is eliminated and the process repeated with variables being eliminated one at a time until it creates a statistically significant decrease in the regression sum of squares. At that point the process ceases and the regression model so-defined is assumed to be the optimum. Generally this method is no more or less efficient than that of forward inclusion, though there might be something to be said in favour of a strategy that starts with all the variables as this provides an overview of the situation that may assist at the stage of interpretation. However, it must not be expected that these two methods will necessarily produce the same optimum solutions, especially where many variables may be involved. There may on occasion even be a case for the researcher to make a non-statistical decision on the best regression model for their own immediate purposes. Draper and Smith (1966) have written extensively on the various methods and conclude that there may be circumstances under which there are no unique solutions. If this is the case, then personal judgements may be part of the statistical methodology. It is on this latter basis that we might elect to examine all possible combinations of independent terms. The advantage of such a procedure is that it eliminates the problems of the internal rearrangments and changes in the partial correlations that result from the forwards-inclusion or backwards-elimination methods. These are once-and-for-all strategies; a variable, once included or excluded, cannot be reconsidered and if such changes do occur they cannot be taken into account. The disadvantage of the 'all possible regressions' strategy is that with n variables there are $2^n - 1$ possibilities with consequent demands on the researcher's and the computer's time.

A method that combines the efficiency of forwards-inclusion (and backwards-elimination) methods with the thoroughness of the 'all possible regressions' procedure is *stepwise regression*. In this scheme the independent variables are re-examined at each stage to identify any that have become superfluous following the introduction of subsequent predictors, or to permit use of previously rejected variables. At each step both inclusion and exclusion are possible until neither are possible at the selected significance level. The stepwise method is not, however, foolproof and researchers should be alive to the creation of models (notably where there are many variables in the regression set) that, while being statistically sound, may not stand up to geographical and theoretical scrutiny. There are indeed those who have gone so far as to suggest that variable selection should be based more firmly on theoretical and deductive considerations rather than statistical measures such t-statistics and F-ratios. Such an approach assumes, however, that we already have a good understanding of the causal structures that we are examining.

These difficulties depend often on why the researcher is engaged in regression analysis: is it principally for investigation or in order to produce a predictive model? The latter is less concerned with the delicacies of theoretically sound models than with reducing the statistical error terms and raising the highest possible R^2 values. Researchers engaged in investigation are less fortunate and may have to dispense with the objectivity offered by statistical criteria and concentrate on models that are theoretically more sound.

11.7 Stepwise regression using SPSS

We have already seen in section 11.5 how SPSS can be used to develop an optimum regression model using an intuitive strategy based on variable selection by reference to zero-order coefficients and the *t*-statistics of each of the partial regression coefficients. This forced entry of pre-selected variables was carried out by using the **ENTER** command with the **METHOD** option. But with SPSS we can make alternative specifications at that point such as **BACKWARD** (for the backwards elimination method), **FORWARD** (for the forwards-inclusion method) or **STEPWISE** (if we want to use the stepwise procedure as outlined in section 11.6). Let us see what model is derived using the same data set as in section 11.5, but by specifying that the stepwise procedure should be used. Figure 11.5 shows the form of the command lines and the nature of the screen results that are produced. The command structure differs little from that used in Figure 11.4 though the screen listing has been edited to show only the final step's results.

Notice from Figure 11.5 how the keyword **STEPWISE** replaces **ENTER** in the **METHOD** specification and is followed by all the variables in the file which are to considered for possible inclusion in the final model. In this case the stepwise procedure concludes with the same optimum model, with only PROTEIN and POPINCR identified as the significant predictor variables, as we found in section 11.5, though such happy agreement between formal and informal procedures cannot be guaranteed, especially where a larger number of closely-correlated variables is being used.

Although the use of the stepwise method leaves the final choice of predictor variables to the objective scrutiny of the program, the researcher must make some choices with respect to the way in which it runs. The significance level at which the regression model is to be accepted or rejected is based on the *F*-ratio of regression to residual variance, and is an obvious area where only the researcher can make a choice. Equally the significance level for the inclusion of new variables within stepwise methods is one that must be decided upon. The default significance level in SPSS for the inclusion of a new variable is 0.05, but is 0.10 for the exclusion of an item already accepted at an earlier stage. These defaults can be redefined using the **CRITERIA** option. This command line must appear before the specification of the dependent variable. Where several models are run within the same program the criteria can be redefined, perhaps for example going back to the default, by placing the necessary choices immediately before each of the lines identifying the dependent variable. Figure 11.6 shows how we might have specified one model using an inclusion (**PIN**) level of 0.01 and an exclusion (**POUT**) significance level of 0.05. The second model uses the same variables but the **CRITERIA** line now instructs the program to revert to the default significance level. The results are not shown but are the same as in Figure 11.5. Because the improvement in the degree of explanation upon inclusion of the second

```
GET /FILE 'world3.dat'.
The SPSS/PC+ sytem file is read from
    file world3.dat

REGRESSION / VARIABLES (COLLECT)
 /STATISTICS R, ANOVA, COEFF, ZPP
 /DEPENDENT LIFE
 /METHOD STEPWISE PROTEIN, POPINCR, ILLITS, DOCTORS, TPERHEAD.
```

```
  * * * *  M U L T I P L E   R E G R E S S I O N  * * * *

Variable(s) Entered on Step Number
    2..     POPINCR

Multiple R              .87510
R Square                .76580
Adjusted R Square       .74965
Standard Error         4.19244

Analysis of Variance
                        DF     Sum of Squares      Mean Square
Regression               2       1666.71265        833.35633
Residual                29        509.71897         17.57652

F =    47.410305           Signif F = .0000

---------------------- Variables in the Equation ----------------------

Variable              B        SE B       Beta     Correl  Part Cor    Partial

PROTEIN         .225626     .053393    .546335    .830767   .379753    .617332
POPINCR       -3.572096    1.167292   -.395636   -.788409  -.275003   -.494059
(Constant)    52.514998    5.576759

---------- in ----------

Variable          T   Sig.T

PROTEIN       4.226   .0002
POPINCR      -3.060   .0047
(Constant)    9.417   .0000
```

Figure 11.5 Regression analysis instructions and results in SPSS using the stepwise method.

variable (POPINCR) is also significant at the 0.01 level, the program would again have stopped at the second step when working to this more rigorous **PIN/POUT** specification.

To understand more clearly how this procedure operates we can

```
GET /FILE 'world3.dat'.
The SPSS/PC+ system file is read from
    file world3.dat

REGRESSION /VARIABLES (COLLECT)
 /CRITERIA PIN(0.01) POUT(0.05)
 /DEPENDENT LIFE
 /METHOD STEPWISE PROTEIN, POPINCR, ILLITS, DOCTORS, TPERHEAD
 /CRITERIA DEFAULT
 /DEPENDENT LIFE
 /METHOD STEPWISE PROTEIN, POPINCR, ILLITS, DOCTORS, TPERHEAD.
```

Figure 11.6 Regression analysis instructions in SPSS for overriding default specifications for inclusion/exclusion levels in stepwise methods.

abstract the results from Figure 9.2 (where the single predictor is PROTEIN) and Figure 11.4 (where the results of the optimum PROTEIN + POPINCR model are listed). Table 11.6 shows that the inclusion of the second variable increases the regression (explained) sum of squares from 1502.1 to 1666.7. Meanwhile the corresponding residual sum of squares has fallen from 674.3 to 509.7. Our task is to determine the significance of that change. To do so we must calculate the F-ratio derived from the change in the regression sum of squares (164.6) against the residual sum of squares after the inclusion of the new variable (509.7). Because only one variable is added, the degrees of freedom for the former are 1, while the degrees of freedom for the latter are $n-k-1$ (where n is sample size and k is the number of variables in the new model). The two variances are hence 164.7 and 17.56 respectively. From Appendices VIb and VIc we find the consequent F-ratio (9.37) to be significant at both 0.05 and 0.01 significance levels. In this example the form of the final model does not depend upon the choice between the two most widely used significance levels. This may not, however, always be the case and

Table 11.6 Analysis of variance table for regression model improvement obtained by adding one further variable to the simple model

Source of variation	Sum of squares	Degrees of freedom	Variance	F-ratio
REGRESSION (k = 1)	1502.1	1		
REGRESSION (k = 2)	1666.7	2		
CHANGE	164.6	2 − 1	164.6	9.37
RESIDUAL (k = 2)	509.7	32 − 2 − 1	17.56	

circumstances will arise in which a variable is included using the 0.05 significance level but excluded in the more stringent 0.01 case where higher F-ratios (and improvements in the degree of 'explanation') are required. More than in most applications this demonstrates the need for care in the selection of significance levels.

11.8 Residuals in multiple regression

Chapter 9 discussed the conditions of variable and residual behaviour that had to be complied with before the simple regression model could be accepted as valid. All of those conditions, though they are not repeated here, apply with equal force to multiple regression models. Once again, however, a study of the residuals not only determines the reliability of the data and the model, it may also direct future developments in the research topic. For example the presence of a high degree of autocorrelation of the residuals may initially suggest the data to be inappropriate though we might more constructively conclude that it suggests the presence of a hitherto unsuspected variable. The subsequent inclusion of such a variable would eliminate the autocorrelation and also, almost certainly, increase the level of explanation offered by the model. Referring to the life expectancy example, the Durbin–Watson d statistic of residual autocorrelation for the simple regression model was, at 2.36, already close enough to 2.00 for the null hypothesis of zero autocorrelation to be rejected (see section 9.5). In the case of the optimum multiple model described in the earlier sections the d-statistic converges yet more closely to 2.00 with a value now of 2.11. The means by which SPSS can be requested, by stipulating the **RESIDUALS** option with the keyword **DURBIN**, to list this information is shown in Figure 11.7. The d-statistic is calculated, as described in section 9.5, by reference to the ordered values of $\hat{Y}$. In this respect the use of two or more predictors has no effect on the method of calculation. The **RESIDUALS** option can also be requested to print a histogram of the standardised residuals (**HISTOGRAM(ZRESID)**) and to identify the ten most extreme residuals (**OUTLIERS(ZRESID)**). The former acts as a quick check on the normality of the residuals, the latter may help to identify hitherto overlooked controls on the dependent variable's behaviour.

The listing includes summaries of the predicted values of the dependent term (PRED), the absolute residuals (RESID) and the same two items expressed in their standardised form (ZPRED and ZRESID). The ten most extreme residuals are listed giving their case number (row on the data matrix) and their standardised value. The histogram is a simple graph, somewhat limited in this case by the small sample size that leaves few items in the classes but suggests, nevertheless, that the distribution of the residuals is normal. In figure 11.7 N is the number of observations in each class, Exp N is the number expected under a perfect normal distribution and Out is the standard error of the class.

There are other means by which residuals can fail to comply with the model's requirements. But with such failures it does not necessarily follow that they must be excluded from further study. It might be necessary to apply transformations to one or more of the independent variables, but to do so will, as outlined in Chapter 10, have an important effect on the regression model illustrated by equations 11.6 and 11.7. The former expresses the regression equation in which the variables have been logged.

$$\log Y = \log a + b_1 \log X_1 + b_2 \log X_2 \ldots b_n \log X_n \tag{11.6}$$

$$Y = aX_1^{b_1} \quad X_2^{b2} \ldots X_n^{bn} \tag{11.7}$$

```
GET /FILE 'world3.dat'.
The SPSS/PC+ system file is read from
    file world3.dat

REGRESSION /VARIABLES (COLLECT)
 /DEPENDENT LIFE
 /METHOD ENTER PROTEIN, POPINCR
 /RESIDUALS DURBIN, HISTOGRAM(ZRESID), OUTLIERS(ZRESID).
```

```
Residuals Statistics:

                Min          Max         Mean     Std Dev     N

*PRED        49.0377      75.5478      58.9084     7.3325    32
*RESID       -8.4077      10.1820        .0000     4.0549    32
*ZPRED       -1.3462       2.2693        .0000     1.0000    32
*ZRESID      -2.0054       2.4287        .0000      .9672    32

Total Cases =     32

Durbin-Watson Test =  2.11246

Outliers - Standardized Residual

  Case #      *ZRESID
      20      2.42867
      25     -2.00545
       9      1.49697
       3      1.43111
       2     -1.41885
       1      1.24743
      30     -1.22668
      13     -1.12949
       4     -1.03599
      15      1.02245
```

```
Histogram — Standardised Residual

NExp N        (* = 1 Cases,    . : = Normal Curve)
0   .02   Out
0   .05  3.00
0   .13  2.67
1   .29  2.33 *
0   .58  2.00 .
0 1.07  1.67 .
3 1.76  1.33 *:*
5 2.58  1.00 **:**
0 3.40   .67   .
1 4.01   .33 *  .
9 4.24   .00 ***:*****
3 4.01  -.33 ***.
4 3.40  -.67 **:*
3 2.58 -1.00 **:
2 1.76 -1.33 *:
0 1.07 -1.67 .
1  .58 -2.00 :
0  .29 -2.33
0  .13 -2.67
0  .05 -3.00
0  .02   Out
```

Figure 11.7 Regression residual analysis in SPSS showing instructions and screen display of results.

The consequence of these changes is that the new variables are, as shown in equation 11.7, multiplicative and not linearly additive in their control on *Y*. In addition, as Chapter 10 made clear, some variables that are non-linearly related to the dependent variable can be made arithmetically linear through the judicious selection from the many possible transformations. Use of the **COMPUTE** option in SPSS allows these changes to be easily made though caution is again counselled, particularly in investigative rather than predictive modelling. Should we be concerned only with the predictive capacity of the regression model these arithmetic implications may be of little account. On the other hand, researchers with a stronger interpretational bias might find such models much more difficult to work with.

If heteroscedasticity or autocorrelation, especially positive auto-correlation, are suspected with respect to any of the independent terms SPSS can be instructed to prepare a *partial plot*. This shows the scatter of the standardised residuals of *Y* against the chosen independent variable but, importantly, with the effect of the other variables in the model included but held constant. Figure 11.8 shows the instructions and screen display of this operation. The example used specifies the PROTEIN/POPINCR model and requests a PARTIALPLOT of the

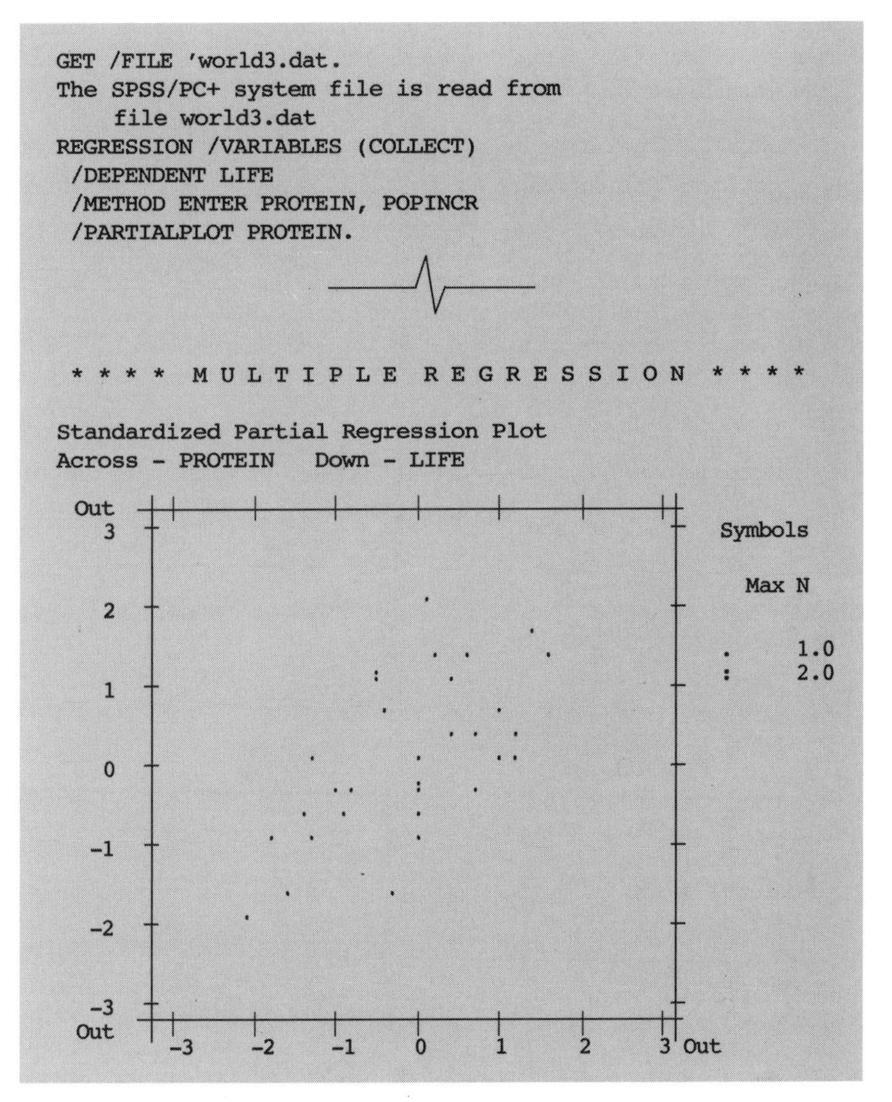

Figure 11.8 Instructions for plotting partial regression residuals in SPSS.

standardised residuals of PROTEIN against LIFE. The results indicate no tendency towards heteroscedasticity. Notice that where two points occur in close proximity they are plotted as a vertically arranged pairs of dots.

Thus far our attention has focused on the behaviour of the residuals as a group, but the character of the individuals cannot be overlooked especially if they constitute *outliers*. When dealing with standardised residuals, outliers might be all those observations that differ from the best-fit model by more than 1.96 or 2.58 standard errors depending upon whether the 0.05 or 0.01 significance level is chosen as the limit for this purpose. Individuals that differ by such degrees often repay attention. At the most practical level, they might be observational or coding errors and should be excluded from the data set or corrected. More probably, if care has been taken at earlier stages in the study, they will reflect random variation about the regression line. Remember, for example, that 5 per cent of the dependent data must by definition differ from their predicted counterparts by more than 1.96 standard errors. In this case there can be no causal explanation of their behaviour other than that offered by inevitable random variation. But this may not always be the case. On closer inspection some outliers may reveal particular characteristics in common that identify potentially new independent variables. The life expectancy example used in this chapter reveals only two significant residuals (out of 32), these being cases 20 and 25. The former is Malaysia, where the observed male life expectancy is far greater than might be anticipated from the nation's average protein intake and natural population increase. Case 25 is that of Guinea where the life expectancy falls far short of what is expected. Future studies might well focus on these extremes to determine what variables, if any, account for their unusual levels of life expectancy and to establish if they exercise a wider influence over the dependent variable.

11.9 Problems and assumptions

Geographers are now more aware than ever of the difficulties inherent not only in regression analysis but also in the indiscriminant use of powerful statistics packages and their options for data manipulation. Several thought-provoking studies have appeared in past years on this topic from Poole and O'Farrell (1971), Mather and Openshaw (1974) and Gould (1970). Problems have been identified in two areas. First, important assumptions necessary for the validity of regression models may not always be fulfilled. Second, unquestioning dependence on the results of pre-written computer packages can be ill-advised, especially if insufficient attention is given to the substantive suitability of the raw data and the sampling frame within which they were gathered. Longley (1975) and Wempler (1970) have written on this particular problem and their comments remain largely valid despite the more recent advances in computer software.

It is becoming clear that the often-quoted need for data normality is subordinate, in regression analysis, to more pressing requirements. Indeed, as Mather and Openshaw have indicated, the urge to normalise and transform data can lead to all manner of interpretational difficulties. Regression coefficient reliability is more responsive to other preconditions and the best estimates are achieved only when these are met. We have already touched upon most of these needs in this and earlier chapters. At the most fundamental level, data accuracy is critical because the regression equations permit errors in Y but assume that X is perfectly measured. Rarely is this the case in primary research. Even if measured to high degrees of accuracy variables must also be studied in terms of how faithfully they reflect the phenomena they represent. The life expectancy model used in this chapter is a good example. We must, for example, ask ourselves how representative protein intake is of the overall nutritional value of the diet. In a similar fashion we took the natural population increase as a surrogate measure of the demographic condition of each nation. It may not be the best measure of this important characteristic. But, even if we harbour reservations about some of our variables, practical impositions may mean that no other data are available and a compromise might be needed between the practical need for data and the more purely academic demands of the exercise.

Enough has been written to indicate the problems that can result from multicollinearity. Only partial correlation analysis can assist in such matters but problems may yet linger if the variables have a high degree of zero-order intercorrelation.

Lastly, it is important that observations used in the forms of regression introduced in this text should be random and independent of one another. This stipulation is perhaps the most difficult for geographers to comply with as their data are often spatially organised with an underlying tendency to spatial autocorrelation. Much has been written on this subject (for example, Cliff and Ord, 1970) and many of the problems it raises have yet to be fully resolved. Happily it is within the compass of most researchers to at least check the reliability of their data and, in most cases, to take steps to counter any non-compliances.

References

Chatfield, C. and Collins, A.J. (1980) *Introduction to Multivariate Analysis*, Chapman and Hall, London.

Cliff, A.D. and Ord, K. (1970) 'Spatial autocorrelation: a review of existing and new measures with applications', *Econ. Geog.*, 46, 269–292.

Draper, N.R. and Smith, H. (1966) *Applied Regression Analysis*, Wiley, New York.

Gould, P. (1970) 'Is *statistix inferens* the geographical name for a wild goose?', *Econ. Geog.*, 46, 439–448.

Johnston, R.J. (1978) *Multivariate Statistical Analysis in Geography*, Longman, London.

Kendall, M. (1975) *Multivariate Analysis*, Griffin, London.

Longley, J.W. (1975) 'An appraisal of least squares programs for the electronic computer from the point of view of the user', *J. Am. Stat. Assoc.*, 62, 819–841.

Mather, P.M. and Openshaw, S. (1974) 'Multivariate methods and geographical data', *The Statistician*, 23, 283–308.

Poole, M.A. and O'Farrell, P.N. (1971) 'The assumptions of the linear regression model', *Trans. Inst. Br. Geogrs*, 52, 145–158.

Wempler, R.H. (1970) 'A report on the accuracy of some widely used least squares computer programs', *J. Am. Stat. Assoc.*, 60, 549–563.

Chapter 12

Factor Analysis and Related Techniques

12.1 Introduction

The aim of this chapter is to introduce the practical aspects associated with the application of factor analytical methods, rather than dealing in detail with any theoretical problems. A full consideration of the theory of factor analysis would take us far beyond the scope of this book and, therefore, our attention will be confined largely to the way in which the SPSS system deals with the practicalities of using the technique and how geographers can understand and interpret the results. We shall start by examining the possible uses of factor analysis, before discussing the variety of techniques that are available to the researcher.

As we suggested in Chapter 11, the most important feature of factor techniques is their ability to reduce a large data set to a smaller number of factors or components. This reduction process may be necessary for a number of reasons, although two main ones can be identified. First we may wish to produce combinations of the original data, which may then be used as new variables in some further analysis. For example, we may use factor analysis to combine so-called independent variables in multiple regression to reduce the effect of multicollinearity. By this means a poorly-understood but large body of data covering a range of variables may be rendered more manageable. Second, we may use it for exploratory purposes, in an attempt to detect and identify groups of functionally-interrelated variables. We may, or may not, be guided in this approach by an established degree of understanding of the variables' interrelationships, and the strategy may be merely one of exploratory investigation or of testing an hypothesis which predicts a number of factors or components. As we shall see the two approaches require rather different methods of factor analysis. It is within the area of exploratory analysis that many geographical uses of factor techniques are to be found, though as such work advances the studies become less investigative and more analytical. This is partly the case with factorial ecological studies of urban areas. Thus, the earlier exploratory studies revealed three major groupings of variables – those concerned with socio-economic status, with stage in the life cycle and with segregation. These have formed the basis of further work, with factor analysis being used to confirm or to reject a range of hypotheses. The factors that are isolated by these methods, though of a purely statistical character in terms of their derivation, are not mere numerical devices when set within the wider interpretational framework. Factoring methods in fact spring from psychological researches in the 1930s in which subjectively well-understood concepts, such as intelligence, could be measured only on the basis of several parameters and no single practical measure could be relied upon. These several parameters combined to form an 'intelligence factor'. In the same way, geographers may wish to measure attributes of selected areas, such as urban decay, social status or social well-being, all of which are the combined effects of several measurable variables none of which individually and adequately encompasses our meaning of the terms. Again factoring methods can take these several variables and identify possibly useful combinations of measurable attributes.

It should be stressed that the term factor analysis does not refer to a single technique, but covers a variety of approaches, though they all share important characteristics. Initially, we can recognise three main stages in their application, each of which may offer the researcher a choice of alternative methods (Table 12.1). In the first stage most factor analyses require product-moment correlation coefficients as the basic data input. At this introductory stage the alternatives are between that of using information measuring correlations between different variables for a group of observations, or of taking a matrix of correlation coefficients measuring the degree of similarity between a set of individuals based on a number of variables or attributes. In the first instance we may examine the behaviour of a set of variables sampled across different areas of a city. This is very much the approach taken in Chapters 9 and 11. As a result we produce a correlation matrix of the selected variables – this is called *R*-mode factoring. We might note that the focus of attention consists of the variables, the geographical or spatial dimension, if present at all, appears only as part of the sampling strategy. But, as geographers, we might be more concerned to look at the degrees of similarity between the areas and to distinguish spatially-based, rather than variable-based, arrangements and groupings. We might therefore produce a correlation matrix not of the variables but of the areas. In effect the roles of variables and cases are exchanged. Pairs of areas which have similar attributes will display a strong positive correlation (though we might prefer to call it now a similarity index), and those pairs of urban areas with contrasting values across the variables will have a strong negative correlation. This approach is referred to as *Q*-mode factoring. Though *Q*-mode factoring is often valuable to geographers it requires a large number of variables with which to measure each case. Ideally, at least as many variables as cases are required for successful *Q*-mode factoring.

The second stage is to explore the possibilities of data reduction by

Table 12.1 A simplified view of options in factoring methods

Stage	Option types	Terminology
Correlation	Between variables Between individuals	*R*-mode factoring *Q*-mode factoring
Extraction of initial orthogonal factors	Variance assumed to be common	Principal components
	Variance apportioned to common and unique categories	Factor analysis
Rotation to final factors	Uncorrelated	Orthogonal
	Correlated	Oblique

constructing a new, and smaller, set of variables based on the interrelationships in the correlation matrix. There are two basic approaches at this level – principal components analysis and factor analysis (Table 12.1).

In both these approaches, new variables are defined as mathematical transformations and combinations of the original data. However, in the factor analysis model the assumption is made that the observed correlations are largely the result of some underlying regularity in the basic data. Specifically, it assumes that the behaviour of the original variable is partly – indeed it would be hoped – largely, influenced by the various factors that are common to all the variables. The degree to which this is the case is expressed through what is termed the *common variance*. A further element is known as the *unique variance* and is an expression of the variance which is specific to the variable itself and to errors in its measurement. This latter effect clearly delimits the utility of the factors. In contrast, principal components analysis makes no such assumptions about the structure of the original variables. It does not presuppose any element of unique variance to exist within each variable and the variances are held to be entirely common in character (Figure 12.1). In essence we use principal components analysis as a purely exploratory tool when we have little understanding or knowledge of the data with which we might be working. Factor analysis, on the other hand, presupposes that we have sufficient understanding to know how many significant factors are to be expected and the degree of common variance for which they account. These differences will be examined in more detail in section 12.3.

The final stage in which variations in method are possible is in the search for interpretable factors. This is the point at which we analyse the character of the factors and qualities that they represent. To help us in this process we might choose to refine or clarify the factor model. To achieve this a numbr of solutions are availble. Up to this point the

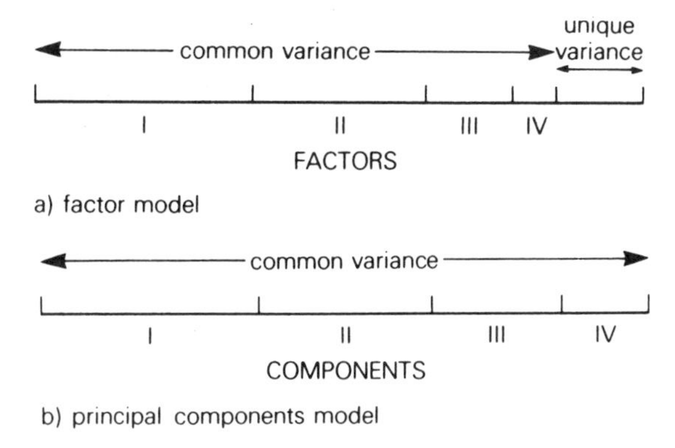

Figure 12.1 Apportionment of common and unique variance in (a) the factor model and (b) the principal components model.

various factors can be thought of as being geometrically orthogonal, i.e. at right-angles to each other in n-dimensional space. In the statistical sense this ensures that they are uncorrelated with one another. We may, however, rotate these axes as a fixed set through n-dimensional space in order possibly to account for a greater degree of the variance of the original data. In this case the factors remain orthogonal and uncorrelated. But we may also rotate the axes independently so that they become oblique to one another. This may further increase the utility of the factoring model but a degree of correlation will now exist between pairs of axes as their orthogonal geometry will have been compromised. These points are discussed more fully and exemplified in section 12.6.

12.2 Theoretical considerations and terminology

Before we examine the operational aspects of the group of techniques collectively known as factor analysis we must consider a little of its theoretical background. There are two basic approaches, one based on algebraic solutions and the second on geometric interpretations. It is the latter upon which we shall focus; those readers interested in more detailed explanations will find many texts, such as Harman (1967), to assist them.

Let us start by considering the case of simple correlation. In section 8.2 we saw how the relationship between two variables can be expressed by means of a scatter diagram with X and Y axes defining the two variables in question. Figure 9.2 takes the example of life expectancy against protein intake. Had the points fallen perfectly along a straight line the correlation would have been perfect with a value of $+1.0$ (perfect negative correlations would also plot as straight lines). On the other hand, had there been no correlation between the two variables the points would have formed a vaguely circular scatter on the graph with no

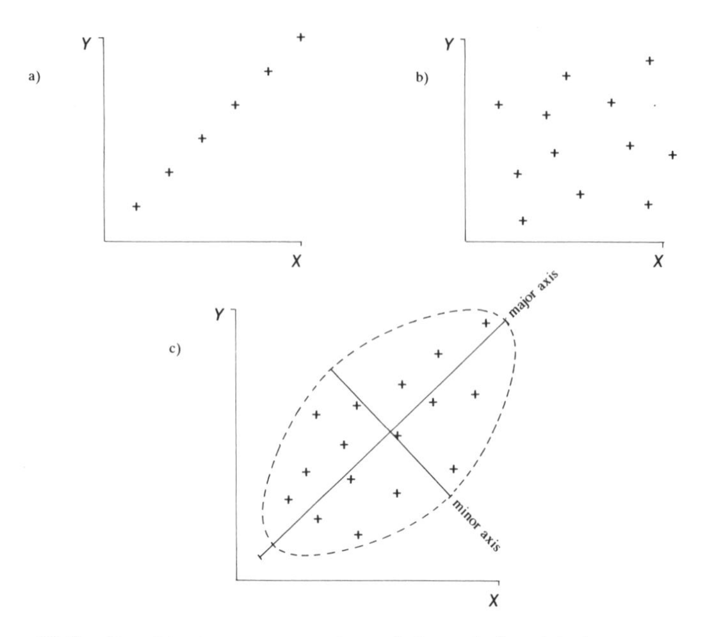

Figure 12.2 Graphical representation of the relationship between pairs of variables. Diagram (a) shows a perfect correlation, (b) zero-correlated variables and (c) the intermediate state in which the principal components can be viewed as orthogonal axes.

apparent trend. From the point of view of the geometry of such graphs, these two extremes – the perfect (ignoring for the moment the question of the sign) and the zero correlation – represent the two limiting conditions between which all other states can be described in the form of ellipses with varying degrees of eccentricity or elongation. This point is illustrated in Figure 12.2 which shows two extreme cases together with an intermediate condition. The latter in particular draws attention to the important point that all such ellipses can be described by reference to a major and a minor axis. The former runs the length of the ellipse. The minor axis lies at right angles to the major.

Within the current scope of our study, the major axis can be viewed as representing the common variance between the two variables and is in some ways comparable to a regression line. The minor axis is more akin to the residual variance, i.e. that which is not accounted for by the major axis. Thus, when we have a perfect relationship the major axis is at its greatest length and the minor axis disappears. As the scatter of points spreads and departs from a perfect relationship, so too does the major axis shorten and the minor axis lengthen until there is zero correlation when the two are of the same length. These two axes represent the only two possible components of a two-variable factor model. Where correlations are strong the two variables can be reduced to the one dominant component (major axis) of the plot and the minor axis, being relatively trivial, can be overlooked. Notice that at most we can only have as many components as we have variables. This is as true in n-dimensional space as it is in two.

Thus far we have done little more than consider the simple two-variable case. The great virtue of factor analytical methods is that they can perform this task of estimating components on data sets containing many variables or cases (depending upon whether we use a R- or Q-mode approach). The problem is that such n-variable models require a similar number of geometric dimensions and cannot be represented in graphical form. Hence the principal axis of the hyperellipsoid (an ellipse in n-dimensional space) may represent the combined effects of not two, but several, inter-correlated variables. At the same time we are now no longer restricted to just two components or axes and can have as many as we have variables. The second (orthogonal) component might no longer, as above, represent the residual element, but might indicate the combined effects of another group of variables. This axis will be 'shorter' than the principal axis but may nevertheless account for a significant proportion of the variability of the data set not accounted for by the latter. A third axis or component may take up a little more of the remaining variability and represent the effect of a third group of variables and so on until all the variability is accounted for. As with multiple regression, however, there will come a time when the addition of further components doesn't significantly help our attempts to explain the data set's variability. The lengths of the axes, or components, are most easily determined by computer programs and are more properly termed *eigenvectors*, and they have a statistical length measured by their *eigenvalue*. A simple example will illustrate the character of the newly-defined components.

Let us consider the relationships between three variables given by the product-moment correlations in Table 12.2. From this information it is possible to calculate each variable's correlation with the first component or factor. This measure is a function of the square root of the sum of all the possible correlations entries in the appropriate column of the correlation matrix. This sum includes the diagonal elements which, as the measure of the variable's association with itself, is 1.0, and the duplicate correlation entries either side of the principal diagonal. The column sums are given in Table 12.2 and these in turn add to 7.2, the square root of which is 2.68. The correlation between variable X_1 and the first factor is found by dividing the appropriate column sum of correlations by the square root of the sum of all the correlations in the matrix. Thus, for example, the correlation between the first factor and variable X_1 is given by $2.3/2.68 = 0.86$. This quantity is more correctly termed the *loading* of the variable on the factor. As they are a measure of the correlation between the variable and the first factor, their respective squares are the proportions of the variance of that variable accounted for by the factor. When the squares of the loadings of all the variables on the one factor are summed they provide a measure of the variance of the whole data

Table 12.2 Hypothetical correlation matrix for factor loading calculations

	X_1	X_2	X_3	
X_1	1.0	0.6	0.7	
X_2	0.6	1.0	0.8	
X_3	0.7	0.8	1.0	
Sum of correlations	2.3	2.4	2.5	total sum = 7.2
Factor loading	0.86	0.89	0.93	

Note: factor loading = column sum/√(total sum of correlations)

Table 12.3 Factor loadings and eigenvalue for hypothetical example

Variable	Loading		Squared loading
X_1	0.86		0.74
X_2	0.89		0.79
X_3	0.93		0.86
		eigenvalue = sum =	2.39

accounted for by that factor. This important quantity is the eigenvalue of the factor. Table 12.3 shows how the eigenvalue of factor 1 is derived from our simple example.

To establish the importance of the first component we can relate it to the total variance of the original data by simply dividing it by the number of variables (n) in the data set (equation 12.1). The answer is usually expressed as a percentage.

$$\text{Percentage of variance} = (\lambda_1/n) \times 100\% \qquad (12.1)$$

Equation 12.1

λ_1 = eigenvalue of factor 1
n = number of original variables

In this case the eigenvalue of 2.39 is divided by 3; indicating that 79.7 per cent of the total variance is accounted for by the first component.

The next step would normally be to estimate the eigenvalue for the second component. This would be extracted from what is termed the residual correlation matrix. The maximum percentage of explanation possible by this second factor is limited to that not accounted for by the first; in this example, 100 – 79.7 or 20.3 per cent. In statistical terms this second component will be wholly uncorrelated with the first and, in geometrical terms, the two will be orthogonal to each other in *n*-dimensional space. In principal components analysis the process could continue until we had as many factors as variables, by which stage all the correlations would have been accounted for and, theoretically, the sum of the eigenvalues should equal to *n* and, therefore provide 100 per cent explanation. In factor analysis we would normally confine ourselves only to those first few factors, by convention those with an eigenvalue of 1.0 or more, that account for the greater part of the data set's variability (see also Cattell, 1978).

Factor analysis differs from principal components analysis in one further important regard. This concerns the concept of *communality*, which is the proportion of variance of each variable accounted for by the common factors, i.e. the common variance identified in section 12.1 and Figure 12.1. It is important to recognise that the communality of each variable is the quantity which appears in the principal diagonal of the correlation matrix used to initiate all factoring procedures. In principal components analysis the behaviour of each variable is assumed to be wholly explained by the factors or components that are common to all the variables. By inference, therefore, there is no variation that is unique to each variable and not accounted for by the cumulative effect of all possible factors or components. On the other hand, the technique known specifically as factor analysis makes no such assumptions. The correlation matrix in the case of factor analysis will have principal diagonal elements (communalities) that are less than 1.0. The degree to which each variable's communality differs from 1.0 is a measure of its unique variance and this cannot be accounted for by the common factors. The difficulty in this approach is that we do not know the communalities to be entered in the principal diagonal. The problem is most easily resolved in computer packages that use iterative techniques in which initial estimates are made and the program then run to give a final estimate. If initial and final communalities differ to a great degree, adjustments are made to the initial estimates, the program rerun and so on until they converge upon consistent figures. This final bound is generally guided by the multiple coefficient of explanation (R^2) of the regression model using all of the variables in the data set, one of which is a dependent term. Higher R^2 values will generally indicate smaller degrees of unique variance.

The initial estimates are often derived from a preliminary principal components analysis. Attention is usually confined only to those principal components with eigenvalues greater than 1.0; on purely qualitative grounds these might be regarded as 'significant' components. Each variable will, as we saw in Table 12.3, have a different loading on the components. A generally reliable estimate of communality of any one variable is provided by the squares of its loadings summed across all the 'significant' components. This total represents the collective percentage variance of the variable accounted for by the common components. These quantities, one for each variable in the data set, must be less than 1.0, though their magnitudes are determined by the efficacy of the 'significant' components. If the latter account for a large proportion of the total variance, then the derived communality estimates will be close to 1.0, otherwise they will be much smaller.

12.3 Factor analysis and principal components analysis: a comparison

As we explained in section 12.1, principal components and factor analysis represent the two important approaches in factoring methods. Geographers have used both techniques. From our point of view the main points of interest are, firstly, in the distinctions between them and, secondly, how to decide on which is the more appropriate in any given case.

Looking first at the differences between the techniques, we have already drawn attention to some of the main contrasts connected with the statistical procedures. The principal component models assume a closed system in which all statistical variation is accounted for by the variables themselves. Hence the communality values are set to 1.0 and, ultimately, we would expect to account for the whole variance of the data set. As a minimum requirement we would want very high correlations between the variables, thereby minimising, ideally eliminating, the unique variance. This approach is favoured because of its simplicity and it solves the communality problem and many texts prefer it to common factor analysis (Blakith and Reyment, 1971; Chatfield and Collins, 1980).

Factor analysis does present difficulties with regard to communality estimates, though the advent of high-speed computers has helped to overcome some of the practical difficulties. However, it might be regarded as a more realistic approach to research problems as it does not presuppose a closed system in which the variables wholly explain each other's behaviour. Thus, in most studies we would acknowledge that we have not collected data on all possible variables and that measurement terms (error variation) would be attached to each of them. This allows for a residual and unexplained variance element unique to each variable.

Given these differences we can go on to consider the criteria for selecting one model in preference to the other. First, it should be pointed out that under certain conditions both models will give similar results, for example, when all the variables are highly intercorrelated. If, however, some of the correlations are low the use of 1.0 as communality entries will be an overestimate through which the principal components and the factor analysis methods will produce divergent results.

We will not concern ourselves too much with the theoretical differences but, rather, focus on the practical issues that determine the models' suitabilities. In general, the factor analysis approach is the more realistic. It acknowledges that errors in measurement will occur, though it also requires at least a general knowledge of the underlying factorial structure of the data. In contrast, principal components models deal with more limiting cases; no underlying structure is assumed and the search is one based largely on data reduction. The general characteristics of the two methods are listed in Table 12.4. Clearly from what we have said, factor analysis provides a more comprehensive model for most geographers and, for this reason, most of what follows is aimed more specifically at this approach.

Table 12.4 Essential characteristics of principal components and of factor analysis

	Principal components	Factor analysis
Assumptions	A closed system, with no assumptions about the underlying variable structure; identifies only common variance	Realistic assumptions concerning errors in measurement; identifies common and unique variance
Best conditions	High correlations between a large number of variables and the need is for simple data reduction	As for principal components, but it will also deal with a small correlation matrix and permits a wider range of analysis and interpretation

12.4 Factor analysis: data input

Any factor analysis model, when applied to a specific problem, can be broken down into a series of operational steps (Table 12.1). These start with the data matrix, from which a correlation matrix can be derived. The next steps would normally consist of deriving the initial orthogonal factors, then rotating them to achieve a better degree of explanation of the original data's variance and, finally, the listing of the factor scores for each variable on each of the factors. Other possibilities may be open but these options are explored in the later sections of this chapter. For the moment we will concentrate on the first stages, those where the raw data and its correlation matrix are the focus of activity. The correlation matrix may not be the only principal data input for the computer program, the raw statistics can also be used by some programs. Nevertheless the correlation matrix might be preferred if only because it allows for the easier identification of possible groupings of variables upon which the later factors may converge.

The main decisions to be taken with regard to the raw data surround the question of transformations. In common with other parametric methods it is assumed that the data are normally distributed and, because the procedures rely so heavily upon correlation coefficients, that the relationships are linear. Distributional transformations may be necessary if these conditions are not met. But, as in so many of these cases, the final interpretation may be hampered through transforming the data and the user should weigh carefully the requirements of the procedures against the problems of interpretation at a later stage. The published literature is ambivalent with some papers using transformed data and others offering cogent arguments against it. If no statistical inferences are being made then the questions of normality and linearity are less important.

The questions posed by the units used to measure the raw data are easily resolved. In any study the different variables will have not only a number of scales upon which they can each be measured (temperatures in degrees F or C, income in sterling or dollars, etc.) but the magnitudes will also differ between the variables. All variables could be given equal statistical weight by expressing them in standardised form. Fortunately this is rarely necessary as the correlation coefficients are, as shown in Chapter 8, scale-independent.

Of greater importance is the selection of variables to be used in factoring methods. The selection will, in part, depend upon data availability. Nevertheless the derived structure of the factors will be determined by the choice of variables. For example, in urban social studies, if the original variables included many which measure demographic characteristics, clearly this would be reflected in the composition and character of the factors. It is often advisable therefore to achieve some kind of balance across the selected variables with regard to what they measure. The decision as to which variables to include is obviously difficult and no clear guide lines exist; although when using factor analysis, rather than principal components analysis, the prerequisite of a working hypothesis may well help.

Finally there is the question of the number of variables. The effect of including a large number of variables that measure similar characteristics and are highly correlated is merely to increase the importance of the factor through which they are represented, without necessarily changing the overall factor structure. Again, some balance should be achieved between the variables. The overriding consideration, in *R*-mode analysis, is that there should be many more cases than variables. In *Q*-mode analysis the reverse requirement prevails and there should be many more variables than cases.

12.5 Factor analysis using SPSS

The SPSS system allows all the techniques of factor analysis thus far described to be employed. Once again we will show how to provide instructions by which the system can be made to carry out a number of tasks, but emphasis will also be placed upon the statistical and the geographical interpretation of the results. We can use SPSS to obtain the following information: a variable list, the correlation matrix, communalities, factor loadings, eigenvalues, the percentage of variance explained by each factor and the factor scores. The results can be also be based on orthogonal and oblique rotations of the derived factors. These too will be explored. In common with all the methods that we have used, there is a great deal more that can be done using the SPSS factor analysis option. We will, however, confine ourselves to the essentials of data and results manipulation, and the interested reader should consult the SPSS handbooks for further information. For the most part we will allow the program to run on the default values of the many options that are available. For example, SPSS allows correlation matrices produced at an earlier stage of analysis to be read directly by **FACTOR**, this is not done here and the raw data are provided from which the program produces its own correlation matrix.

A number of variables were selected to reflect socio-economic conditions in the United States (see Table 12.5). Collectively the variables measure aspects of prosperity and affluence throughout the 48 contiguous states (Alaska, Hawaii and the District of Columbia were excluded). The data were assembled into an ASCII code file (**USA.DAT**) which was then read into the SPSS system using the **DATA LIST FILE** instruction. Some data manipulation was required beforehand and the **COMPUTE** option was used to convert the absolute mileage of roads per state into miles per person (**ROADRATE**). In order to avoid having to perform this calculation and data file retrieval on every subsequent occasion, the **SAVE** instruction was used to store the data and auxiliary instructions in an SPSS system file called **USASTUDY** (see section 2.9). Thereafter the instruction **GET /FILE 'USASTUDY'** was all that was needed to set up the data for factor, or indeed any other, analysis.

The first step was to perform a simple principal components analysis of the data. The program rotates the components as a default option and in this case that was overruled by using **/ROTATION NOROTATE** subcommand. This follows the only other command necessary at this preliminary stage, that specifying the **FACTOR** option, followed by a list of variables for analysis. In this case the original data set consisted of many more variables but, as is sometimes the case in factor analysis, the use of many, perhaps poorly-correlated variables creates computational difficulties and leads to uncertain results. As a consequence of this risk the number of variables was trimmed to 13, focusing on those that were generally well-correlated but embraced simultaneously a range of social, demographic and economic aspects. Table 12.5 has already listed the selected variables and provides a brief account of what they represent. Figure 12.3 contains the SPSS instructions and the subsequent print-out. The former should need little elaboration, but the latter deserve closer attention.

The results begin with the section on 'Initial Statistics'. The initial

Table 12.5 List of variables used in factor and principal components analysis

SPSS code name	Description of variable
POPDENS	Population in numbers of people per square mile
BRATE	Birth rate as live births per thousand population
URBAN	Percentage of each state's population living within metropolitan districts
DOCTORS	Number of physicians per thousand population
CHARGE	Average daily hospital room charges
EDUC1	State expenditure on primary and secondary schools in dollars per head
CRIME	Crime rate as reported crimes per 100,000 population
UNEMPLY	Unemployment rate as a percentage of the civilian labour force
INCOME	Average annual income per person in dollars
SALES	Retail sales per person in dollars
PAPERS	Average number of newspapers sold per person per day
ROADRATE	Density of major roads in miles of road per square mile
POVERTY	Percentage of people living below the official poverty line

Source: Statistical Abstract of the United States 1987, US Department of Commerce, Washington DC (1987)

communality estimates were set to 1.0 for each variable. Adjacent to these communalities, and beginning under the column headed 'Factor', are details of the derived factors including their eigenvalues, the percentage of variation of the data set for which they individually account and the cumulative percentages of the latter. Of the 13 factors only the first three have eigenvalues in excess of 1.0. They collectively account for 70.7 per cent of the variance in the data set and it is these three that the program identifies for further study. The subsequent factor matrix shows how each variable loads on the three factors. The variables INCOME and DOCTORS load highly on factor 1 but poorly on the others – as should be the case. On the other hand UNEMPLY and ROADRATE are the dominant variables on factor 2. Factor 3 reveals BRATE and CRIME to be dominant in that dimension. Should we wish to attribute broad characteristics to these derived factors, on the basis of the variable loadings, we might suggest factor 1 to be the 'wealth' factor, 2 to be one of 'economic activity' and 3 more social and demographic in character. The final section of the results shows the communality estimates for each variable produced at the final stage. These are not set to 1.0 because they are based only on the three extracted factors.

Thus far we have been able to reduce, with little effort, the original 13 variables to just three factors. However close attention to the factor matrix reveals a degree of ambiguity in this form of interpretation. Variables such as SALES and POVERTY appear to score similarly, though with different signs, on the first two factors. We will see in the

```
GET /FILE 'USASTUDY'.
The SPSS/PC+ system file is read from
    file USASTUDY

FACTOR /VARIABLES POPDENS BRATE URBAN DOCTORS CHARGE EDUC1
CRIME UNEMPLY INCOME SALES PAPERS ROADRATE POVERTY
/ROTATION NOROTATE.
```

```
        - - - -  F A C T O R   A N A L Y S I S - - - -

Extraction 1 for Analysis 1, Principle-Components Analysis (PC)

Initial Statistics:

Variable    Communality  *  Factor  Eigenvalue  Pct of Var   Cum Pct
                         *
POPDENS         1.00000  *     1      5.81133        44.7       44.7
BRATE           1.00000  *     2      1.81460        14.0       58.7
URBAN           1.00000  *     3      1.57052        12.1       70.7
DOCTORS         1.00000  *     4       .98418         7.6       78.3
CHARGE          1.00000  *     5       .66891         5.1       83.5
EDUC1           1.00000  *     6       .51351         4.0       87.4
CRIME           1.00000  *     7       .46514         3.6       91.0
UNEMPLY         1.00000  *     8       .37259         2.9       93.9
INCOME          1.00000  *     9       .24613         1.9       95.7
SALES           1.00000  *    10       .22036         1.7       97.4
PAPERS          1.00000  *    11       .14665         1.1       98.6
ROADRATE        1.00000  *    12       .11082          .9       99.4
POVERTY         1.00000  *    13       .07525          .6      100.0

PC Extracted  3 factors.

Factor Matrix:

              FACTOR 1      FACTOR 2      FACTOR 3

POPDENS         .72342        .17075       -.39779
BRATE          -.39902       -.23664        .74913
URBAN           .77550        .46160        .20650
DOCTORS         .86569        .13557       -.05269
CHARGE          .65919        .06366        .27727
EDUC1           .77532       -.21469       -.16512
CRIME           .40535        .32888        .73613
UNEMPLY        -.49183        .56675        .00501
INCOME          .93651       -.11704        .05119
SALES           .63494       -.47984        .20674
PAPERS          .59416       -.14684       -.29037
ROADRATE       -.47458       -.76397        .00966
POVERTY        -.69112        .41149       -.17274
```

```
Final Statistics:

Variable    Communality  *  Factor  Eigenvalue  Pct of Var    Cum Pct
                         *
POPDENS         .71073   *     1      5.81133       44.7        44.7
BRATE           .77641   *     2      1.81460       14.0        58.7
URBAN           .85712   *     3      1.57052       12.1        70.7
DOCTORS         .77058   *
CHARGE          .51547   *
EDUC1           .67449   *
CRIME           .81436   *
UNEMPLY         .56313   *
INCOME          .89336   *
SALES           .67613   *
PAPERS          .45890   *
ROADRATE        .80897   *
POVERTY         .67680   *
```

Figure 12.3 SPSS instructions and screen display for principal components analysis with a specification that the components should not be rotated. For details on the variables consult Table 12.5.

following section how such ambiguities may be resolved and improve the efficiency of the factor model.

12.6 Rotational solutions in factor analysis

To a large extent the aim of factor analysis is to define new variables or factors that adequately and clearly describe the original variables. The ideal is to search for a 'simple factor structure' whereby each original variable loads high on one factor and low (close to zero) on the others. Unfortunately, what often happens is that the initial solution derived by the factor analysis program does not provide such an unambiguous factor structure. Some indication of this has already been noted in the preceding section. Variables may load highly on more than one factor. In such cases the factors are not clearly describing and accounting for the original variables and their interpretation becomes correspondingly more difficult. Rotational factor analysis often provides an alternative solution that overcomes these difficulties. Factor rotation therefore aims to simplify further the factor structure by more clearly identifying groups of variables. In any such solution the axes that describe each factor can be fixed, by rotation about their common origin, in an infinite number of positions. Some of these rotations will provide a better account of the original data set than others, though there may be little to choose between several possibilities. Finally, before examining the current example, it should be noted that different methods exist to define the orthogonal

```
GET /FILE 'USASTUDY'.
The SPSS/PC+ system file is read from
    file USASTUDY

FACTOR /VARIABLES POPDENS BRATE URBAN DOCTORS CHARGE EDUC1
CRIME UNEMPLY INCOME SALES PAPERS ROADRATE POVERTY.
```

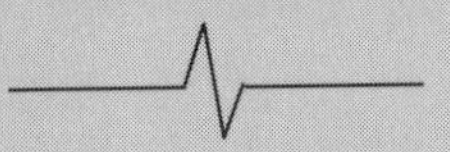

```
    - - - -  F A C T O R   A N A L Y S I S - - - -

Final Statistics:

Variable    Communality  *  Factor  Eigenvalue  Pct of Var   Cum Pct
                         *
POPDENS          .71073  *    1      5.81133       44.7         44.7
BRATE            .77641  *    2      1.81460       14.0         58.7
URBAN            .85712  *    3      1.57052       12.1         70.7
DOCTORS          .77058  *
CHARGE           .51547  *
EDUC1            .67449  *
CRIME            .81436  *
UNEMPLY          .56313  *
INCOME           .89336  *
SALES            .67613  *
PAPERS           .45890  *
ROADRATE         .80897  *
POVERTY          .67680  *

Varimax  Rotation 1, Extraction 1, Analysis 1 - Kaiser Normalization.

  Varimax converged in   5 iterations.

Rotated Factor Matrix:

              FACTOR 1      FACTOR 2      FACTOR 3

POPDENS         .31185        .29820        .72426
BRATE           .00916        .02662       -.88069
URBAN           .22957        .83501        .32736
DOCTORS         .48569        .53840        .49478
CHARGE          .44561        .55375        .10128
EDUC1           .65791        .20440        .44706
CRIME           .15124        .82715       -.32758
UNEMPLY        -.74041        .09258       -.07972
INCOME          .73032        .47141        .37116
SALES           .80750        .15513       -.00403
PAPERS          .47602        .08067        .48403
ROADRATE        .22444       -.74648       -.44874
POVERTY        -.79182       -.21186       -.07026
```

Figure 12.4 SPSS instructions and screen display for principal components analysis allowing the system to run on the default of orthogonal rotation of the components. The initial estimates are the same as those in Figure 12.3 and are omitted.

rotation model. These are known as the quartimax, equimax and the varimax solutions. The latter is the most widely used and depends upon a simplification of the columns of the factor matrix thereby maximising the sum of the variance of the squared loadings in each column; hence the name varimax. We need not concern ourselves too much with the mathematical background of the various rotational methods, rather we should focus on how they may help to provide more clearly understandable factors.

Figure 12.4 shows the results of our analysis when the program is allowed to default to the VARIMAX rotation option. The preliminary results are exactly as in Figure 12.3 and are not repeated here. The figure shows only the instructions and the final rotated factor loadings. The picture has indeed become less ambigous with variables now being clearly distinguishable in their loadings on the three factors. SALES and POVERTY, the two variables observed to weight similarly on factors 1 and 2 in the non-rotated model, are now clearly distinguishable in terms of the factor to which they can be attributed and both weight preferentially on factor 1 and only lightly on factor 2. Most important, the factor structure is now clearer and there is little ambiguity as most variables loading heavily on one of the factors are low on the other two. Factor 1 appears to be more decidedly 'economic' with SALES, POVERTY and UNEMPLY dominating the loadings. Factor 2, with URBAN and CRIME, is more 'social' and factor 3, with POPDENS and BRATE is 'demographic' in character.

Section 12.1 has already indicated that solutions can be sought by rotating the axis as a single orthogonal set. A further option that is available is to select an oblique solution. In such cases the factors are rotated independently of one another. In doing so they lose their mutually orthogonal relationship and, depending on the angle between the factors (or axes, if we retain a more geometrical terminology), will become correlated between themselves. This may not be an entirely unrealistic representation of the data as groups of variables, while possessing distinctive characteristics, may also be linked and are not necessarily orthogonal (and uncorrelated). From the results of our example thus far we might anticipate that the 'economic' and the 'social' factors may not themselves be wholly uncorrelated or causally unconnected. On the other hand we must be careful, in rotating factors, not to merely reproduce our original data set! A further difficulty with oblique rotations is that there are no unique solutions and many decisions, especially that connected with the freedom to rotate the factors, are in the hands of the researcher and consequently subjective in character.

The oblique rotation approach also alters a fundamental assumption in orthogonal factoring which is that any variable will, ideally, load as ±1.0 on one axis only, and as zero on all others. This is no longer true and the consequences of this change further complicate matters by producing two sets of loadings, the essential geometry of which is

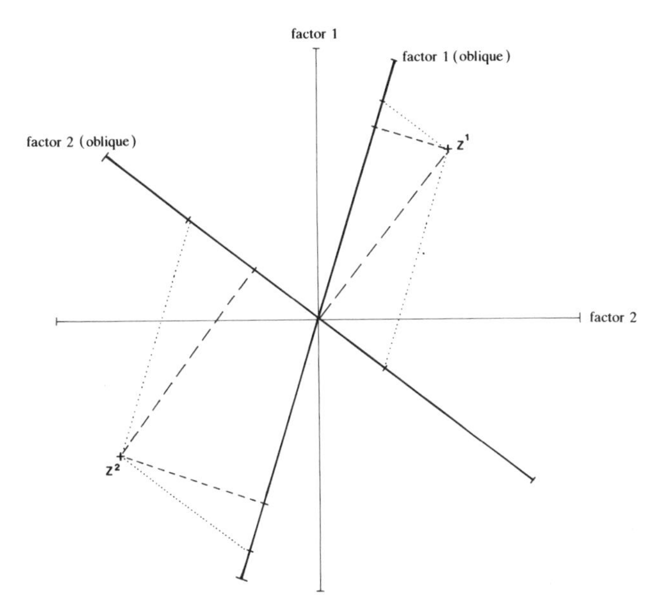

Figure 12.5 Figure illustrating the differences between structure and pattern loadings in factor analysis. The geometry of the structure loadings is shown by chain lines, that of the pattern loadings by dotted lines.

illustrated in simplified form in Figure 12.5. The first, the so-called *structure loadings* are analogous to the loadings of variables on orthogonal models and represent, in geometrical terms, the independent perpendicular projections of each variable onto each factor; these projections are indicated by the chain lines in Figure 12.5. Such perpendicular projections, or loadings, overlook any non-orthogonal relationship between the factors. The second set, the *pattern loadings*, again represent the loadings of each variable on each factor once again, but now that loading takes into account the fact that orthogonality no longer exists. In simple geometrical terms these are not the perpendicular projections of a variable onto any one factor but a projection whose line is parallel to the other factor. These projections are shown as dotted lines in the figure. The degree to which that projection departs from the perpendicular depends upon the angle between the two factors. The whole issue becomes of course more complicated when three or more axes are involved, but the same basic geometric principles apply no matter how many factors are extracted.

In Figure 12.6 we have gone two steps further in our analysis. First, the **ROTATION** subcommand requests the oblique rotation **OBLIMIN**. Second, the analysis is genuinely factor, rather than principal components, analysis as the **EXTRACTION** subcommand overrides the default with **PA2** which requires the program to factorise using initial communality estimates based here on each variable's squared multiple correlation coefficient with other variables in the set. In this respect it

```
GET /FILE 'USASTUDY'.
The SPSS/PC+ system file is read from
    file USASTUDY

FACTOR /VARIABLES POPDENS BRATE URBAN DOCTORS CHARGE EDUC1
CRIME UNEMPLY INCOME SALES PAPERS ROADRATE POVERTY
/EXTRACTION PA2
/ROTATION OBLIMIN.
```

---- F A C T O R   A N A L Y S I S ----

Extraction 1 for Analysis 1, Principle Axis Factoring (PAF)

Initial Statistics:

Variable	Communality	*	Factor	Eigenvalue	Pct of Var	Cum Pct
		*				
POPDENS	.69821	*	1	5.81133	44.7	44.7
BRATE	.60888	*	2	1.81460	14.0	58.7
URBAN	.84156	*	3	1.57052	12.1	70.7
DOCTORS	.77082	*	4	.98418	7.6	78.3
CHARGE	.58376	*	5	.66891	5.1	83.5
EDUC1	.73774	*	6	.51351	4.0	87.4
CRIME	.67358	*	7	.46514	3.6	91.0
UNEMPLY	.54975	*	8	.37259	2.9	93.9
INCOME	.88393	*	9	.24613	1.9	95.7
SALES	.67810	*	10	.22036	1.7	97.4
PAPERS	.49950	*	11	.14665	1.1	98.6
ROADRATE	.71335	*	12	.11082	.9	99.4
POVERTY	.68067	*	13	.07525	.6	100.0

PAF Extracted 3 factors. 9 Iterations required.

Factor Matrix:

	FACTOR 1	FACTOR 2	FACTOR 3
POPDENS	.69938	.10985	-.36741
BRATE	-.38240	-.14496	.66210
URBAN	.78430	.46662	.16739
DOCTORS	.85179	.10192	-.07121
CHARGE	.60985	.03150	.19330
EDUC1	.74426	-.23244	-.12736
CRIME	.40031	.34323	.65454
UNEMPLY	-.45093	.42540	-.02871
INCOME	.95242	-.14960	.05690
SALES	.60275	-.40680	.19801
PAPERS	.54093	-.14881	-.19086
ROADRATE	-.47301	-.71123	.08367
POVERTY	-.65909	.37185	-.16802

Figure 12.6 SPSS instructions and screen display for factor analysis of US data. The commands specify factor analysis with oblique rotation of the significant factors. The output lists both the pattern and structure matrices and the degree of correlation between the factors.

Final Statistics:

Variable	Communality	*	Factor	Eigenvalue	Pct of Var	Cum Pct
		*				
POPDENS	.63619	*	1	5.48661	42.2	42.2
BRATE	.60562	*	2	1.46912	11.3	53.5
URBAN	.86087	*	3	1.20340	9.3	62.8
DOCTORS	.74100	*				
CHARGE	.41027	*				
EDUC1	.62417	*				
CRIME	.70648	*				
UNEMPLY	.38512	*				
INCOME	.93273	*				
SALES	.56800	*				
PAPERS	.35118	*				
ROADRATE	.73659	*				
POVERTY	.60090	*				

Factor Matrix:

	FACTOR 1	FACTOR 2	FACTOR 3
POPDENS	.32339	.22906	-.55395
BRATE	-.02663	.00609	.77340
URBAN	.18955	.81348	-.12406
DOCTORS	.48633	.41865	-.29861
CHARGE	.42493	.37235	.03468
EDUC1	.67178	.05710	-.25776
CRIME	.11341	.75983	.47346
UNEMPLY	-.65456	.16590	-.00718
INCOME	.77990	.29954	-.14484
SALES	.77324	-.00993	.13253
PAPERS	.45461	.01527	-.28804
ROADRATE	.26397	-.78244	.34580
POVERTY	-.77824	-.03105	-.08208

Structure Matrix:

	FACTOR 1	FACTOR 2	FACTOR 3
POPDENS	.51324	.44043	-.66673
BRATE	-.18756	-.15326	.77782
URBAN	.47559	.89824	-.32265
DOCTORS	.68298	.63228	-.48269
CHARGE	.53659	.50135	-.12743
EDUC1	.74430	.32201	-.41037
CRIME	.25647	.70372	.30137
UNEMPLY	-.60005	-.04184	.09831
INCOME	.90611	.57698	-.36752
SALES	.74215	.21127	-.02838
PAPERS	.52015	.21671	-.38676
ROADRATE	-.05885	-.76555	.44282
POVERTY	-.77088	-.26370	.08788

Factor Correlation Matrix:

	Factor 1	Factor 2	Factor 3
FACTOR 1	1.00000		
FACTOR 2	.31951	1.00000	
FACTOR 3	-.21060	-.19505	1.00000

should be noted that the degrees of explanation offered by the final three factors differs from those found using the default principal components approach.

Figure 12.6 lists also the final communality estimates together with the structure and pattern loadings of each variable on the extracted factors. In this case the results differ only marginally from those derived by principal components methods. Only three factors can again be identified as important, they account for a similar degree of variability of the original data set and again they emerge as respectively 'economic', 'social' and 'demographic' in character. The similarity of the pattern and the structure loadings suggests that little orthogonality has been lost between the factors, and this point is reinforced by the factor correlation matrix which shows very little correlation between the three. Based on our earlier studies, any hypothesis utilising the notion of 'economic/affluence', and 'social' and a 'demographic' factors is again borne out by the loadings on factors 1, 2 and 3 respectively.

The above strategy is one of hypothesis-checking and can be contrasted with that of hypothesis-creation. Strictly speaking we should only use factor, as opposed to principal components, analysis when we are working within the framework of an existing hypothesis or hypotheses. In this context we are rotating the factors towards 'targets' that accommodate our hypotheses. How far the results converge upon our targets will, naturally, depend upon the validity of our hypotheses. Where coincidence is not achieved we may move towards their modification, a more hypothesis-creation form of approach to the problem. Hence we should make no distinction at this practical level between the two strategies.

12.7 The use of factor scores

Thus far we have concentrated on the relationships between the variables and the factors, but one important part of the output from factor analysis can be the matrix of factor scores which provides a measure of the relationship between each observation, as opposed to each variable, and the factors. As we shall see this has proved to be a valuable means by which geographers can search for spatial order in their data. The factor scores consist of a matrix of values for each observation, or US state in this example, on each factor. Here too we can see the advantages of simplifying the original variables into a smaller number of factors as each case's factor score on each of the factors summarises the effect of all variables on that particular case. However it is important to remember that, as we have shown, each variable itself scores more or less heavily on each factor. More important yet, groups of variables dominate on each factor and we might find them to be loosely associated in terms of interpretation. Joshi (1972) has already discussed some of the problems in interpreting factor scores. In the above examples factor 1 was described as 'economic' in character on which the variables such as SALES, POVERTY and INCOME loaded heavily. As a result the scores of each state on that factor can be regarded as an 'economic' score determined by the collective effect of those same variables. Similarly factor 2 was determined to be more social in character with CRIME and URBAN dominating the loadings. The scores of the states on that factor will, consequently, reflect the combined influence of those high-scoring variables. Other variables are not excluded in any case-score rating but their contributions are minimised as a result of their low loadings. By this means we might compare one case with another in terms of several variables simultaneously. If the cases represented a spatial sample we could, for example, plot a map of scores on any one of the factors, thereby identifying groupings or regions with similar characteristics. We might also plot a graph of factor scores on the two most important factors, by which means statistical groupings of spatial or of non-spatial data might be recognised. It is by such simple methods that we might take full advantage of factoring methods.

The role and character of case-based factor scores becomes clearer when we note the manner of their derivation, which is very much like that of deriving an estimate from a multiple regression model. But instead of each variable having a regression coefficient it now has, for each variable in turn, a factor score. These scores may be obtained by including the instruction /**PRINT FSCORES** at the conclusion of the factor analysis. It has the effect of suppressing all other listings and is best used only when interest lies exclusively with the case factor scores. The results of the analysis are given in Figure 12.7. Exact factor scores can be derived from principal components only and not from factors. SPSS can provide factor scores derived from the latter, but they are estimates only and subject to a degree of error dependent upon the factor structure. The following example is based on principal components analysis with orthogonal rotation (Figure 12.4). From Figure 12.7 it can be seen that, for factor 1 the 'regression equation' for each state's factor score would be:

$$\text{SCORE} = -0.02029(\text{POPDENS}) + 0.11782(\text{BRATE}) - 0.07136(\text{URBAN})$$

$$+ 0.04314(\text{DOCTORS}) \ldots\ldots\ldots\ldots -0.25987(\text{POVERTY})$$

This calculation, using the standardised values of each individual observation, would be repeated for each state, the final score being a standarised expression of the effect of all variables but, as indicated above, giving greatest weight to those variables that load heavily on that factor. In this latter respect note that the 'regression coefficients' will change from factor to factor. Notice also that they are not directly related to the derived factor loadings as shown in Figure 12.4. Thus, for factor 2, the expression would be:

```
GET /FILE 'USASTUDY'.
The SPSS/PC+ system file is read from
    file USASTUDY

FACTOR /VARIABLES POPDENS BRATE URBAN DOCTORS CHARGE EDUC1 CRIME UNEMPLY
INCOME SALES PAPERS ROADRATE POVERTY
 /PRINT FSCORES.

        - - - - F A C T O R   A N A L Y S I S - - - -

Analysis Number 1  Listwise deletion of cases with missing values

Extraction 1 for Analysis 1, Principal-Components Analysis (PC)

   PC Extracted  3 factors.

Factor Score Coefficient Matrix:

                FACTOR 1      FACTOR 2      FACTOR 3

POPDENS         -.02029       -.00577        .29675
BRATE            .11782        .13266       -.46666
URBAN           -.07136        .30670        .02550
DOCTORS          .04314        .11247        .11995
CHARGE           .07868        .17935       -.08312
EDUC1            .15998       -.05775        .11801
CRIME           -.01204        .40428       -.30631
UNEMPLY         -.28079        .15456        .04461
INCOME           .16098        .06542        .03161
SALES            .28347       -.03844       -.13174
PAPERS           .10002       -.09359        .18006
ROADRATE         .24652       -.31125       -.16220
POVERTY         -.25987        .02053        .09852
```

Figure 12.7 SPSS instructions and screen display requesting the factor score coefficient matrix. By default no other output is included.

$$\text{SCORE} = -0.00577(\text{POPDENS}) + 0.13266(\text{BRATE}) + 0.30670(\text{URBAN}) + 0.11247(\text{DOCTORS}) \ldots\ldots\ldots\ 0.02053(\text{POVERTY})$$

SPSS does not automatically carry out task of estimating the case-based factor scores and has to be instructed to do so. The results are then stored in the system but, unusually for SPSS, will be lost upon conclusion of the session unless they are displayed on the screen using the **LIST** instruction and consequently contained within the workfile SPSS.LIS which can be printed out at a later stage (see section 2.9). These tasks are accomplished (Figure 12.8) by specifying the **SAVE** option after the completion of the factor analysis. This is followed by the keyword **REGRESSION** which instructs the system to prepare case factor scores using the regression method outlined above (there are other options but they are not discussed here). We need also to specify over how many factors the scores must be calculated. There is neither need nor virtue in specifying any more than the number that are 'significant', i.e. have eigenvalues greater than 1.0.

```
GET /FILE 'USASTUDY'.
The SPSS/PC+ system file is read from
    file USASTUDY

FACTOR /VARIABLES POPDENS BRATE URBAN DOCTORS CHARGE EDUC1
CRIME UNEMLY INCOME SALES PAPERS ROADRATE POVERTY
SAVE /REGRESSION (3 FAC).
```

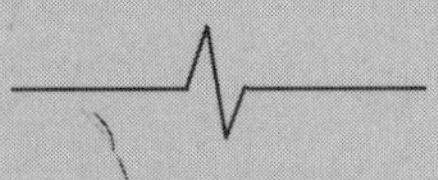

```
3 PC EXACT SCORES WILL BE SAVED WITH ROOTNAME: FAC

FOLLOWING FACTOR SCORES WILL BE ADDED TO THE ACTIVE FILE:

NAME        LABEL

FAC1        REGR FACTOR SCORE 1  FOR ANALYSIS 1
FAC2        REGR FACTOR SCORE 2  FOR ANALYSIS 1
FAC3        REGR FACTOR SCORE 3  FOR ANALYSIS 1
```

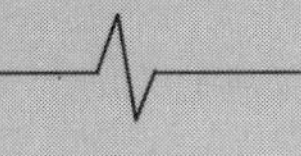

```
LIST /VARIABLES = STATE, FAC1, FAC2, FAC3.
```

STATE	FAC1	FAC2	FAC3
ME	.22384	-.93413	.18975
NH	1.32016	-.69151	-.06476
VT	.87803	-.90463	-.11085
MA	1.65985	.41663	1.93584
AZ	-.28744	1.40091	-1.31385
..	....	....	....
..	....	....	....
UT	-.26438	.98430	-2.52495
NV	.87401	.92318	-1.28888
WA	.04513	1.24346	-.46438
OR	-.11285	.89742	-.37481
CA	.52744	2.04894	-.78332

Figure 12.8 Sequence of commands and screen displays needed to obtain the case factor scores of each US state. Only a sample of the latter are listed in the figure. The states are denoted by their standard post codes, this variable being included in original data file USA.DAT under the variable name STATE. The LIST specification allows this 'variable' to be printed.

In this case that number is 3. We need also to provide a name for the three new 'variables' which will contain the three sets of scores and here we have named them as **FAC**. Both the latter items of information appear within parenthesis after the **REGRESSION** specification. The three variables can then be displayed on the screen using the **LIST /VARIABLES** instruction as shown in Figure 12.8. Notice that in order to identify each state's factor score the **LIST** specification includes the hitherto unused column of the original data matrix which contains the state's two-letter code.

Armed with this information we may now plot or map the results in the search for geographical or statistical groupings. SPSS does not create such graphical displays, though some other programs dedicated to plotting and mapping may do so. The following figures were prepared manually and show the spatial patterns of state scores on all three principal factors. In interpreting these maps it must not be forgotten that the factors have different eigenvalues. Attention should be focused on the map of factor 1 as it accounts for over 40 per cent of the variability of the data set; the other two (Figure 12.4) provide a far lower degree of statistical explanation. For a case to achieve a high score on any factor it must have a high value for variables with positive loadings, but low values for variables with negative loadings. Conversely for a case to gain a low score it must have a low value on variables with a positive loading, but a high value on variables with a negative loading.

Figures 12.9 to 12.11 map the factor scores for each of the 48 US states. All three reveal interesting spatial groupings. Factor 1, the 'economic' factor, identifies the southern state grouping of Louisiana, Mississippi, Arkansas and Alabama one region that rates at the lower end of this factor. To understand what this means we must refer back to Figure 12.4 to remind ourselves of the variables that load heavily on this factor and, importantly, of the sign of the loading. These states are characterised by high unemployment and high poverty levels (negatively-weighted variables on this factor) and by low income and low retail sales (positively-weighted variables). Northern states generally do much better especially North Dakota and Wyoming in the Mid-West as do several of the smaller New England states where unemployment and poverty are generally lower and retail sales and income higher. Factor 2, the 'social' dimension, presents a different geographic picture with the southern–western region, led by California and Arizona doing well. That is to say, in terms of the study variables that load heavily on this factor, they are highly urbanised, but have high crime rates. The distant states of Florida and Michigan also fall into this category. On the other hand North Dakota and Wyoming, which rated high on the economic factor, form with South Dakota a region of low case-scores on the social scale. This is not, however, to assume them to be socially deprived, indeed we may conclude the opposite as these states, though poorly urbanised, have low crime rates! Lastly factor 3 shows another clear

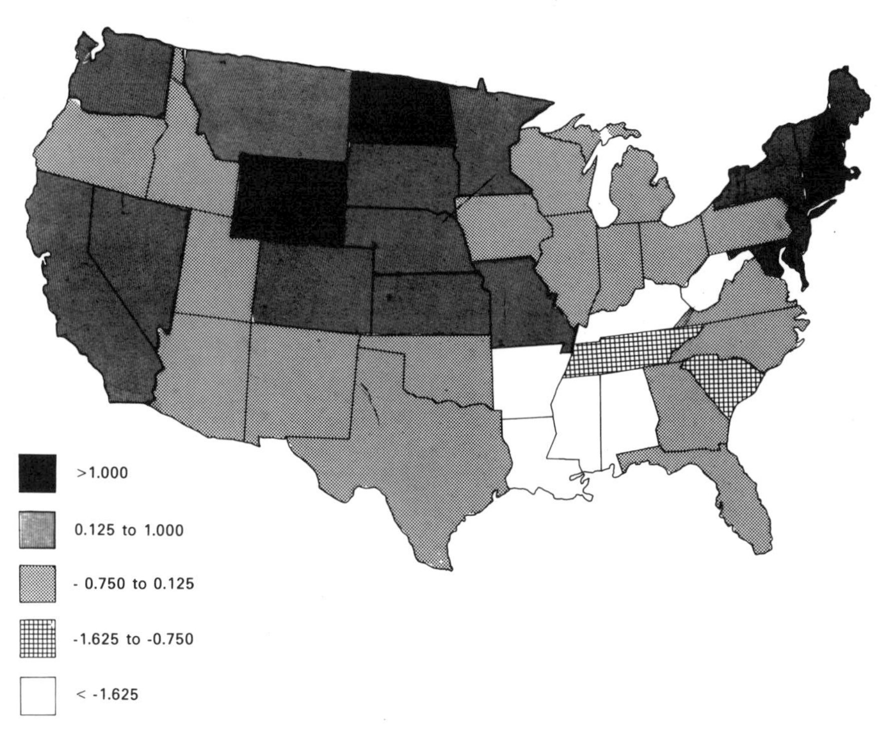

Figure 12.9 Map of scores on factor 1 based on the output displayed in Figure 12.8. This factor accounts for 44.7% of the data variance.

geographic grouping. Remembering that POPDENS loads at 0.72426 and BRATE as −0.88069 on this factor (Figure 12.4), the contiguous states of the New England region, Ohio and West Virginia secure high scores indicating that they have very high population densities, but very low birth rates. Conversely Wyoming, Utah and New Mexico are states with low scores suggesting low population densities but high birth rates. This is not the place to enquire more closely into the socio-economic geography of the United States, but it should be clear that factor and principal components analysis offers wide scope to assist in such studies.

12.8 Problems in the application of factor analysis

Like most other multivariate statistical techniques applied to geographical problems, factor analysis has attracted a degree of criticism. Among statisticians there are those who dismiss the method as 'an elaborate way of doing something which can only ever be crude, namely the picking out of clusters of interrelated variables'. In these debates such people favour the mathematical clarity of principal components analysis. However, countering such criticism is the work of the pro-factor analysts, whose studies are mainly rooted within the social sciences

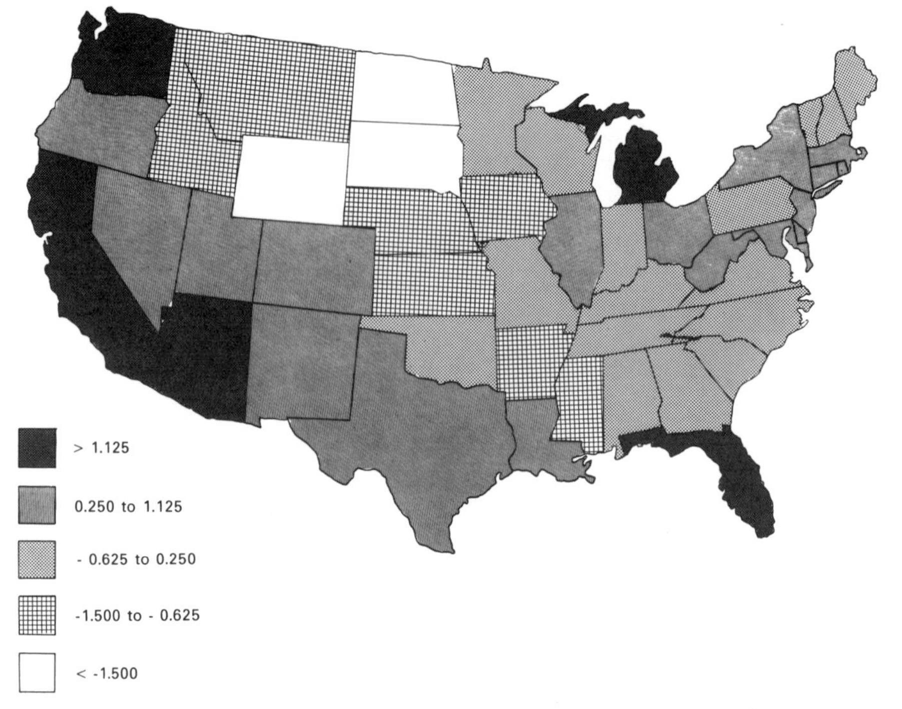

Figure 12.10 Map of scores on factor 2 based on the output displayed in Figure 12.8. This factor accounts for 14.0% of the data variance.

(Cattell, 1978). To these people the advantages of factor analysis are the objectivity of the method and its more realistic assumptions of the data structures.

These debates have encouraged geographers to question the methods and two main issues have emerged. The first concerns how factor analysis should be used by geographers, a problem that to a large extent is created by the very flexibility of the methods. Without perhaps fully appreciating it to be the case, geographers have used factoring methods in one of three ways: (1) to attempt to create some order in a large group of poorly-understood data; (2) to explore working hypotheses and to measure the related dimensions of sets of variables; and (3) to analyse patterns of factor loadings. These uses may be further complicated by the variety of methods that exist to solve each problem and the question is quickly raised of which is the best method. The problem becomes yet more vexing when different programs, using different methods, produce different results from the same data. Such problems cannot be fully resolved until geographers explore the advantages and limitations of different factor models in the light of their own clearly-defined needs.

The second problem is that of interpretation. One way in which this problem arises is through the study of factor loadings of the variables. These represent the square roots of the proportions of standardised

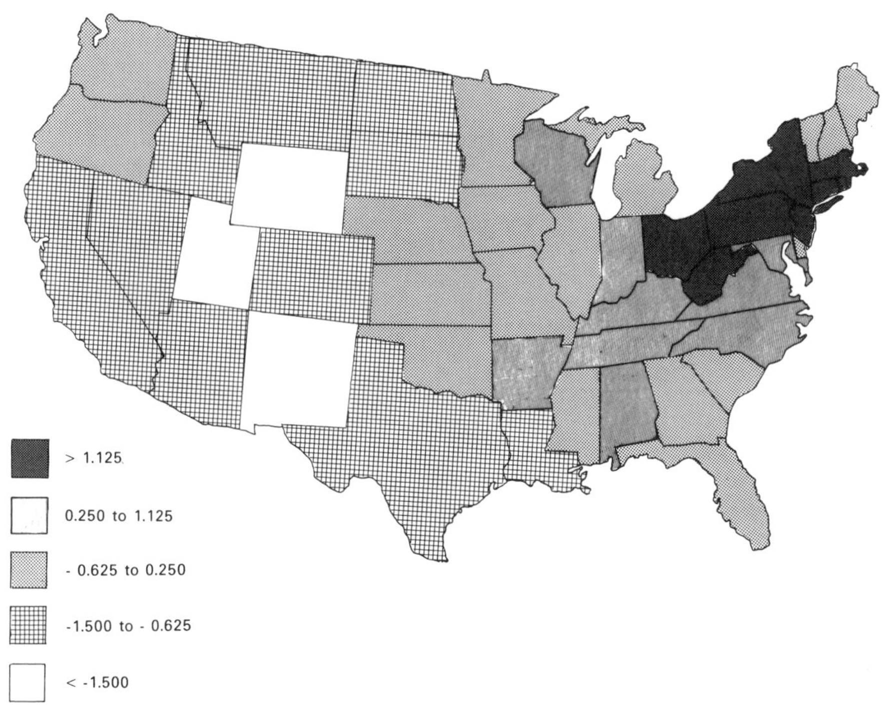

Figure 12.11 Map of scores on factor 3 based on the output displayed in Figure 12.8. This factor account for 12.1% of the data variance.

variation of each variable on each factor. Thus these factors are not as substantive as their numerical size might indicate. We might, in this sense, overestimate the importance of groups of variables. In particular we might lay too great a stress on pairs of variables that are poorly correlated. Johnston (1976) has observed that this problem stems partly from geographers' use of imprecise goals.

Variable selection is a perennial problem and one which has already been reviewed in section 12.4. Some workers have gone so far as to suggest the selective removal of variables that score low on all derived factors, thereby simplifying the data structure. But the necessities of subjective assessment would normally make this inadvisable. Other problems in urban studies result from the mismatch between census enumeration districts (the usual unit in British studies) and social groups. This may, owing to unfortunate boundary locations, introduce spurious internal heterogeneity into each area making it thereby more difficult to abstract useful factors. In factorial ecological studies of urban areas such problems have contributed to a more cautious use of factoring methods, often replacing them by less mechanistic approaches (Herbert and Thomas, 1982).

References

Blakith, R.E. and Reyment, R.A. (1971) *Multivariate Morphometrics*, Academic Press, London.
Cattell, R.B. (1978) *The Scientific Use of Factor Analysis*, Plenum Press, New York.
Chatfield, C. and Collins, A.J. (1980) *Introduction to Multivariate Analysis*, Chapman and Hall, London.
Harman, M.H. (1967) *Modern Factor Analysis*, Chicago University Press, Chicago.
Herbert, D. and Thomas, C. (1982) *Urban Geography: a first approach*, Wiley, Chichester.
Johnston, R.J. (1976) 'Residential area characteristics: research methods for identifying urban sub-areas. Social area analysis and factorial ecology', in D.T. Herbert and R.J. Johnston (eds) *Social Areas in Cities*, Wiley, Chichester.
Joshi, T.R. (1972) 'Towards computing factor scores', in W.P. Adams and F. Helleiner (eds) *International Geography*, University of Toronto Press, Toronto.

Chapter 13
Spatial Indices and Pattern Analysis

13.1 Introduction

Many of the statistical techniques that geographers commonly use tend to be of a non-spatial type, that can just as easily be applied to any other scientific discipline. Indeed, the majority of the techniques mentioned so far in this book fall into this category, where location is but one of many variables under examination. However, there also exists a set of spatial statistics which allow the geographer to summarise and describe numerically a variety of spatial patterns ranging from simple dot distributions through to contour maps. As Unwin (1981) demonstrates, it is possible to recognise a typology of maps that loosely correspond to the nominal, ordinal, interval and ratio scales of measurement that we discussed in Chapter 4. In this way a picture can be built up of the types of data we are likely to want to analyse using different spatial statistics. To carry out such spatial analysis three major groups of techniques are available, namely: centrographic techniques, related spatial indices and point pattern analysis.

13.2 Centrographic techniques

Centrographic techniques are an extension of the descriptive statistics discussed in Chapter 4, but applied to data in two-dimensional space. Such techniques, also termed geostatistics, have been used for some considerable time, with the concept of the mean centre being introduced in the USA census of population as early as 1870. Similarly, a school of so-called centrography, based on these methods, was developed in Russia in the early part of the twentieth century (Taylor, 1977). The main measures used within these studies were those concerned with central tendency, in particular the arithmetic mean centre. Geographers have used such spatial statistics since the 1950s (Hart, 1954), and have extended the techniques through the work of Warntz and Neft (1960).

The mean centre can be most easily used to summarise a spatial distribution of point patterns. In this case the data must first be identified by grid co-ordinates. If the mean centre of a point distribution is to be calculated, then the first step is to derive the mean of the X and Y co-ordinates for the point distribution, using equation 13.1:

$$\bar{X} = \frac{\Sigma X_i}{N} \qquad \bar{Y} = \frac{\Sigma Y_i}{N} \tag{13.1}$$

Equation 13.1

X_i, Y_i = co-ordinates of individual points
N = total number of points

These calculations are demonstrated for a simple, hypothetical example in Table 13.1; while Figure 13.1 shows how the mean centre of this dot distribution is represented by the intersection of $\bar{X}$ and $\bar{Y}$.

Table 13.1 Calculation of mean centre

X	Y	Point
1	5	1
2	6	2
2	4	3
2	3	4
3	7	5
3	5	6
3	4	7
4	6	8
4	3	9
5	5	10
—	—	—
29	48	10

$\bar{X}$ = 29/10 = 2.9; $\bar{Y}$ = 48/10 = 4.8

It is also possible to calculate the weighted mean centre of a distribution where each dot represents a different value. For example, each dot on the map may represent a factory of different size, and we may want to work out the mean centre of factory floorspace. In this case we could weight out calculations in terms of the size of each factory. The mean centre is then found by multiplying the weights of each occurrence by its X and Y co-ordinate values, and dividing by the sum of the weights, as in equation 13.2:

$$\bar{X}_w = \frac{\Sigma(X_i W_i)}{\Sigma W_i} \qquad \bar{Y}_w = \frac{\Sigma(Y_i W_i)}{\Sigma W_i} \tag{13.2}$$

where W are the weighted values.

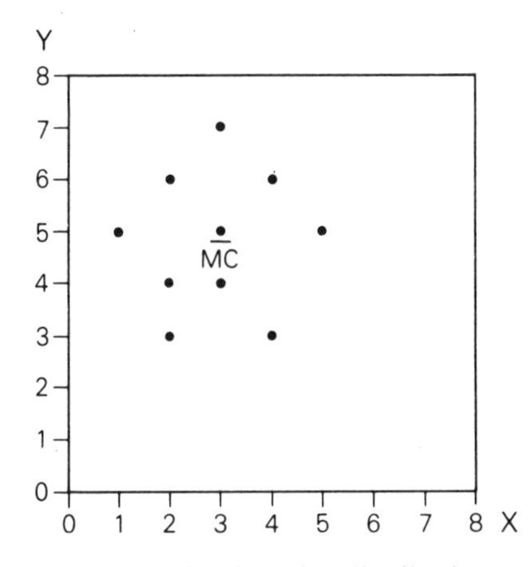

Figure 13.1 Mean centre for a simple point distribution.

As with conventional measures of central tendency, geostatistics can also make use of the median centre and the modal centre. The former, as Ebdon (1977) points out, is unfortunately described in quite different ways by a number of standard statistical texts. For example, Cole and King (1968) and Hammond and McCullagh (1974) define the median centre as the point of intersection of two orthogonal lines from the *X* and *Y* axes, which have an equal number of points on each side (Figure 13.2(a)). In contrast, Neft (1966), King (1969) and Smith (1975) refer to the median centre as the point in a distribution at which the sum of the absolute deviations of each point is minimised. That is, distances between the median centre and each point are at a minimum. It therefore represents the point of theoretical 'minimum aggregate travel', and its position can be found by the use of grid overlays and an interactive procedure as outlined by Seymour (1965). There are three basic steps involved, in what can be a fairly lengthy process without the aid of a computer. The first stage is to overlay a co-ordinate grid on the map, the limits of which are set by the four most extreme points in the pattern. Second, for each new grid co-ordinate point (X_0, Y_0), the square root of the sum of the squared distances to the *n* points (X_i, Y_i), of the original pattern is calculated (Figure 13.2(b)). The point having the lowest value, using equation 13.3, is then identified:

$$M_c = \sqrt{}\Sigma[(X_0 - X_i)^2 + (Y_0 - Y_i)^2] \qquad (13.3)$$

Equation 13.1

X_0, Y_0 = grid co-ordinates
X_i, Y_i = original points
M_c = median centre

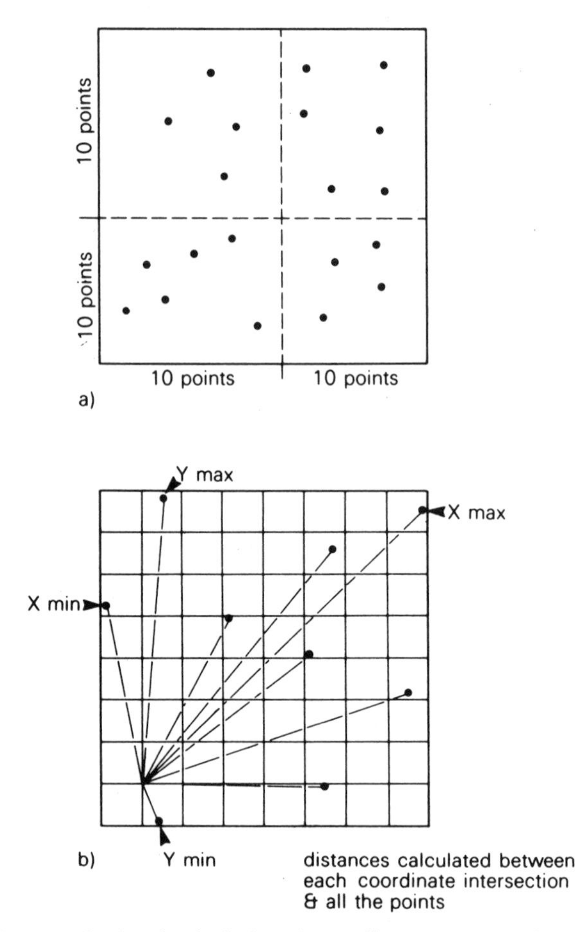

Figure 13.2 Two methods of calculating the median centre: (a) the equal frequency approach, (b) the minimum aggregate distance approach.

In the third stage, this point of minimum value is now taken as the centre of a new, finer grid overlay determined in a subjective manner. At this stage, step two is repeated using equation 13.3. A new, more accurate, point of minimum distance is thus established. Such iterations can be carried out as deemed necessary to determine more accurately the 'median centre'.

The concept of the median centre as the point of minimum aggregate travel is of considerable use within the study of economic geography (Smith, 1975). Furthermore this measure, together with a wider set of distance concepts developed by Warntz and Neft (1960), has extended such analysis into the area of spatial modelling.

At a simpler, but nevertheless effective, level of analysis the concept of the mean centre has proved extremely useful in studying the changing pattern of distributions over time. These changes can be described by calculating the mean centre of a distribution at different time periods and

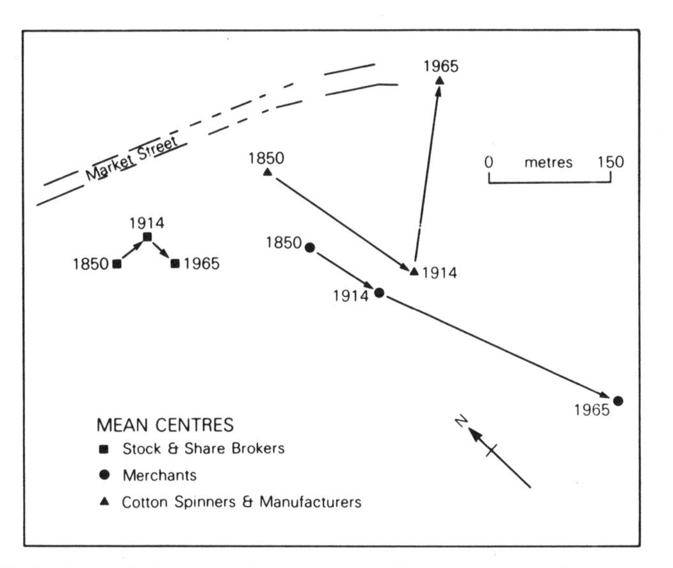

Figure 13.3 Use of the spatial mean to indicate functional change in central Manchester (modified from Varley, 1968).

by plotting such changes, as shown in Figure 13.3. Considerable work has also been carried out by Russian economic geographers during the 1920s and 1930s; when they calculated the mean centre for a variety of economic activities and constructed so-called centrograms (Sviatlovsky and Eells, 1937).

As was shown in Chapter 4, measures of central tendency were only one way of describing a distribution, since use could also be made of statistics measuring dispersion. In geostatistics a commonly used measure of dispersion is the standard distance, which is analogous to the standard deviation in simple, descriptive statistics. The simplest method of calculating the standard distance of a distribution is shown in equation 13.4:

$$SD = \sqrt{\left[\frac{\Sigma(X - \bar{X})^2}{N} + \frac{\Sigma(Y - \bar{Y})^2}{N}\right]} \tag{13.4}$$

Equation 13.4

N = number of points
$\bar{X}, \bar{Y}$ = means of X and Y
SD = standard distance

This is based on Pythagoras' theorem, where linear distances between points can be calculated from their X and Y co-ordinates (Figure 13.4). The method of calculation is shown in Table 13.2 for the same data as in Table 13.1.

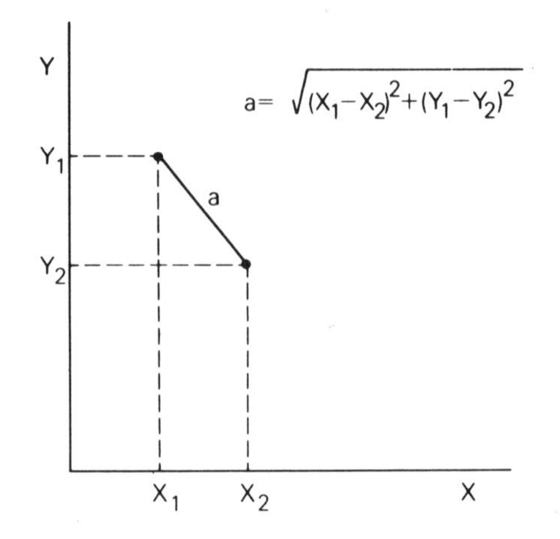

Figure 13.4 Calculation of point to point distances using Pythagoras' theorem.

Very often it is useful to use the mean centre and the standard distance together in order to describe adequately a spatial distribution. However, a glance through the literature will show that few studies have used both measures.

Table 13.2 Calculation of standard distance

X	Y	$X-\bar{X}$	$(X-\bar{X})^2$	$Y-\bar{Y}$	$(Y-\bar{Y})^2$
1	5	−1.9	3.61	0.2	0.04
2	6	−0.09	0.81	1.2	1.44
2	4	−0.09	0.81	−0.8	0.64
2	3	−0.09	0.81	−1.8	3.24
3	7	0.1	0.01	2.2	4.84
3	5	0.1	0.01	0.2	0.04
3	4	0.1	0.01	−0.8	0.64
4	6	1.1	1.21	1.2	1.44
4	3	1.1	1.21	−1.8	3.24
5	5	2.1	4.41	0.2	0.04
			12.90		15.60

$\bar{X}$ = 2.9; $\bar{Y}$ = 4.8 (from Table 13.1); standard distance on X=12.9/10=1.29; standard distances on Y=15.66/10=1.55; combined standard distances=2.87.

Shachar's (1964) analysis of the dispersion of commercial functions in the cities of Tel Aviv, Jerusalem and Rome adequately demonstrated the usefulness of standard distance, while also drawing attention to the problem of using absolute measures of dispersion in a comparative study. Thus, absolute standard distance is affected by the shape of the area under study and also by the size of the study area. For example, in Shachar's results the standard distances of commercial functions in Rome were usually larger than those found in Tel Aviv or Jerusalem, mainly because it was a larger city. To overcome this problem use needs to be made of relative standard distance measures. There are a number of ways of calculating such relative measures, depending on the phenomena under investigation. In Shachar's study the commercial functions were related to the population distribution and the relative dispersion was calculated by dividing the standard distance of a commercial distribution by the standard distance of the population. In contrast Neft (1966), studying population dispersion in different countries, used a relative measure based on the radius of the area of each country, which was assumed to be circular.

13.3 Spatial indices and the Lorenz curve

Apart from the centrographic techniques discussed in the previous section, distributions can also be compared using spatial indices that relate to the Lorenz curve. The latter is a simple, but effective, means of illustrating graphically the difference between spatial patterns, and in many texts it is introduced as a technique of map comparison (Unwin, 1981). A number of stages are involved in the calculation and construction of the Lorenz curve, as can be illustrated in a simple example which compares the distributions of black people and white people in the USA ((Table 13.3). First, the ratio between blacks and whites needs to be calculated for each area (in our example census regions). Second, these areal units are then ranked on the basis of these ratios, from the smallest to the largest. Next, each variable is converted into a percentage of the total for its own area. Thus, in our example, for each census region of the USA the numbers of black and white people are expressed as percentages. Finally, as Table 13.3 shows, these percentage values are accumulated, maintaining the ranks, from 1

Table 13.3 Calculation of Lorenz curve for ethnic segregation in the USA (1981)

Regions	Number (000) of: white	black	X/Y	Rank	X (%)	Y (%)	Cumulation X(%)	Y(%)
N England	11586	475	24.4	8	6.2	1.8	94.7	99.0
Mid Atlantic	30743	4374	7.0	4	16.3	16.5	47.6	69.5
EN Central	36139	4548	7.9	5	19.2	17.2	66.8	86.7
WN Central	16045	789	20.3	7	8.5	3.0	88.5	97.2
S Atlantic	28648	7648	3.7	1	15.2	28.9	15.2	28.9
EC Central	11700	2868	4.1	2	6.2	10.8	21.4	39.7
WS Central	18597	3525	5.3	3	9.9	13.3	31.3	53.0
Mountain	9957	269	37.0	9	5.3	1.0	100.0	100.0
Pacific	24926	1993	12.5	6	13.2	7.5	80.0	94.2
	188341	26489						

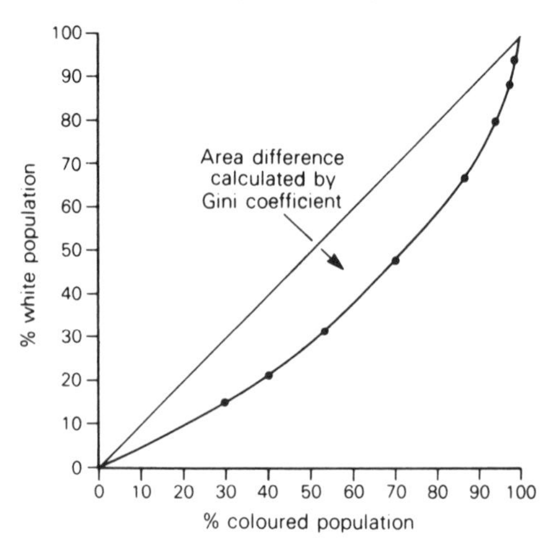

Figure 13.5 The Lorenz curve, based on the data in Table 13.3.

upwards. These values can then be plotted out as a graph, or Lorenz curve, as shown in Figure 13.5.

In general terms the Lorenz curve has a number of obvious features; one being that if the distributions are proportionally identical in each area then the plot will be a straight line (Figure 13.5). Differences between the distributions will be shown in the form of deviations away from this diagonal line. The extreme case is one of complete separation between two distributions, where the line would follow the X axis and then the vertical boundary of the graph when the value of X reached 100 per cent. In the example given in Figure 13.5 it can be seen that a degree of difference does exist between the two distributions, representing in this case some measure of racial separation at a fairly broad, regional scale within the USA.

One method of measuring the differences revealed by the Lorenz curve is to calculate the index of dissimilarity (D_s). This may be defined as the maximum vertical distances between the diagonal line and the Lorenz curve (Figure 13.5). In fact there are three different ways of obtaining this index. First, it can be calculated from the data, as the maximum difference in the cumulative percentages of the two distributions. Thus, in Table 13.3 D_s would be given by 69.5 − 47.6, and has the value 21.9. A second method of calculating the index is by measuring it from the Lorenz curve in Figure 13.5, since the area between the two curves is a measure of how poor the fit is. Finally, D_s can be calculated using equation 13.5, which has the advantage that it deals with uncumulated percentages and therefore avoids some of the work in Table 13.3:

$$D_s = \frac{\Sigma |X_i - Y_i|}{2} \qquad (13.5)$$

where X_i and Y_i are the individual percentages of each variable. D_s has a range of 0 to 100. This statistic is also known as the Gini coefficient.

A glance through the literature shows that owing to the flexibility of the index of dissimilarity, it has been applied to a variety of studies. Thus, in our example the index was used as a measure of ethnic segregation, indicating that at a regional scale segregation was not particularly high, with an index of 21.9. Indeed, it is within the context of segregation studies that the index has been widely used. Tauber and Tauber (1965), for example, applied it to over 200 cities in the USA and found that the index varied between 60.4 and 98.1, illustrating high levels of segregation. In Northern Ireland Poole and Boal (1973) applied the technique to a study of segregation tendencies in Belfast. Finally, the more general application of the Lorenz curve to segregation studies was reviewed by Jones and Eyles (1976).

A second type of use for Lorenz curve analysis is within the area of economic geography. In this instance the measure of dissimilarity from the Lorenz curve is known as the 'coefficient of geographical association'. This produces a measure of the extent to which economic activities are concentrated spatially relative to some other form of activity. Smith (1975), for example, presents coefficients of localization for a number of different industries, showing the relative degrees of concentration within the UK.

A further use of the index would be to compare population or employment with land area. Thus, population would be plotted on the Y axis and the size of each areal unit as values of X. A perfectly even distribution of population would therefore have an index of zero, with values approaching 100 (or 1, depending on whether percentages are used) indicating population concentration. Such measures of dispersal or concentration based on the Lorenz curve can obviously be applied to a variety of activities other than population. For example, Wild and Shaw (1974) used a locational index based on the Lorenz curve to rank shop types in order of their locational behaviour and whether they were dispersed or concentrated.

In addition to the measures based on the Lorenz curve, geographers have also used a variety of spatial indices, or coefficients (Isard, 1960). The most extensively used is probably the 'location quotient', which measures the extent to which different areas depart from some norm; for example the national average. The quotient can be calculated using equation 13.6, or if the data are expressed as percentages, equation 13.7:

$$LQ = \frac{(X_i/X)}{(Y_i/Y)} \tag{13.6}$$

Equation 13.6

X_i = employment in a given activity i, in an area
X = total employment in an area
Y_i = national employment in activity i
Y = total national employment

$$LQ_j = \frac{X_j}{k} \qquad (13.7)$$

Equation 13.7

X_j = percentage of an activity in j
k = national percentage

In equation 13.6 the location quotient indicates the degree of concentration, with higher values of *LQ* representing high concentrations, and values of 1 indicating equal distributions. Table 13.4 illustrates how such measures can be derived for two different types of economic activity. Geographical variations in employment concentrations based on the location quotient can be illustrated by mapping the quotients (Figure 13.6).

Table 13.4 Calculation of location quotients for employment in engineering and textile manufacturing in England (1976), thousands.

Standards	Enginering	*LQ*	Textiles	*LQ*	Total manufacturing
North	192	0.94	52	0.97	438
Yorks and Humberside	242	0.74	149	1.71	711
East Midlands	210	0.77	171	2.38	587
East Anglia	82	0.91	14	0.58	195
South East	943	1.10	108	0.47	1851
South West	215	1.10	36	0.70	420
West Midlands	577	1.28	42	0.36	976
North West	399	0.86	190	1.53	1005
	2860		762		6183

From equation 13.6, *LQ* for the North (engineering) = (192/2860)/(438/6183) = 0.067/0.071 = 0.94.

One of the problems with the location quotient, and indeed with a number of these spatial indices, is that statistically, nothing is known about their sampling distributions (King, 1969). Furthermore, as the scale for the quotient is arranged around unity, values below the national norm are compressed between 0 and 1; but above unity the quotient can rise to any value. One coefficient that does not suffer from these disadvantages is the Gini-coefficient of concentration which can be calculated using equation 13.5, and which varies on a scale of 0 to 100. Both Glasser (1962) and King (1969) illustrate how the sampling distribution of the coefficient can be derived; and that if the sampling is done without replacement from an infinite population then the sampling distribution will approximate to normal.

Finally, some mention must be made of the problems associated with

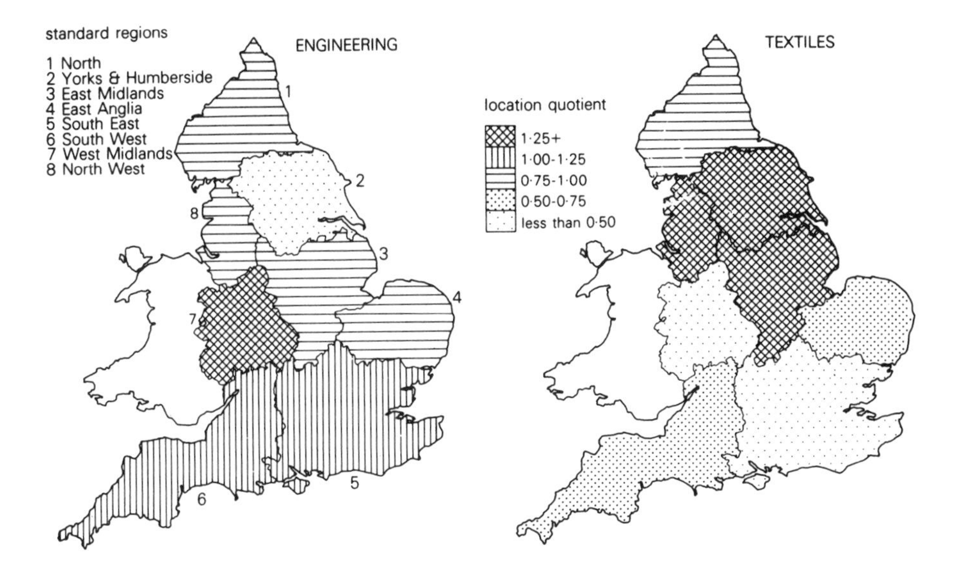

Figure 13.6 Use of location quotients to describe engineering and textile employment in England (1976).

using the Lorenz curve. First, there are data restrictions since variables must be expressed as frequencies for each areal unit, and negative values cannot be included. This therefore makes it rather more difficult to apply the technique to a study of a continuous variable. Second, the index is affected by changes in the spatial boundaries of the study units, and changes in spatial scale. This can be partly illustrated by the example of ethnic segregation, used in Table 13.2, which at a regional scale showed little evidence of segregation. However, the results from the analysis of individual cities by Tauber and Tauber (1965) showed very high levels of segregation at this smaller spatial scale. Indeed there are predictable variations in the index D_s with changes in spatial scale, and lower values of D_s are to be expected in a study with a few areal units. A more sensitive technique in the study of ethnic segregation is Lieberson's isolation index (P^*), which was fully discussed by Robinson (1980).

Leading on from these difficulties, associated with changing boundaries and scale, is a third problem relating to the fact that the Lorenz curve is insensitive to spatial arrangement, or pattern. Thus, the Lorenz curve and its related indices are an effective way of describing the relationships between distributions, but offer no indication or measure of spatial pattern.

13.4 Pattern analysis

In the earlier sections of this chapter, the techniques we discussed were concerned with either measuring spatial distributions or providing summary statistics to describe such distributions. However, in many circumstances geographers may be interested in the locations of individuals relative to each other, often expressed as points on a map. Such work is concerned with the techniques of point pattern analysis. Early approaches to this type of study were subjective and merely involved the mapping of individuals to produce a simple dot map (Monkhouse and Wilkinson, 1971). Very often the interpretation of these patterns lacked objectivity, making accurate description difficult and comparitive studies impossible.

The first attempts to study point patterns in an objective fashion were made by plant ecologists in the analysis of plant communities (Clark and Evans, 1954). These studies were soon adapted and used by geographers, initially by Dacey (1960) and King (1962) and then by many others (see Rogers, 1974; Cliff and Ord 1981 for a review of this work). All these studies have one important feature in common – they all make use of some type of probability distribution as a means of describing spatial patterns. Thus, a link is established between an observed distribution of points on a map and probability theory as explored in Chapter 5.

One of the most widely used measures is the Poisson probability distribution, which has the following assumptions. First, it postulates the condition of equal probability, which in the context of point patterns refers to the situation where any location on a map has an equal probability of receiving a point. For example, if we had a map of a particular woodland, the Poisson theory suggests that any part of the area would have an equal opportunity of having a tree. We may infer from this that the process producing such patterns is therefore a random one. Second, the theory assumes a condition of independence, whereby each of the points located on a map would be independent of one another. Thus, in our woodland example the assumption is made that the location of one tree would neither repel nor attract another. Based on these assumptions and the use of the Poisson probability function, as discussed in section 5.17, it is possible to model a distribution of points on a map. These expected patterns can then be compared with observed or real patterns, and thus used as a standard yardstick, from which we can measure deviations. By applying the Poisson distribution, we are therefore using the concept of randomness as our basic measure.

In the study of point patterns we can recognise two important violations of the assumptions made by the Poisson model, each of which produce non-random processes. The first of these concerns patterns that result from competitive processes. This may be illustrated by the example of food stores competing with each other in a city. Over time, those located close to other food stores may be driven out of business by more powerful competitors, thereby reducing store clusters and producing a regular pattern. The best example of such competitive processes in the geographical literature is that of the evolution of market centres and described by central place theory. A second deviation from the Poisson model is the situation where the locations of existing activities attract others. This is termed a contagious process, and tends to produce a clustered pattern. Harvey (1966) has used the idea of contagious processes in the context of spatial diffusion studies as presented by Hägerstrand (1953).

Given this background it is possible to recognise three basic types of point patterns, namely regular, random, and clustered, each of which can be modelled by a particular probability distribution (Figure 13.7). These distributions provide the geographer with some basic yardsticks with which to compare observed point patterns, and also provide a conceptual framework to account for some of the possible variations from randomness.

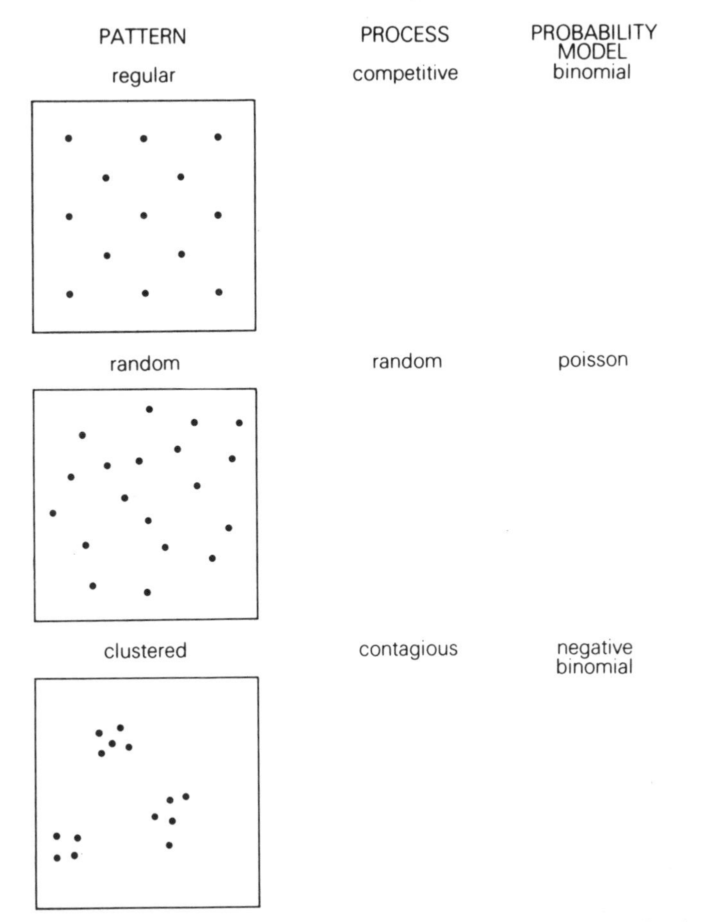

Figure 13.7 Types of point patterns and their probability distributions.
Note: the binomial and Poisson distributions have already been discussed in Chapter 5, but no formal examination of the negative binomial distribution is attempted here.

13.5 Measurement of point patterns

At the start of this section it is important to make clear the distinction that statisticians draw between 'pattern' and 'dispersion'. Both Dacey (1973) and Sibley (1976) have urged caution in our use of such terms. By their operational definitions 'pattern' should be taken to mean the distances between and arrangements of points in space, and 'dispersion' refers, by contrast, to the areal extent of a collection of points. Sibley has observed that failure to bear these distinctions in mind can lead to meaningless and misleading expressions, such as, 'random dispersion'. Attention in this section is devoted to patterns in the sense that they are defined above.

There are a number of ways in which geographers have studied point patterns. Later in this section we shall look at 'nearest neighbour analysis', but we shall start with 'quadrat analysis'. The latter has been popular with both geographers and botanists but has recently been less widely adopted as an appreciation of some of its problems has grown. Nevertheless it remains an important part of geographical methodology.

In quadrat analysis the study area is overlaid by a grid of lines forming units of equal size, and the number of points in each cell are counted. We have already used this method in section 5.16 when the spatial distribution of grocers' shops in Sunderland was examined. Traditionally the grid systems are based on squares, although other shapes, such as hexagons, could be used provided that they combine to form a complete cover. Irregular shapes, rectangles or units of different areas should not be used. Squares are clearly the easiest units to construct and measure and will form the basis of what follows, although the theoretical implications apply to all suitable shapes.

The points that can be studied and counted using the quadrat method can consist of any spatially distributed 'point' phenomenon. Shops have already been cited, and there is a substantial literature on their study in this way (Rogers, 1974). But species of plant, cases of disease, industrial locations and even settlement patterns are equally amenable. The points, or events, must be spatially discrete, and continuous variables such as rainfall or altitude are not suitable. We hope that by counting the numbers of point events within each square we can derive a measure of the points' pattern. The guiding principle underlying these attempts is that point patterns can be described according to their location along a continuum which varies from perfectly regular (all points equidistant) at one end to perfectly clustered (all points touching) at the other. The random distribution lies mid-way between these extremes. Figure 13.7 conveys something of the form of these patterns, and later in this section we shall see how the distances between points can be used to describe the patterns: but another method of description is to compare the quadrat counts with a hypothesised probability distribution. The Poisson distribution has already been examined and has a particular, though limited, application to this problem.

In the earlier example of Sunderland shops we hypothesised that their spatial distribution was random. If that hypothesis was correct then the observed distribution should approximate to the Poisson which is, as we explained in Chapter 5, a random distribution in which events are independent and located without spatial preference. Section 7.6 demonstrated how the differences between observed and hypothesised observations were tested. The distributions are described by the number of quadrats which contain 0, 1, 2, 3, etc. points, and in the latter example the observed data differed so little from that expected from a Poisson distribution that the pattern was concluded to be random.

Another approach to this problem of pattern analysis would have been to use the variance–mean ratio of the observed distribution. The mean, denoted by λ, is given by the observed density of points, which in the earlier example was 0.4 shops per quadrat. The variance of the observed distribution is given by:

$$\sigma^2 = \frac{\Sigma X^2}{\Sigma X} - \frac{\Sigma X}{N} \tag{13.8}$$

where X is the number of points in all quadrats and N the number of quadrats.

One of the hallmarks of the Poisson distribution is the equality of mean and variance. Thus, if the variance–mean ratio is unity or close to it, we may conclude the distribution to be Poisson and, consequently, spatially random.

Departures from unity in the ratio reflect tendencies towards either clustering or regularity. Regularly located points yield very low variances because most quadrats record a similar number of points, with the result that the variance–mean ratio is less than 1.0. On the other hand clustered point patterns give variances that are very high, because a few quadrats have many points and the majority have very few or none, and the resulting variance–mean ratios are greater than 1.0. The degree of departure from 1.0 can be converted to a z score after calculating the standard error of the difference (SE_X) from:

$$SE_X = \surd[2/(N - 1)] \tag{13.9}$$

in which N is the total number of quadrats, and:

$$z = \frac{\text{observed ratio} - \text{expected ratio}}{\text{standard error}} \tag{13.10}$$

Given this information we can rework the shops example. The mean of the spatial pattern we have already established to be 0.4 shops per quadrat, and its variance we estimate from equation 13.8. The constituent observations (X) consist of the number of shops counted in each of the 135 quadrats. In this case there were 96 quadrats with zero shops, 27 with one, 9 with two, 2 with three and only 1 with four. We may prepare our data for equation 13.8 in the form shown in Table 13.5, taking care to keep the distinction between N (the number of quadrats) and ΣX (the total number of shops) clearly in mind. We find that $\sigma^2 = 1.357$ and the variance–mean ratio is therefore $1.357/0.4 = 3.393$.

Table 13.5 Method of calculating the variance of spatial patterns using quadrat counts

Number of shops per quadrat (n)	Number of quadrats with n shops (q)	Number of shops by quadrats (X)	X^2
0	96	0	0
1	27	27	27
2	9	18	36
3	2	6	18
4	1	4	16
	135	55	97

In their original form the X data would consist of 135 observations corresponding to the number of shops in each of the 135 quadrats. Here they are treated in a simpler, condensed, form in which $X = nq$ and $X^2 = n^2q$.

We can move on to examine the degree to which the variance–mean ratio of the shop pattern differs from the hypothesised value of 1.0. The first step is to calculate the standard error of the difference from equation 13.9:

$$SE_X = \sqrt{[2/(135-1)]}$$
$$= \sqrt{0.01493} = 0.1222$$

From this the z value (t value if ΣX is less than 30) is calculated by dividing the standard error into the difference between the observed and expected variance–mean ratios:

$$z = \frac{3.393 - 1.0}{0.1222} = 19.583$$

Thus, at the 0.01 significance level, for which the critical z values are ± 2.58, the differences are sufficiently great for us to conclude that the distribution is, by this method, non-random. Moreover, if we pay attention to the sign of the z value we can see, because of the nature of equation 13.10, that positive results show a tendency towards clustering while negative values arise from a tendency towards regularity. The positive result here suggests a degree of clustering of the points.

Hence we come, by different routes, to contradictory conclusions concerning this distribution. Such difficulties are not uncommon in point pattern analysis and warn us to be on our guard. There are, unfortunately, further problems that have to be considered. Most importantly, both the observed and hypothesised (Poisson) distributions are density-dependent as by varying the grid-size placed over the study area or its limits the measured density can be drastically altered. Figure 5.22 has already shown the dramatic effect that variations in density (which provides the Poisson mean λ) can have on the character of the distribution. It is unfortunate that pattern is completely independent of density, yet the distributions used here to analyse them are strongly density-dependent. Equally patterns are independent of scale, and those shown in Figure 13.7 are as likely to be encountered on a scale of kilometres as a scale of centimetres. Nevertheless, variations in quadrat size can still lead to unrepresentative contrasts between the observed distributions that they generate.

Clearly we need to select the boundaries of the study areas with care. Less easy to accomplish is the choice of a suitable quadrat size. Greig-Smith (1964) has pointed out that by the judicious selection of quadrat size we can simulate a Poisson distribution where none exists. The effect of quadrat size has been widely examined by plant ecologists (Greig-Smith, 1964; Pielou, 1969). Their work has shown that, in visually clustered patterns such as that in Figure 13.7, very small quadrats usually produce results suggesting randomness; since it is likely that quadrats would only contain small numbers of points and not measure the clusters. Similarly, very large quadrats in such circumstances produce results that seem to indicate a regular pattern, as most quadrats would contain similar numbers of points. If, on the other hand, we are examining a pattern that appears to be visually regular, then this characteristic will be shown as quadrat size increases (Greig-Smith, 1952).

There are two ways in which this problem of quadrat size can be tackled. The first is to derive some method of determining an ideal quadrat size for particular point distributions. Ecologists have examined such a notion and define quadrat size as $2A/N$, where A is the area of study and N is the number of points. i.e. twice the mean area around each point. Taylor (1977), however, suggests that such quadrats are probably too large for geographical studies, particularly where spatial competition is important. In these conditions he recommends the use of quadrats determined by the area of the map divided by the number of points (A/N).

A further, though time-consuming, solution is to test for Poisson randomness over a range of quadrat sizes. If it is present throughout we might be confident in claiming its existence. The problem of appropriate quadrat size has yet to be resolved and the final choice may well be determined by the nature of the project rather than the character of the data. But we should, despite these warnings, note the advice of Harvey (1966) who reminds us that 'the quadrat sampling has considerable potential for testing models of location'.

In addition we can use the effect of quadrat size variations in a positive fashion. Thus, by using quadrats of different sizes to examine the same pattern we can explore variations in scale or 'grain' of point distributions. This has been discussed by Kershaw (1964), who plotted quadrat size against variance of the distributions and argued that peaks of high variance on the resulting graphs could be interpreted as point clusters.

There is, however nothing intrinsically sacred about the Poisson distribution and its applicability of spatial observations. Scatters of points in space can be described by other distributions, for example the negative binomial, whose exclusion from this text should not be interpreted as a relegation of their importance. Most of these distributions make allowances for a 'contagion' effect in which occurrences influence the probability of other events. The Poisson distribution specifically forbids any such effect and requires the events to be wholly independent. Contagion effects can often be very important in describing point patterns. Some services tend, by their nature, to be randomly located over wide areas of towns and cities. On the other hand there are other retail functions, such as banks and department stores, that tend to cluster together in city centres; it would be inappropriate to attempt to describe their patterns in the urban field by the Poisson distribution.

There are some probability distributions that can take specific account of contagion effects, and they can be applied to cases in which the Poisson distribution is theoretically imprecise, i.e. if it is thought that the point pattern is not random. These distributions possess rather exotic titles such as 'Neyman's type A' and the 'Polya–Aeppli distribution'. They are less widely known than those reviewed in Chapter 5, but are particularly useful when studying patterns generated by clustering and contagion. A good example might be the pattern of locations created by car component factories which tend to cluster around the car factories themselves. However, the equations describing these probability distributions are not easy to evaluate. The interested reader is referred to more detailed discussions (e.g. Rogers, 1974).

Despite the mathematical elegance of these distributions they remain, like the simpler Poisson distribution, highly dependent on quadrat size. As a result the failure of a point pattern to conform to one of them does not imply that the pattern is not one of contagion. Neither does a correspondence provide irrefutable proof that contagion is present. Indeed, any point pattern may be approximated by two or three different probability distributions. Here, then, is another reason why quadrat analysis is now less popular; because different theoretical interpretations, based on approximations to different distributions, can be made for each case. Nevertheless the thoughtful use of quadrat analysis for point pattern description or for testing a priori assumptions against real-world observations remains a valuable geographical tool.

An alternative method to quadrat analysis is to use measures based on the spacing between points, by taking the distance of each point to its nearest neighbour. The pioneer work was carried out by plant ecologists Clark and Evans (1954), who developed a measurement index and linked it to the Poisson probability distribution. The test requires a knowledge of the population density, and the analysis involves a comparison between an observed spacing of a point distribution and the spacing expected in a random pattern. The average expected distances are calculated using equation 13.11, the derivation of which was presented by Clark and Evans:

$$\bar{r}_e = 0.5\ \sqrt{(A/N)} \qquad (13.11)$$

Equation 13.11

$\bar{r}_e$ = average expected distance
A = area of study region
N = number of points

Once again, such expected values are described by the Poisson function. The nearest neighbour statistic (R) is derived by dividing the observed distances by the expected, with the results falling within a range of values from 0 to 2.1491:

$$R = \bar{r}_a/\bar{r}_e \qquad (13.12)$$

Equation 13.12

$\bar{r}_a$ = average observed distance
$\bar{r}_e$ = average expected distance

This index shows how more, or less, spaced the observed distribution is compared with a random one.

Under conditions of maximum aggregation all the individuals in a point distribution occupy the same locus and the distance to the nearest neighbour is therefore zero. At the other extreme, conditions of maximum spacing, the individuals will be distributed in an even hexagonal pattern (Figure 13.7). Consequently, every point will be equidistant from six other individuals, so that the mean distance to the nearest neighbour is maximised and $R = 2.1491$. If $R = 1$, then the observed and expected distances are equal, thus indicating a random pattern. Therefore, when values of R are less than 1 this suggests distributions tending toward a clustered pattern, while values above 1 describe tendencies towards dispersion.

The concept of randomness has traditonally been applied as the base measure in point pattern analysis. In theory a random pattern is one in which the location of each point is totally uninfluenced by the remaining points. However, in practice it is much more useful to view patterns as deviating from clustered or regular, with the results falling on a continuum between the two. In reality, locational forces are unlikely to operate randomly, but are more often capable of transforming either of the extreme conditions towards a random pattern. The applicability of this approach can be improved with the application of a test assessing the significance of the R value, which takes into account possible variations in the random processes. Thus, the probability that an R value could have arisen by chance can be established, using the standard error and z scores (see Chapter 6). The standard error of the expected average nearest neighbour distance can be calculated using equation 13.13:

$$SE_{r_e} = \frac{0.26136}{\sqrt{[N(N/A)]}} \tag{13.13}$$

Equation 13.13

N = number of points
A = area of study
SE_{r_e} = standard error of expected average distance

The standard error can be then used in the normal fashion, with 95 per cent of the expected values falling within $\pm 1.96 SE_{\bar{r}_e}$ of the average computed distance. Getis (1964), in some early work on the use of nearest neighbour techniques, extended the use of standard errors to test the significance of patterns by computing the z value using equation 13.14:

$$Z_R = \frac{|\bar{r}_e - \bar{r}_a|}{SE_{\bar{r}_e}} \tag{13.14}$$

Equation 13.14

$\bar{r}_e$ = average expected distance
$\bar{r}_a$ = average observed distance
$SE_{\bar{r}_e}$ = standard error of expected average distance

These inferential statistics can in turn be used to delimit a range of random matching conditions, as suggested by Pinder and Witherick (1972).

The implication of the nearest neighbour technique is that the area under study is an isotropic surface. Attention therefore needs to be given to defining the actual study area in terms of a 'biotope space' (Hudson, 1969); which for example, in an urban environment involves measuring the built-up area. In addition, consideration needs to be given to the definition of any study area, since both the size and shape of the area may influence the results. As a general rule the study area should be defined relative to the problem under investigation. For example, as Getis (1964) showed in his study of urban retail patterns, the most appropriate study area was a circle, enclosing most of the built-up area.

The basic nearest neighbour technique can be extended to take into account the scale elements in a point pattern. This can be achieved by measuring order neighbour distances up to the *n*th value, and by calculating the corresponding values of R. The formula for such measures was derived by Thompson (1956):

$$\text{Expected distance to } n\text{th neighbour} = \frac{1}{\sqrt{M}} \frac{(2n)!n}{(2^n n!)^2} \qquad (13.15)$$

Equation 13.15

M = density of points per unit area
n = order of nearest neighbour

If measurements are taken to the *n*th order neighbour, then clearly some idea of the scale at which point patterns are occurring can be gained. In theory, by plotting rank order of the nearest neighbour against the corresponding R values a measure of the 'grain' or scale of pattern intensity can be achieved. Thus, in terms of mean distance between points, if the pattern were of a 'fine-grained' nature then a plot of order neighbour against mean distance would give a curve of only gradual increase, with a smooth profile. Conversely, in a 'coarse-grained' pattern such a curve would be less smooth with sharp increases, as distance measurements become of an inter-group type. One problem in applying this extension concerns the number of neighbours to which measurements should be taken, for as Cowie (1968) states 'there is no objective method to indicate how many neighbours should be employed'. However, in the few attempts at this type of analysis the number selected has been at least of the order of three (Cowie, 1968; Sibley, 1971). The calculation of the technique and the extension proposed by Thompson are obviously time-consuming by hand, especially for large numbers of points; and in such circumstances the measurement and calculations can be speeded up using a computer program.

The technique can be illustrated by an examination of patterns of retail change, which has seen a frequent use of such spatial statistics (Kivell and Shaw, 1980). The example presented here relates specifically to changes in the pattern of footwear retailers in Kingston-upon-Hull from 1880 to 1950. For the purposes of the analysis two distinct retail organisational types were recognised, multiples and independents, that had different locational requirements. The former refers to those firms having five or more branch shops, and they were often controlled by national companies.

Table 13.6 Trends in the pattern of footwear retailers

	Independents			Multiples		
Time	Nearest neighbour value	Pattern type	Shop	Nearest neighbour value	Pattern type	Shops
1880	1.053	regular	256	nil	nil	nil
1890	1.264	regular	275	0.915	random	19
1900	0.975	random	235	0.630	clustered	16
1910	0.919	random	189	0.684	clustered	22
1920	0.991	random	124	0.873	clustered	25
1930	0.923	random	100	0.640	clustered	32
1940	0.863	clustered	52	0.426	clustered	29
1950	0.781	clustered	22	0.591	clustered	19

An analysis of first-order neighbours using equation 13.12 gives an indication of pattern intensity, and clear differences can be recognised between the two shop types (Table 13.6). Despite differences in basic patterns, the overall trend for both types was towards a clustered state, a condition that took the independents longer to achieve. Such results essentially represent the local variations in the retail pattern, and the spatial scale of the analysis can be extended using higher-order neighbours. In this case, calculations were also made to the tenth-order neighbour to measure the grain of the pattern. An example of a graph, plotting rank order neighbour against corresponding values of R, is illustrated in Figure 13.8.

The independent retailers exhibited a fine-grained pattern and the resultant curves decrease in a monotonic fashion towards a general clustered pattern. At high-order neighbours the R value is low, indicating a high degree of large-scale clustering. In contrast, the multiples had a more complex pattern, with marked spatial variations occurring at different scales of measurement. Indications are that they had a coarser-grained pattern with a much stronger element of localized clustering. The step-like nature of the rank order curves suggests that the multiples occurred in loosely associated clumps of stores, probably at points of relatively high accessibility.

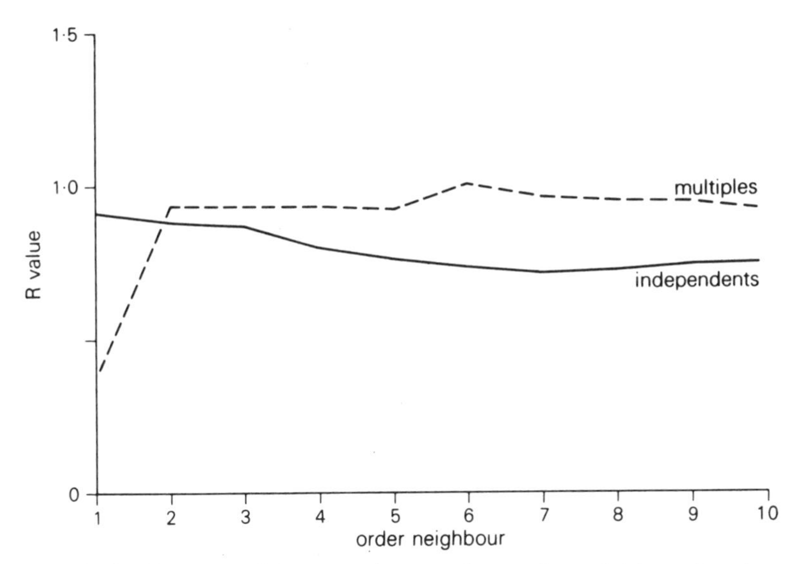

Figure 13.8 Nearest neighbour values against order neighbour for stores in Hull.

13.6 Problems in the interpretation of point patterns

Point pattern analysis, in addition to having a number of methodological problems, also suffers from difficulties in the interpretation of results. From the previous two sections in this chapter it is clear that the analysis of point patterns is based on a comparison of the observed distribution with some known theoretical one. If there is a close fit between the two then we may describe a pattern as either random, clustered or dispersed. However, we should be very cautious about inferring processes from point patterns. First, as Harvey (1966) and Rogers (1974) have shown, there may be a number of different probability models that fit our observed points, and we therefore must in some cases use other background evidence to discriminate between them. Second, even if we only have one clear model that fits our data, this is not sufficient evidence to infer conclusively that a specific process produced the pattern.

It is because of such inferential problems that since the mid-1970s point pattern analysis has rather fallen from favour in many areas of geographical research. Thus, while the technique can help us objectively to describe and classify point patterns, it fails to provide the necessary information about spatial processes. This is especially the case in human geography with its increasing emphasis on the study of behavioural processes.

References

Clark, P. J., and Evans, F. C. (1954) 'Distance to nearest neighbour as a measure of spatial relation in populations', *Ecology*, **35**, 445–453.

Cliff, A. D., and Ord, J. K. (1981). *Spatial Processes: Models and Applications*, Pion, London.

Cole, J. P., and King C. A. M. (1968) *Quantitative Geography*, Wiley, London.

Court, A. (1964). 'The elusive point of minimum travel', *Anns Ass. Am. Geogrs*, **54**, 400–403.

Cowie, S. R. (1968) 'The cumulative frequency nearest neighbour method for the identification of spatial patterns', Seminar Paper Ser. 10, Dept. of Geog., University of Bristol.

Dacey, M. F. (1960) 'A note on the derivation of nearest-neighbour distances', *J. Reg. Sci.*, **2**, 81–87.

Dacey, M. F. (1966) 'A country-seat model for the areal pattern of an urban system', *Geog. Rev.*, **56**, 527–542.

Dacey, M. F. (1973) 'Some questions about spatial distributions', in R. J. Chorley (ed.) *Directions in Geography*, Methuen, London.

Ebdon, D. (1977) *Statistics in Geography: A practical approach*, Blackwell, Oxford.

Getis, A. (1964) 'Temporal land use pattern analysis with the use of nearest neighbour and quadrat methods', *Anns Ass. Am. Geogrs*, **54**, 391–399.

Glasser, G. J. (1962) 'Variance formulas for the mean difference and coefficient of concentration', *J. Am. Stats Assoc.*, **57**, 648–654.

Greig-Smith, P. (1952) 'The use of random and contiguous quadrats in the study of the structure of plant communities', *Anns Botany (N.S.)*, **16**, 293–316.

Greig-Smith, P. (1964) *Quantitative Plant Ecology*, Butterworths, London.

Hägerstrand, T. (1953) 'On Monte Carlo simulation of diffusion', reprinted in W. L. Garrison and D. F. Marble (eds) *Quantitative Geography* (1967), Northwestern Univ., Evanston.

Hammond, R., and McCullagh, P. S. (1974) *Quantitative Techniques in Geography: an Introduction*, 2nd edn, Clarendon, Oxford.

Hart, J. F. (1954) 'Central tendency in areal distributions', *Econ. Geog*, **30**, 48–59.

Harvey, D. W. (1966) 'Geographic processes and the analysis of point patterns', *Trans. Inst. Brit. Geogrs*, **40**, 81–95.

Hudson, J. C. (1969) 'A location theory for settlement', *Anns. Ass. Am. Geogrs*, **59**, 365–381.

Isard, W. (1960) *Methods of Regional Analysis*, Wiley, New York.

Jones, E., and Eyles, J. (1976) *An Introduction to Social Geography*, Oxford Univ. Press, Oxford.

Kershaw, K. A. (1964) *Quantitative and Dynamic Ecology*, Arnold, London.

King, L. J. (1962) 'A quantitative expression of the pattern of urban settlements in the USA', *Tijds. voor Econs. en Sociale Geog.*, **53**, 1–7.

King, L. J. (1969) *Statistical Analysis in Geography*, Prentice-Hall, Englewood Cliffs.

Kivell, P. T., and Shaw, G. (1980) 'The study of retail location', in J. A. Dawson (ed.) *Retail Geography*, Croom Helm, London.

Monkhouse, F. J., and Wilkinson, H. R. (1971) *Maps and Diagrams*, 3rd edn. Methuen, London.

Neft, D. (1966) *Statistical Analysis for Areal Distributions*, Reg. Sci. Inst. Monograph 2, Philadelphia.

Pielou E. C. (1969) *An Introduction to Mathematical Ecology,* Wiley, New York.

Pinder, D. A., and Witherick, M. E. (1972) 'The principles, practice and pitfalls of nearest neighbour analysis', *Geography*, **57**, 277–88.

Poole, M. A., and Boal, F. (1973) 'Segregation in Belfast', in B. D. Clarke and M. B. Cleave (eds) *Social Patterns in Cities*, IBG Special Publication.

Robinson, V. (1980) 'Lieberson's isolation index', *Area*, **12** (4), 307–312.

Rogers, A. (1974) *Statistical Analysis of Spatial Dispersion: the Quadrat Method*, Pion, London.
Seymour, D. R. (1965) 'IBM 7090 program for locating bivariate means and bivariate medians', Tech. Rep. 16, Dept of Geog., Northwestern University.
Shachar, A. (1964) 'Some applications of geo-statistical methods in urban research', *Papers and Proceedings, Reg. Sci. Ass.*, **18**, 197–202.
Sibley, D. (1971) 'A temporal analysis of the distribution of shops in British cities', unpublished PhD thesis, Univ. of Cambridge.
Sibley, D. (1976) 'On pattern and dispersion', *Area*, **8**, 163–165.
Smith D. M. (1975) *Patterns in Human Geography*, Penguin, Harmondsworth.
Sviatlovsky, E. E., and Eells, W. C. (1937) 'The centrographical method and regional analysis', *Geog. Rev.* **27**, 240–254.
Tauber, K. E., and Tauber, A. F. T. (1965) *Negroes in Cities*, Aldine, Chicago.
Taylor, P. J. (1977) *Quantitative Methods of Geography,* Houghton-Mifflin, Boston.
Thompson, H. R. (1956) 'Distribution of distance to Nth neighbour in a population of randomly distributed individuals', *Ecology*, **37**, 391–394.
Unwin, D. (1981) *Introductory Spatial Analysis,* Methuen, London.
Varley, R. (1968) 'Land use analysis in the city centre with special reference to Manchester', Uni. of Wales unpub. M.A.
Warntz, W., and Neft, D. (1960) 'Contributions to a statistical methodology for areal distributions', *J. Reg. Sci.*, **2**, 47–66.
Wild, M. T., and Shaw, G. (1974) 'Locational behaviour of urban retailing during the nineteenth century', *Trans. Inst. Brit. Geogrs*, **61**, 101–118.

APPENDICES

APPENDIX I

Areas beneath the standard normal curve

z	p	z	p	z	p	z	p
0.00	0.00000	0.35	0.13683	0.70	0.25805	1.05	0.35314
0.01	0.00399	0.36	0.14058	0.71	0.26115	1.06	0.35543
0.02	0.00798	0.37	0.14431	0.72	0.26424	1.07	0.35769
0.03	0.01197	0.38	0.14803	0.73	0.26730	1.08	0.35993
0.04	0.01595	0.39	0.15173	0.74	0.27035	1.09	0.36214
0.05	0.01994	0.40	0.15542	0.75	0.27337	1.10	0.36433
0.06	0.02392	0.41	0.15910	0.76	0.27637	1.11	0.36650
0.07	0.02790	0.42	0.16276	0.77	0.27935	1.12	0.36864
0.08	0.03188	0.43	0.16640	0.78	0.28230	1.13	0.37076
0.09	0.03586	0.44	0.17003	0.79	0.28524	1.14	0.37286
0.10	0.03983	0.45	0.17364	0.80	0.28815	1.15	0.37493
0.11	0.04380	0.46	0.17724	0.81	0.29103	1.16	0.37698
0.12	0.04776	0.47	0.18082	0.82	0.29389	1.17	0.37900
0.13	0.05172	0.48	0.18439	0.83	0.29673	1.18	0.38100
0.14	0.05567	0.49	0.18793	0.84	0.29955	1.19	0.38298
0.15	0.05962	0.50	0.19146	0.85	0.30234	1.20	0.38493
0.16	0.06356	0.51	0.19497	0.86	0.30511	1.21	0.38686
0.17	0.06750	0.52	0.19847	0.87	0.30785	1.22	0.38877
0.18	0.07142	0.53	0.20194	0.88	0.31057	1.23	0.39065
0.19	0.07535	0.54	0.20540	0.89	0.31327	1.24	0.39251
0.20	0.07926	0.55	0.20884	0.90	0.31594	1.25	0.39435
0.21	0.08317	0.56	0.21226	0.91	0.31859	1.26	0.39617
0.22	0.08706	0.57	0.21566	0.92	0.32121	1.27	0.39796
0.23	0.09095	0.58	0.21904	0.93	0.32381	1.28	0.39973
0.24	0.09483	0.59	0.22240	0.94	0.32639	1.29	0.40147
0.25	0.09871	0.60	0.22575	0.95	0.32894	1.30	0.40320
0.26	0.10257	0.61	0.22907	0.96	0.33147	1.31	0.40490
0.27	0.10642	0.62	0.23237	0.97	0.33398	1.32	0.40658
0.28	0.11026	0.63	0.23565	0.98	0.33646	1.33	0.40824
0.29	0.11409	0.64	0.23891	0.99	0.33891	1.34	0.40988
0.30	0.11791	0.65	0.24215	1.00	0.34134	1.35	0.41149
0.31	0.12172	0.66	0.24537	1.01	0.34375	1.36	0.41309
0.32	0.12552	0.67	0.24857	1.02	0.34614	1.37	0.41466
0.33	0.12930	0.68	0.25175	1.03	0.34849	1.38	0.41621
0.34	0.13307	0.69	0.26490	1.04	0.35083	1.39	0.41774

Cont.

APPENDIX I (*continued*)

z	p	z	p	z	p	z	p
1.40	0.41924	1.65	0.45053	1.90	0.47128	2.75	0.49702
1.41	0.42073	1.66	0.45154	1.91	0.47193	2.80	0.49744
1.42	0.42220	1.67	0.45254	1.92	0.47257	2.85	0.49781
1.43	0.42364	1.68	0.45352	1.93	0.47320	2.90	0.49813
1.44	0.42507	1.69	0.45449	1.94	0.47381	2.95	0.49841
1.45	0.42647	1.70	0.45543	1.95	0.47441	3.00	0.49865
1.46	0.42785	1.71	0.45637	1.96	0.47500	3.05	0.49886
1.47	0.42922	1.72	0.45728	1.97	0.47558	3.10	0.49903
1.48	0.43056	1.73	0.45818	1.98	0.47615	3.15	0.49918
1.49	0.43189	1.74	0.45907	1.99	0.47670	3.20	0.49931
1.50	0.43319	1.75	0.45994	2.00	0.47725	3.25	0.49942
1.51	0.43448	1.76	0.46080	2.05	0.47982	3.30	0.49952
1.52	0.43574	1.77	0.46164	2.10	0.48214	3.35	0.49960
1.53	0.43699	1.78	0.46246	2.15	0.48422	3.40	0.49966
1.54	0.43822	1.79	0.46327	2.20	0.48610	3.45	0.49972
1.55	0.43943	1.80	0.46407	2.25	0.48778	3.50	0.49977
1.56	0.44062	1.81	0.46485	2.30	0.48928	3.55	0.49981
1.57	0.44179	1.82	0.46562	2.35	0.49061	3.60	0.49984
1.58	0.44295	1.83	0.46638	2.40	0.49180	3.65	0.49987
1.59	0.44408	1.84	0.46712	2.45	0.49286	3.70	0.49989
1.60	0.44520	1.85	0.46784	2.50	0.49379	3.75	0.49991
1.61	0.44630	1.86	0.46856	2.55	0.49461	3.80	0.49993
1.62	0.44738	1.87	0.46936	2.60	0.49534	3.85	0.49994
1.63	0.44835	1.88	0.46995	2.65	0.49598	3.90	0.49995
1.64	0.44950	1.89	0.47062	2.70	0.49653	4.00	0.49997

Columns headed p give the probabilities of an event within the range of $z = 0.0$ to the selected z value. Because the normal distribution is symmetrical about $z = 0.0$ the table applies equally to negative z values.

To estimate the probability of an event exceeding the selected z value, subtract the associated probability from 0.50000.

APPENDIX II

Critical values on Student's *t*-distribution

	Confidence Limits				
	0.90	0.95	0.98	0.99	0.999
	two-tailed significance levels (one-tailed levels in brackets)				
	0.10	0.05	0.02	0.01	0.001
	(0.05)	(0.025)	(0.01)	(0.005)	(0.0005)
υ					
1	6.31	12.71	31.81	63.66	636.6
2	2.92	4.30	6.97	9.93	31.60
3	2.35	3.18	4.54	5.84	12.92
4	2.13	2.78	3.75	4.60	8.61
5	2.02	2.57	3.37	4.03	6.86
6	1.94	2.45	3.14	3.71	5.96
7	1.90	2.37	3.00	3.50	5.41
8	1.86	2.31	2.90	3.36	5.04
9	1.83	2.26	2.82	3.25	4.78
10	1.81	2.25	2.76	3.17	4.59
11	1.80	2.20	2.72	3.11	4.44
12	1.78	2.18	2.68	3.06	4.32
13	1.77	2.16	2.65	3.01	4.23
14	1.76	2.15	2.62	2.98	4.14
15	1.75	2.13	2.60	2.95	4.07
16	1.75	2.12	2.58	2.92	4.02
17	1.74	2.11	2.57	2.90	3.97
18	1.73	2.10	2.55	2.88	3.92
19	1.73	2.09	2.54	2.86	3.88
20	1.73	2.09	2.53	2.85	3.85
21	1.72	2.08	2.52	2.83	3.82
22	1.72	2.07	2.51	2.82	3.79
23	1.71	2.07	2.50	2.81	3.77
24	1.71	2.06	2.49	2.80	3.75
25	1.71	2.06	2.49	2.79	3.73
26	1.71	2.06	2.48	2.79	3.71
27	1.70	2.05	2.47	2.77	3.69
28	1.70	2.05	2.47	2.76	3.67
29	1.70	2.05	2.46	2.76	3.66
30	1.70	2.04	2.46	2.75	3.65
40	1.68	2.02	2.42	2.70	3.55
60	1.67	2.00	2.39	2.66	3.46
over 60		approximates to the normal distribution			
z	1.64	1.96	2.35	2.58	3.29

The critical t value is found by reference to the appropriate degrees of freedom (υ) and the selected significance level or confidence limit. In the former case the values for two-tailed tests should be read as + and $-t$. Equivalent one-tailed critical values are found under the bracketed headings and should be assigned to either + or to $-t$.

APPENDIX III

Critical values on the χ^2 distribution

	significance level				
	0.10	0.05	0.01	0.005	0.001
υ					
1	2.71	3.84	6.64	7.88	10.83
2	4.60	5.99	9.21	10.60	13.82
3	6.25	7.82	11.34	12.84	16.27
4	7.78	9.49	13.28	14.86	18.46
5	9.24	11.07	15.09	16.75	20.52
6	10.64	12.59	16.81	18.55	22.46
7	12.02	14.07	18.48	20.28	24.32
8	13.36	15.51	20.29	21.96	26.12
9	14.68	16.92	21.67	23.59	27.86
10	15.99	18.31	23.21	25.19	29.59
11	17.28	19.68	24.72	26.76	31.26
12	18.55	21.03	26.22	28.30	32.91
13	19.81	22.36	27.69	30.82	34.55
14	21.06	23.68	29.14	31.32	36.12
15	22.31	25.00	30.58	32.80	37.70
16	23.54	26.30	32.00	34.27	39.29
17	24.77	27.59	33.41	35.72	40.75
18	25.99	28.87	34.80	37.16	42.31
19	27.20	30.14	36.19	38.58	43.82
20	28.41	31.41	37.57	40.00	45.32
21	29.62	32.67	38.93	41.40	46.80
22	30.81	33.92	40.29	42.80	48.27
23	32.01	35.17	41.64	44.16	49.73
24	33.20	36.42	42.98	45.56	51.18
25	34.38	37.65	44.31	46.93	52.62
26	35.56	35.88	45.64	48.29	54.05
27	36.74	40.11	46.96	49.65	55.48
28	37.92	41.34	48.28	50.99	56.89
29	39.09	42.56	49.59	52.34	58.30
30	40.26	43.77	50.89	53.67	59.70
40	51.81	55.76	63.69	66.77	73.40
50	63.17	67.51	76.15	79.49	86.66
60	74.40	79.08	88.38	91.95	99.61
70	85.53	90.53	100.43	104.22	112.32
80	96.58	101.88	112.33	116.32	124.84
90	105.57	113.15	124.12	128.30	137.21
100	118.30	124.34	135.81	140.17	149.45

The critical values are determined by reference to the sample degrees of freedom (υ) and the selected signficance level. If the test statistic equals or exceeds the critical (tabled) value, the null hypothesis is rejected.

APPENDIX IV

Critical values of the Kolmogorov–Smirnov statistic (D)

			significance level		
n	0.20	0.15	0.10	0.05	0.01
1	0.900	0.925	0.950	0.975	0.995
2	0.684	0.726	0.776	0.842	0.929
3	0.565	0.597	0.642	0.708	0.828
4	0.494	0.525	0.564	0.624	0.733
5	0.446	0.474	0.510	0.565	0.669
6	0.410	0.436	0.470	0.521	0.618
7	0.381	0.405	0.438	0.486	0.577
8	0.358	0.381	0.411	0.457	0.543
9	0.339	0.360	0.388	0.432	0.514
10	0.322	0.342	0.368	0.410	0.490
11	0.307	0.326	0.352	0.391	0.468
12	0.295	0.313	0.338	0.375	0.450
13	0.284	0.302	0.325	0.361	0.433
14	0.274	0.292	0.314	0.349	0.418
15	0.266	0.283	0.304	0.338	0.404
16	0.258	0.274	0.295	0.328	0.392
17	0.250	0.266	0.286	0.318	0.381
18	0.244	0.259	0.276	0.309	0.371
19	0.237	0.252	0.272	0.301	0.363
20	0.231	0.246	0.264	0.294	0.356
25	0.210	0.220	0.240	0.270	0.320
30	0.190	0.200	0.220	0.240	0.290
35	0.180	0.190	0.210	0.230	0.270
over 35	$\frac{1.07}{\sqrt{n}}$	$\frac{1.14}{\sqrt{n}}$	$\frac{1.22}{\sqrt{n}}$	$\frac{1.36}{\sqrt{n}}$	$\frac{1.63}{\sqrt{n}}$

The null hypothesis is rejected if the observed D statistic exceeds the critical value for that sample size (n) at the selected significance level.

For n of greater than 35 the appropriate equation (determined by the choice of signficance level) should be used to determine the critical value.

APPENDIX Va

Probabilities of the Mann – Whitney statistic (U)

$n_2 = 3$

U \ n_1	1	2	3
0	0.250	0.100	0.050
1	0.500	0.200	0.100
2	0.750	0.400	0.200
3		0.600	0.350
4			0.500
5			0.650

$n_2 = 4$

U \ n_1	1	2	3	4
0	0.200	0.067	0.028	0.014
1	0.400	0.133	0.057	0.029
2	0.600	0.267	0.114	0.057
3		0.400	0.200	0.100
4		0.600	0.314	0.171
5			0.429	0.243
6			0.571	0.343
7				0.443

$n_2 = 5$

U \ n_1	1	2	3	4	5
0	0.167	0.047	0.018	0.008	0.004
1	0.333	0.095	0.036	0.016	0.008
2	0.500	0.190	0.071	0.032	0.016
3	0.667	0.286	0.125	0.056	0.028
4		0.429	0.196	0.095	0.048
5		0.571	0.286	0.143	0.075
6			0.093	0.206	0.111
7			0.500	0.278	0.155
8			0.607	0.365	0.210
9				0.452	0.274
10				0.548	0.345
11					0.421
12					0.500
13					0.579

$n_2 = 6$

U \ n_1	1	2	3	4	5	6
0	0.143	0.036	0.012	0.005	0.002	0.001
1	0.286	0.071	0.024	0.010	0.004	0.002
2	0.428	0.143	0.048	0.019	0.009	0.004
3	0.571	0.214	0.083	0.033	0.015	0.008
4		0.321	0.131	0.057	0.026	0.013
5		0.429	0.190	0.086	0.041	0.021
6		0.571	0.275	0.129	0.063	0.032
7			0.357	0.176	0.089	0.047
8			0.452	0.238	0.123	0.066
9			0.548	0.305	0.165	0.090
10				0.381	0.214	0.170
11				0.457	0.268	0.155
12				0.545	0.331	0.197
13					0.396	0.242
14					0.465	0.294
15					0.535	0.350
16						0.409
17						0.469

The notes at the foot of Appendix Vc provide details on the use of these tables

APPENDIX Vb

Probabilities of the Mann–Whitney statistic (U)

$n_2 = 7$

U \ n_1	1	2	3	4	5	6	7
0	0.125	0.028	0.008	0.003	0.001	0.001	0.000
1	0.250	0.056	0.017	0.006	0.003	0.001	0.001
2	0.375	0.111	0.033	0.012	0.005	0.002	0.001
3	0.500	0.167	0.058	0.021	0.009	0.004	0.002
4	0.625	0.250	0.092	0.036	0.015	0.007	0.003
5		0.333	0.133	0.055	0.024	0.011	0.006
6		0.444	0.192	0.082	0.037	0.017	0.009
7		0.556	0.258	0.115	0.053	0.026	0.013
8			0.333	0.158	0.074	0.037	0.019
9			0.417	0.206	0.101	0.051	0.027
10			0.500	0.264	0.134	0.069	0.036
11			0.583	0.324	0.172	0.090	0.049
12				0.394	0.216	0.117	0.064
13				0.464	0.265	0.147	0.082
14				0.538	0.319	0.183	0.104
15					0.378	0.223	0.130
16					0.438	0.267	0.159
17					0.500	0.314	0.191
18					0.562	0.365	0.228
19						0.418	0.267
20						0.473	0.310
21						0.527	0.355
22							0.402
23							0.451
24							0.500

The notes at the foot of Appendix Vc provide details on the use of these tables.

APPENDIX Vc

Probabilities of the Mann–Whitney statistic (U)

U \ n_1	1	2	3	$n_2=8$ 4	5	6	7	8
0	0.111	0.022	0.006	0.002	0.001	0.000	0.000	0.000
1	0.222	0.044	0.012	0.004	0.002	0.001	0.000	0.000
2	0.333	0.089	0.024	0.008	0.003	0.001	0.001	0.000
3	0.444	0.135	0.042	0.014	0.005	0.002	0.001	0.001
4	0.556	0.200	0.067	0.024	0.009	0.004	0.002	0.001
5		0.257	0.097	0.036	0.015	0.006	0.003	0.001
6		0.356	0.139	0.055	0.023	0.010	0.005	0.002
7		0.444	0.188	0.077	0.033	0.015	0.007	0.003
8		0.556	0.248	0.107	0.047	0.021	0.010	0.005
9			0.315	0.141	0.064	0.030	0.014	0.007
10			0.387	0.184	0.085	0.041	0.020	0.010
11			0.461	0.230	0.111	0.054	0.027	0.014
12			0.539	0.285	0.142	0.071	0.036	0.019
13				0.341	0.177	0.091	0.047	0.025
14				0.404	0.217	0.114	0.060	0.032
15				0.467	0.262	0.141	0.076	0.041
16				0.533	0.311	0.172	0.095	0.052
17					0.362	0.207	0.116	0.065
18					0.416	0.245	0.140	0.080
19					0.472	0.286	0.168	0.097
20					0.528	0.331	0.198	0.117
21						0.377	0.232	0.139
22						0.426	0.268	0.164
23						0.475	0.306	0.191
24						0.525	0.347	0.221
25							0.389	0.253
26							0.433	0.287
27							0.478	0.323
28							0.522	0.360
29								0.399
30								0.439
31								0.480
32								0.520

Appendices Va to Vc are arranged to give probabilities for each U statistic determined by reference to the larger and smaller group sizes (n_1 and n_2 respectively). The null hypothesis is rejected if the tabled probability of U is less than the selected significance level.

APPENDIX Vd

Probabilities of the Mann–Whitney statistic (U)

significance level = 0.10 (two-tailed) or 0.05 (one-tailed)

n_2 \ n_1	9	10	11	12	13	14	15	16	17	18	19	20
2	1	1	1	2	2	2	3	3	4	4	4	4
3	3	4	5	5	6	7	7	8	9	9	10	11
4	6	7	8	9	10	11	12	14	15	16	17	18
5	9	11	12	13	15	16	18	19	20	22	23	25
6	12	14	16	17	19	21	23	25	26	28	30	32
7	15	17	19	21	24	26	28	30	33	35	37	39
8	18	20	23	26	28	31	33	36	39	41	44	47
9	21	24	27	30	33	36	39	42	45	48	51	54
10	24	27	31	34	37	41	44	48	51	55	58	62
11	27	31	34	38	42	46	50	54	57	61	65	69
12	30	34	38	42	47	51	55	60	64	68	72	77
13	33	37	42	47	51	56	61	65	70	75	80	84
14	36	41	46	51	56	61	66	71	77	82	87	92
15	39	44	50	55	61	66	72	77	83	88	94	100
16	42	48	54	60	65	71	77	83	89	93	101	107
17	45	51	57	64	70	77	83	89	96	102	109	115
18	48	55	61	68	75	82	88	93	102	109	116	123
19	51	58	65	72	80	87	94	101	109	116	123	130
20	54	62	69	77	84	92	100	107	115	123	130	138

The null hypothesis is rejected if the test statistic (U) is LESS than or equal to the tabled critical value for the larger and smaller group sizes (n_1 and n_2 respectively) at the selected significance level.

APPENDIX Ve

Probabilities of the Mann–Whitney statistic (U)

significance level = 0.02 (two-tailed) or 0.01 (one-tailed)

n_2 \ n_1	9	10	11	12	13	14	15	16	17	18	19	20
2					0	0	0	0	0	0	1	1
3	1	1	1	2	2	2	3	3	4	4	4	5
4	3	3	4	5	5	6	7	7	8	9	9	10
5	5	6	7	8	9	10	11	12	13	14	15	16
6	7	8	9	11	12	13	15	16	18	19	20	22
7	9	11	12	14	16	17	19	21	23	24	26	28
8	11	13	15	17	20	22	24	26	28	30	32	34
9	14	16	18	21	23	26	28	31	33	36	38	40
10	16	19	22	24	27	30	33	36	38	41	44	47
11	18	22	25	28	31	34	37	41	44	47	50	53
12	21	24	28	31	35	38	42	46	49	53	56	60
13	23	27	31	35	39	43	47	51	55	59	63	67
14	26	30	34	38	43	47	51	56	60	65	69	75
15	28	33	37	42	47	51	56	61	66	70	75	80
16	31	36	41	46	51	56	61	66	71	76	82	87
17	33	38	44	49	55	60	66	71	77	82	88	93
18	36	41	47	53	59	65	70	76	82	88	94	100
19	38	44	50	56	63	69	75	82	88	94	101	107
20	40	47	53	60	67	73	80	87	93	100	107	114

The null hypothesis is rejected if the test statistic (U) is LESS than or equal to the tabled critical value for the larger and smaller group sizes (n_1 and n_2 respectively) at the selected significance level.

APPENDIX Vf

Probabilities of the Mann–Whitney statistic (U)

significance level = 0.05 (two-tailed) or 0.025 (one-tailed)

n_2 \ n_1	9	10	11	12	13	14	15	16	17	18	19	20
2	0	0	0	1	1	1	1	1	2	2	2	2
3	2	3	3	4	4	5	5	6	6	7	7	8
4	4	5	6	7	8	9	10	11	11	12	13	13
5	7	8	9	11	12	13	14	15	17	18	19	20
6	10	11	13	14	16	17	19	21	22	24	25	27
7	12	14	16	18	20	22	24	26	28	30	32	34
8	15	17	19	22	24	26	29	31	34	36	38	41
9	17	20	23	26	28	31	34	37	39	42	45	48
10	20	23	26	29	33	36	39	42	45	48	52	55
11	23	26	30	33	37	40	44	47	51	55	58	62
12	26	29	33	37	41	45	49	53	57	61	65	69
13	28	33	37	41	45	50	54	59	63	67	72	76
14	31	36	40	45	50	55	59	64	67	74	78	83
15	34	39	44	49	54	59	64	70	75	80	85	90
16	37	42	47	53	59	64	70	75	81	86	92	98
17	39	45	51	57	63	67	75	81	87	93	99	105
18	42	48	55	61	67	74	80	86	93	99	106	112
19	45	52	58	65	72	78	85	92	99	106	113	119
20	48	55	62	69	76	85	90	98	105	112	119	127

The null hypothesis is rejected if the test statistic (U) is LESS than or equal to the tabled critical value for the larger and smaller group sizes (n_1 and n_2 respectively) at the selected significance level.

APPENDIX VIa

Critical values on the *F*-distribution

significance level = 0.10

υ_2 \ υ_1	1	2	3	4	5	6	8	12	24	inf
1	39.86	49.50	53.59	55.83	57.24	58.20	59.44	60.71	62.00	63.33
2	8.52	9.00	9.16	9.24	9.29	9.33	9.37	9.41	9.45	9.49
3	5.54	5.46	5.39	5.34	5.31	5.28	5.25	5.22	5.18	5.13
4	4.54	4.32	4.19	4.11	4.05	4.01	3.95	3.90	3.83	3.76
5	4.06	3.78	3.62	3.52	3.45	3.40	3.34	3.27	3.19	3.10
6	3.78	3.46	3.29	3.18	3.11	3.05	2.98	2.90	2.80	2.72
7	3.59	3.26	3.07	2.96	2.88	2.83	2.75	2.67	2.58	2.47
8	3.46	3.11	2.92	2.81	2.73	2.67	2.59	2.50	2.40	2.29
9	3.36	3.01	2.81	2.69	2.61	2.55	2.47	2.38	2.28	2.16
10	3.29	2.92	2.73	2.61	2.52	2.46	2.38	2.28	2.18	2.06
11	3.23	2.86	2.66	2.54	2.45	2.39	2.30	2.21	2.10	1.97
12	3.18	2.81	2.61	2.48	2.39	2.33	2.24	2.15	2.04	1.90
13	3.14	2.76	2.56	2.43	2.35	2.28	2.20	2.10	1.98	1.85
14	3.10	2.73	2.52	2.39	2.31	2.24	2.15	2.05	1.94	1.80
15	3.07	2.70	2.49	2.36	2.27	2.21	2.12	2.02	1.90	1.76
16	3.05	2.67	2.46	2.33	2.24	2.18	2.09	1.99	1.87	1.72
17	3.03	2.64	2.44	2.31	2.22	2.15	2.06	1.96	1.84	1.69
18	3.01	2.62	2.42	2.29	2.20	2.13	2.04	1.93	1.81	1.66
19	2.99	2.61	2.40	2.27	2.18	2.11	2.02	1.91	1.79	1.63
20	2.97	2.59	2.38	2.25	2.16	2.09	2.00	1.89	1.77	1.61
21	2.96	2.57	2.36	2.23	2.14	2.08	1.98	1.87	1.75	1.59
22	2.95	2.56	2.35	2.22	2.13	2.06	1.97	1.86	1.75	1.57
23	2.94	2.55	2.34	2.21	2.11	2.05	1.95	1.84	1.72	1.55
24	2.93	2.54	2.33	2.19	2.10	2.04	1.94	1.83	1.70	1.53
25	2.92	2.53	2.32	2.18	2.09	2.02	1.93	1.82	1.69	1.52
26	2.91	2.52	2.31	2.17	2.08	2.01	1.92	1.81	1.68	1.50
27	2.90	2.51	2.30	2.17	2.07	2.00	1.91	1.80	1.67	1.49
28	2.89	2.50	2.29	2.16	2.06	2.00	1.90	1.79	1.67	1.48
29	2.89	2.50	2.28	2.15	2.06	1.99	1.89	1.78	1.65	1.47
30	2.88	2.49	2.28	2.14	2.05	1.98	1.88	1.77	1.64	1.46
40	2.84	2.44	2.23	2.09	2.00	1.93	1.83	1.71	1.57	1.38
60	2.79	2.39	2.18	2.04	1.95	1.87	1.77	1.66	1.51	1.29
120	2.75	2.35	2.13	1.99	1.90	1.82	1.72	1.60	1.45	1.19
inf	2.71	2.30	2.08	1.94	1.85	1.77	1.67	1.55	1.38	1.00

The critical F value is determined by reference to the degrees of freedom associated with the greater and the lesser variances (υ_1 and υ_2 respectively). The observed variance ratio is significant if it exceeds the critical value. Intermediate values should be estimated by interpolation.

APPENDIX VIb

Critical values on the *F*-distribution

significance level = 0.05

v_2 \ v_1	1	2	3	4	5	6	8	12	24	inf
1	161.4	199.7	215.7	224.6	230.2	234.0	238.9	243.9	249.0	254.3
2	18.51	19.00	19.16	19.25	19.30	19.33	19.37	19.41	19.45	19.50
3	10.13	9.55	9.28	9.12	9.01	8.94	8.84	8.74	8.64	8.53
4	7.71	6.94	6.59	6.39	6.26	6.16	6.04	5.91	5.77	5.65
5	6.61	5.79	5.41	5.19	5.05	4.95	4.81	4.68	4.53	4.36
6	5.99	5.14	4.76	4.53	4.39	4.28	4.15	4.00	3.84	3.67
7	5.59	4.74	4.35	4.12	3.97	3.87	3.73	3.57	3.41	3.23
8	5.32	4.46	4.07	3.84	3.69	3.58	3.44	3.28	3.12	2.93
9	5.12	4.26	3.86	3.63	3.48	3.37	3.23	3.07	2.90	2.71
10	4.96	4.10	3.71	3.48	3.33	3.22	3.07	2.91	2.74	2.54
11	4.84	3.98	3.59	3.36	3.20	3.09	2.95	2.79	2.61	2.40
12	4.75	3.88	3.49	3.26	3.11	3.00	2.85	2.69	2.50	2.30
13	4.67	3.80	3.41	3.18	3.02	2.92	2.77	2.60	2.42	2.21
14	4.60	3.74	3.34	3.11	2.96	2.85	2.70	2.53	2.35	2.13
15	4.54	3.68	3.29	3.06	2.90	2.79	2.64	2.48	2.29	2.07
16	4.49	3.63	3.24	3.01	2.85	2.74	2.59	2.42	2.24	2.01
17	4.45	3.59	3.20	2.96	2.81	2.70	2.55	2.38	2.19	1.96
18	4.41	3.55	3.16	2.93	2.77	2.66	2.51	2.34	2.15	1.92
19	4.38	3.52	3.13	2.90	2.74	2.63	2.48	2.31	2.11	1.88
20	4.35	3.49	3.10	2.87	2.71	2.60	2.45	2.28	2.08	1.84
21	4.32	3.47	3.07	2.84	2.68	2.57	2.42	2.25	2.05	1.81
22	4.30	3.44	3.05	2.82	2.66	2.55	2.40	2.23	2.03	1.78
23	4.28	3.42	3.03	2.80	2.64	2.53	2.38	2.20	2.00	1.76
24	4.26	3.40	3.01	2.78	2.62	2.51	2.36	2.18	1.98	1.73
25	4.24	3.38	2.99	2.76	2.60	2.49	2.34	2.16	1.96	1.71
26	4.22	3.37	2.98	2.74	2.59	2.47	2.32	2.15	1.95	1.69
27	4.21	3.35	2.96	2.72	2.57	2.46	2.30	2.13	1.93	1.67
28	4.20	3.34	2.95	2.71	2.56	2.44	2.29	2.12	1.91	1.65
29	4.18	3.33	2.93	2.70	2.54	2.43	2.28	2.10	1.90	1.64
30	4.17	3.32	2.92	2.69	2.53	2.42	2.27	2.09	1.89	1.62
40	4.08	3.23	2.84	2.61	2.45	2.34	2.18	2.00	1.79	1.51
60	4.00	3.15	2.76	2.52	2.37	2.25	2.10	1.92	1.70	1.39
120	3.93	3.07	2.68	2.45	2.29	2.17	2.02	1.83	1.61	1.25
inf	3.84	2.99	2.60	2.37	2.21	2.09	1.94	1.75	1.52	1.00

The critical F value is determined by reference to the degrees of freedom associated with the greater and the lesser variances (v_1 and v_2 respectively). The observed variance ratio is significant if it exceeds the critical value. Intermediate values should be estimated by interpolation.

APPENDIX VIc

Critical values on the *F*-distribution

significance level = 0.01

v_1	1	2	3	4	5	6	8	12	24	inf
v_2	4052	4999	5403	5625	5764	5859	5981	6106	6234	6366
2	98.49	99.01	99.17	99.25	99.30	99.33	99.36	99.42	99.46	99.50
3	34.12	30.81	29.46	28.71	28.24	27.91	27.49	27.05	26.60	26.12
4	21.20	18.00	16.69	15.98	15.52	15.21	14.80	14.37	13.93	13.46
5	16.26	13.27	12.06	11.39	10.97	10.67	10.27	9.89	9.47	9.02
6	13.74	10.92	9.78	9.15	8.75	8.47	8.10	7.72	7.31	6.88
7	12.25	9.55	8.45	7.85	7.46	7.19	6.84	6.47	6.07	5.65
8	11.26	8.65	7.59	7.01	6.63	6.37	6.03	5.67	5.28	4.86
9	10.56	8.02	6.99	6.42	6.06	5.80	5.47	5.11	4.73	4.31
10	10.04	7.56	6.55	5.99	5.64	5.39	5.06	4.71	4.33	3.91
11	9.65	7.20	6.22	5.67	5.32	5.07	4.74	4.40	4.02	3.60
12	9.33	6.93	5.95	5.41	5.06	4.82	4.50	4.16	3.78	3.36
13	9.07	6.70	5.74	5.20	4.86	4.62	4.30	3.96	3.59	3.16
14	8.86	6.51	5.56	5.03	4.69	4.46	4.14	3.80	3.43	3.00
15	8.68	6.36	5.42	4.89	4.56	4.32	4.00	3.67	3.29	2.87
16	8.53	6.23	5.29	4.77	4.44	4.20	3.89	3.55	3.18	2.75
17	8.40	6.11	5.18	4.67	4.34	4.10	3.79	3.45	3.08	2.65
18	8.28	6.01	5.09	4.58	4.25	4.01	3.71	3.37	3.00	2.57
19	8.18	5.93	5.01	4.50	4.17	3.94	3.63	3.30	2.92	2.49
20	8.10	5.85	4.94	4.43	4.10	3.87	3.56	3.23	2.86	2.42
21	8.02	5.78	4.87	4.37	4.04	3.81	3.51	3.17	2.80	2.36
22	7.94	5.72	4.82	4.31	3.99	3.76	3.45	3.12	2.75	2.31
23	7.88	5.66	4.76	4.26	3.94	3.71	3.41	3.07	2.70	2.26
24	7.82	5.61	4.72	4.22	3.90	3.67	3.36	3.03	2.66	2.21
25	7.77	5.57	4.68	4.18	3.86	3.63	3.32	2.99	2.62	2.17
26	7.72	5.53	4.64	4.14	3.82	3.59	3.29	2.96	2.58	2.13
27	7.68	5.49	4.60	4.11	3.78	3.56	3.26	2.93	2.55	2.10
28	7.64	5.45	4.57	4.07	3.75	3.53	3.23	2.90	2.52	2.06
29	7.60	5.42	4.54	4.04	3.73	3.50	3.20	2.87	2.49	2.03
30	7.56	5.39	4.51	4.02	3.70	3.47	3.17	2.84	2.47	2.01
40	7.31	5.18	4.31	3.83	3.51	3.29	2.99	2.66	2.29	1.80
60	7.08	4.98	4.13	3.65	3.34	3.12	2.82	2.50	2.12	1.60
120	6.85	4.79	3.95	3.48	3.17	2.96	2.66	2.34	1.95	1.38
inf	6.64	4.60	3.78	3.32	3.02	2.80	2.51	2.18	1.79	1.00

The critical F value is determined by reference to the degrees of freedom associated with the greater and the lesser variances (v_1 and v_2 respectively). The observed variance ratio is significant if it exceeds the critical value. Intermediate values should be estimated by interpolation.

APPENDIX VII

Critical values of the Kruskal – Wallis statistic (H)

sample sizes			significance levels		
n_1	n_2	n_3	0.10	0.05	0.01
2	2	2	4.571		
3	2	1	4.286		
3	2	2	4.470	4.714	
3	3	1	4.571	5.143	
3	3	2	4.556	5.710	
3	3	3	4.622	5.600	6.489
4	2	1	4.199	4.822	
4	2	2	4.458	5.125	6.000
4	3	2	4.444	5.400	6.300
4	3	3	4.700	5.727	6.746
4	4	1	4.066	4.966	6.667
4	4	2	4.555	5.300	6.875
4	4	3	4.477	5.586	7.144
4	4	4	4.581	5.692	7.490
5	2	1	4.170	5.000	
5	2	2	4.342	5.071	6.373
5	3	1	5.982	4.915	6.400
5	3	2	4.495	5.205	6.822
5	3	3	4.503	5.580	7.030
5	4	1	3.974	4.923	6.885
5	4	2	4.522	5.268	7.118
5	4	3	4.542	5.631	7.440
5	4	4	4.619	5.618	7.752
5	5	1	4.056	5.018	7.073
5	5	2	4.508	5.339	7.269
5	5	3	4.545	5.642	7.543
5	5	4	4.521	5.643	7.791
5	5	5	4.560	5.720	7.980

The null hypothesis is rejected if the calculated H value equals or exceeds the critical value on the tables. The latter is determined by reference to the three sample sizes, n_1, n_2 and n_3, though for some combinations there are no reliable values within the selected significance levels and those entries are left blank.

APPENDIX VIII

Critical values of the Pearson product-moment correlation coefficient

	two-tailed significance levels (one-tailed in brackets)			
n	0.10(0.05)	0.05(0.025)	0.02(0.01)	0.01(0.005)
3	0.988	0.997	1.000	1.000
4	0.900	0.950	0.980	0.990
5	0.805	0.878	0.934	0.959
6	0.729	0.811	0.882	0.917
7	0.669	0.754	0.833	0.875
8	0.621	0.707	0.789	0.834
9	0.582	0.666	0.750	0.798
10	0.549	0.632	0.715	0.765
11	0.521	0.602	0.685	0.735
12	0.497	0.576	0.658	0.708
13	0.476	0.553	0.634	0.684
14	0.458	0.532	0.612	0.661
15	0.441	0.514	0.592	0.641
16	0.426	0.497	0.574	0.623
17	0.412	0.482	0.558	0.606
18	0.400	0.468	0.543	0.590
19	0.389	0.456	0.529	0.575
20	0.378	0.444	0.516	0.561
21	0.369	0.433	0.503	0.529
22	0.360	0.423	0.492	0.537
23	0.352	0.413	0.482	0.526
24	0.344	0.404	0.472	0.515
25	0.337	0.396	0.462	0.505
26	0.330	0.388	0.453	0.496
27	0.323	0.381	0.445	0.487
28	0.317	0.374	0.437	0.479
29	0.311	0.367	0.430	0.471
30	0.306	0.361	0.423	0.463
31	0.301	0.355	0.416	0.456
32	0.296	0.349	0.409	0.449
33	0.291	0.344	0.403	0.443
34	0.287	0.339	0.397	0.436
35	0.283	0.334	0.391	0.430
40	0.264	0.312	0.367	0.403
45	0.249	0.294	0.346	0.380
50	0.235	0.279	0.328	0.361
55	0.224	0.266	0.313	0.345
60	0.214	0.254	0.300	0.330
65	0.206	0.244	0.288	0.317
70	0.198	0.235	0.278	0.306
75	0.191	0.227	0.268	0.296
80	0.185	0.220	0.260	0.286
85	0.180	0.213	0.252	0.278
90	0.174	0.207	0.245	0.270
95	0.170	0.202	0.238	0.263
100	0.165	0.197	0.232	0.256
150	0.135	0.160	0.190	0.210
200	0.117	0.139	0.164	0.182

The observed correlation coefficient is significant if it exceeds the tabled value at the selected significance level with the sample size *n*.

APPENDIX IX

Critical values of the Spearman rank correlation coefficient

	two-tailed significance levels (one-tailed in brackets)			
n	0.10(0.05)	0.05(0.025)	0.02(0.01)	0.01(0.005)
5	0.900	1.000	1.000	1.000
6	0.829	0.886	0.943	1.000
7	0.714	0.786	0.893	0.929
8	0.643	0.738	0.833	0.881
9	0.600	0.700	0.783	0.833
10	0.563	0.648	0.745	0.794
11	0.536	0.618	0.709	0.755
12	0.503	0.587	0.678	0.727
13	0.484	0.560	0.648	0.703
14	0.464	0.538	0.626	0.679
15	0.446	0.521	0.604	0.654
16	0.429	0.503	0.582	0.635
17	0.414	0.488	0.566	0.618
18	0.401	0.472	0.550	0.600
19	0.391	0.460	0.535	0.584
20	0.380	0.447	0.522	0.570
21	0.370	0.436	0.509	0.556
22	0.361	0.425	0.497	0.544
23	0.353	0.416	0.486	0.532
24	0.344	0.407	0.476	0.521
25	0.337	0.398	0.466	0.511
26	0.331	0.390	0.457	0.501
27	0.324	0.383	0.449	0.492
28	0.318	0.375	0.441	0.483
29	0.312	0.368	0.433	0.475
30	0.306	0.362	0.425	0.467
31	0.301	0.356	0.419	0.459
32	0.296	0.350	0.412	0.452
33	0.291	0.345	0.405	0.446
34	0.287	0.340	0.400	0.439
35	0.283	0.335	0.394	0.433
40	0.264	0.313	0.368	0.405
45	0.248	0.294	0.347	0.382
50	0.235	0.279	0.329	0.363
55	0.224	0.266	0.314	0.346
60	0.214	0.255	0.301	0.331
65	0.206	0.245	0.291	0.322
70	0.198	0.236	0.280	0.310
80	0.185	0.221	0.262	0.290
90	0.174	0.208	0.247	0.273
100	0.165	0.197	0.234	0.259

The observed correlation coefficient is significant if, for sample size n, it exceeds the critical value at the selected significance level.

APPENDIX Xa

Critical bounds of the Durbin–Watson statistic (d)

significance level = 0.05

n	$k = 1$ d_l	$k = 1$ d_u	$k = 2$ d_l	$k = 2$ d_u	$k = 3$ d_l	$k = 3$ d_u	$k = 4$ d_l	$k = 4$ d_u	$k = 5$ d_l	$k = 5$ d_u
15	1.077	1.361	0.946	1.543	0.814	1.750	0.685	1.977	0.562	2.220
16	1.106	1.371	0.982	1.539	0.857	1.728	0.734	1.935	0.615	2.157
17	1.133	1.381	1.015	1.536	0.897	1.710	0.779	1.900	0.664	2.104
18	1.158	1.391	1.046	1.535	0.933	1.696	0.820	1.872	0.710	2.060
19	1.180	1.401	1.074	1.536	0.967	1.685	0.859	1.848	0.752	2.023
20	1.201	1.411	1.100	1.537	0.998	1.676	0.894	1.828	0.792	1.991
21	1.221	1.420	1.125	1.538	1.026	1.669	0.927	1.812	0.829	1.964
22	1.239	1.429	1.147	1.541	1.053	1.664	0.958	1.797	0.863	1.940
23	1.257	1.437	1.168	1.543	1.078	1.660	0.986	1.785	0.895	1.920
24	1.273	1.446	1.188	1.546	1.101	1.656	1.013	1.775	0.925	1.902
25	1.288	1.454	1.206	1.550	1.123	1.654	1.038	1.767	0.953	1.886
26	1.302	1.461	1.224	1.553	1.143	1.652	1.062	1.759	0.979	1.873
27	1.316	1.469	1.240	1.556	1.162	1.651	1.084	1.753	1.004	1.861
28	1.328	1.476	1.255	1.560	1.181	1.650	1.104	1.747	1.028	1.850
29	1.341	1.483	1.270	1.563	1.198	1.650	1.124	1.743	1.050	1.841
30	1.352	1.489	1.284	1.567	1.214	1.650	1.143	1.739	1.071	1.833
31	1.363	1.496	1.297	1.570	1.229	1.650	1.160	1.735	1.090	1.825
32	1.373	1.502	1.309	1.574	1.244	1.650	1.177	1.732	1.109	1.819
33	1.383	1.508	1.321	1.577	1.258	1.651	1.193	1.730	1.127	1.813
34	1.393	1.514	1.333	1.580	1.271	1.652	1.208	1.728	1.144	1.808
35	1.402	1.519	1.343	1.584	1.283	1.653	1.222	1.726	1.160	1.803
36	1.411	1.525	1.354	1.587	1.295	1.654	1.236	1.724	1.175	1.799
37	1.419	1.530	1.364	1.590	1.307	1.655	1.249	1.723	1.190	1.795
38	1.427	1.535	1.373	1.594	1.318	1.656	1.261	1.722	1.204	1.792
39	1.435	1.540	1.382	1.597	1.328	1.658	1.273	1.722	1.218	1.789
40	1.442	1.544	1.391	1.600	1.338	1.659	1.285	1.721	1.230	1.786
45	1.475	1.566	1.430	1.615	1.383	1.666	1.336	1.720	1.287	1.776
50	1.503	1.585	1.462	1.628	1.421	1.674	1.378	1.721	1.335	1.771
55	1.528	1.601	1.490	1.641	1.452	1.681	1.414	1.724	1.374	1.768
60	1.549	1.616	1.514	1.652	1.480	1.689	1.444	1.727	1.408	1.767
65	1.567	1.629	1.536	1.662	1.503	1.696	1.471	1.731	1.438	1.767
70	1.583	1.641	1.554	1.672	1.525	1.703	1.494	1.735	1.464	1.768
75	1.598	1.652	1.571	1.680	1.543	1.709	1.515	1.739	1.487	1.770
80	1.611	1.662	1.586	1.688	1.560	1.715	1.534	1.743	1.507	1.772
85	1.624	1.671	1.600	1.696	1.575	1.721	1.550	1.747	1.525	1.774
90	1.635	1.679	1.612	1.703	1.589	1.726	1.566	1.751	1.542	1.776
95	1.645	1.687	1.623	1.709	1.602	1.732	1.579	1.755	1.557	1.778
100	1.654	1.694	1.634	1.715	1.613	1.736	1.592	1.758	1.571	1.780

The upper (d_u) and the lower (d_l) bounds are determined by reference to the number of observations (n) and the number of independent terms (k).

APPENDIX Xb

Critical bounds of the Durbin–Watson statistic (d)

significance level = 0.01

n	$k=1$ d_l	$k=1$ d_u	$k=2$ d_l	$k=2$ d_u	$k=3$ d_l	$k=3$ d_u	$k=4$ d_l	$k=4$ d_u	$k=5$ d_l	$k=5$ d_u
15	0.811	1.070	0.700	1.252	0.591	1.464	0.488	1.704	0.391	1.967
16	0.844	1.086	0.737	1.252	0.633	1.446	0.532	1.663	0.437	1.900
17	0.874	1.102	0.772	1.255	0.672	1.432	0.574	1.630	0.480	1.847
18	0.902	1.118	0.805	1.259	0.708	1.422	0.613	1.604	0.522	1.803
19	0.928	1.132	0.835	1.265	0.742	1.415	0.650	1.584	0.561	1.767
20	0.952	1.147	0.863	1.271	0.773	1.411	0.685	1.567	0.598	1.737
21	0.957	1.161	0.890	1.277	0.803	1.408	0.718	1.554	0.633	1.712
22	0.997	1.174	0.914	1.284	0.831	1.407	0.748	1.543	0.667	1.691
23	1.018	1.187	0.938	1.291	0.858	1.407	0.777	1.534	0.698	1.673
24	1.037	1.199	0.960	1.298	0.882	1.407	0.805	1.528	0.728	1.658
25	1.055	1.211	0.981	1.305	0.906	1.409	0.831	1.523	0.756	1.645
26	1.072	1.222	1.001	1.312	0.928	1.411	0.855	1.518	0.783	1.635
27	1.089	1.233	1.019	1.319	0.949	1.413	0.878	1.515	0.808	1.626
28	1.104	1.244	1.037	1.325	0.969	1.415	0.900	1.513	0.832	1.618
29	1.119	1.254	1.054	1.332	0.988	1.418	0.921	1.512	0.855	1.611
30	1.133	1.263	1.070	1.339	1.006	1.421	0.941	1.511	0.877	1.606
31	1.147	1.273	1.085	1.345	1.023	1.425	0.960	1.510	0.897	1.601
32	1.160	1.282	1.100	1.352	1.040	1.428	0.979	1.510	0.917	1.597
33	1.172	1.291	1.114	1.358	1.055	1.432	0.996	1.510	0.936	1.594
34	1.184	1.299	1.128	1.364	1.070	1.435	1.012	1.511	0.954	1.591
35	1.195	1.307	1.140	1.370	1.085	1.439	1.028	1.512	0.971	1.589
36	1.206	1.315	1.153	1.376	1.098	1.442	1.043	1.513	0.988	1.588
37	1.217	1.323	1.165	1.382	1.112	1.446	1.058	1.514	1.004	1.586
38	1.227	1.330	1.176	1.388	1.124	1.449	1.072	1.515	1.019	1.585
39	1.237	1.337	1.187	1.393	1.137	1.453	1.085	1.517	1.034	1.584
40	1.246	1.344	1.198	1.398	1.148	1.457	1.098	1.518	1.048	1.584
45	1.288	1.376	1.245	1.423	1.201	1.474	1.156	1.528	1.111	1.584
50	1.324	1.403	1.285	1.446	1.245	1.491	1.205	1.528	1.164	1.587
55	1.356	1.427	1.320	1.466	1.284	1.506	1.247	1.548	1.209	1.592
60	1.383	1.449	1.350	1.484	1.317	1.520	1.283	1.558	1.249	1.598
65	1.407	1.468	1.377	1.500	1.346	1.534	1.315	1.568	1.283	1.604
70	1.429	1.485	1.400	1.515	1.372	1.546	1.343	1.578	1.313	1.611
75	1.448	1.501	1.422	1.529	1.395	1.557	1.368	1.587	1.340	1.617
80	1.466	1.515	1.441	1.541	1.416	1.568	1.390	1.595	1.364	1.624
85	1.482	1.528	1.458	1.553	1.435	1.578	1.411	1.603	1.386	1.630
90	1.496	1.540	1.474	1.563	1.452	1.587	1.429	1.611	1.406	1.636
95	1.510	1.552	1.489	1.573	1.468	1.596	1.446	1.618	1.425	1.642
100	1.522	1.562	1.503	1.583	1.482	1.604	1.462	1.625	1.441	1.647

The upper (d_u) and the lower (d_l) bounds are determined by reference to the number of observations (n) and the number of independent terms (k).

APPENDIX XI

Table of random numbers

19223	95034	05756	27813	96409	12351	42544	82853
73676	47150	98400	01927	27764	42468	84225	36290
45467	71709	77588	00095	32863	29485	82226	90056
52711	38889	93074	60227	40011	85848	48767	52573
95592	94007	69971	91481	60779	53971	17297	59335
68417	35013	15829	72765	85089	56067	50211	47487
82739	57890	20807	47511	81767	55330	94383	14983
60940	72042	17868	24943	61790	90565	87964	18885
36009	19365	15412	39638	85453	46816	84385	41979
39448	49789	18338	24697	39364	42006	76688	08708
81486	69487	60513	09297	00412	71238	27499	39950
59636	88804	04643	71197	18352	73089	84898	45785
62568	70206	40235	03699	71080	22553	11486	11776
45159	32992	73750	66280	03819	56202	02938	70915
61041	77684	94322	24709	73689	14526	31893	32593
38162	98532	61283	70632	23417	26185	41448	75532
73190	32533	04470	29669	84407	90785	65496	86382
95857	07718	87664	92099	58806	66979	98624	84826
35476	55975	39421	65850	04266	35435	43742	11937
71487	09984	29077	14863	61683	47052	62224	51025
13875	81598	95052	90908	73592	75186	87136	95761
54580	81507	27102	56027	55893	33063	41842	81868
71035	09001	43367	49497	72719	96758	27611	91965
96927	19931	36089	74192	77567	88741	48409	41903
43909	99477	25330	64359	40085	16925	85117	36071
15689	14227	06565	14374	13352	49367	81982	87209
36759	58984	68288	22913	18638	54303	00795	07827
69051	64817	87174	09751	84534	06489	87201	97245
05007	16332	81194	14873	04197	85576	45195	96505
68732	55259	84292	08796	43456	93739	31865	97150
45740	41807	65561	33302	07051	93623	18132	09547
66925	55685	39100	78458	11206	19876	87151	31260
08421	44753	77377	28744	75592	08563	79140	94254
53645	66812	61421	47836	12609	15374	98481	14592
66831	68908	40772	21558	47781	33586	79117	06928
55588	99404	70708	41098	43563	56934	48394	51719
12975	13258	13048	45114	72321	81940	00360	02428
96767	35964	23822	96012	94591	65194	50842	53372
72829	50232	97892	63408	77919	44575	24870	04178
88565	42628	17787	49376	61762	16593	88604	12724
62974	88145	83083	69543	46109	59505	69680	00900
19687	12633	57857	95806	09931	02150	43163	58636
37609	59057	66979	83401	60705	02384	90597	93600
54973	86278	88737	74351	47500	84552	19909	67181
00694	05977	19664	65441	20903	62371	22725	53340

APPENDIX XII

Requirements for use of MINITAB and SPSS PC+

The MINITAB examples used in this book were obtained using Release 8 of the package. Nevertheless the instructions and their syntax are equally applicable to earlier Release 7 of this package. As explained in Chapter 2, Release 8 contains several improvements on earlier versions including the pull-down menu system, though this was not specifically referred to in any of the examples. First-time users of MINITAB would be advised to get the latest version. This being the case the minimum machine specifications for the successful and easy running of the package should be noted. These are:

1. An IBM or compatible computer with an Intel processor 80286 or higher.
2. The machine system should be DOS version 3.0 or later.
3. The computer's hard disc should have a minimum of 5Mb of available storage.
4. Although not referred to in the text, there exists the facility for high resolution screen graphics on MINITAB. If this facility is to be employed a further 600K of available hard disc space will be needed with a suitable graphics board.
5. At least 1Mb of RAM including 475K of available conventional memory and 150K of available expanded memory.
6. A maths coprocessor and a mouse are strongly advised.

Teachers with those machines should also note that a version of Release 8 MINITAB is available for use on the Apple Mackintosh series of desktop computers. Further details on these packages may be obtained from either:

MINITAB Inc	or	CLECOM Ltd
3081 Enterprise Drive		The Research Park
State College		Vincent Drive
Pennsylvania 16801-2756		Edgbaston
USA		Birmingham B15 2SQ, UK

The SPSS examples in the book were prepared using SPSS PC+ 4.0. Unlike the MINITAB system which is delivered as one complete package, the SPSS system consists of a number of modules. All the examples cited in this text are based on options found within the Base Module (which is the minimum requirement) and the Statistics Module. The full list of system modules available within SPSS and an outline of their functions is as follows:

The Base System consists of the following:
Menu-driven split-screen command editor; reads and writes data in ASCII, SPSS/PC+ Data Entry, Lotus, dBASE, Symphony and Multiplan format files and from and to SPSS on other platforms. The data manipulation functions include sorting, selecting, merging and matching data. The exploratory data analysis techniques include means, standard deviations, frequencies, and crosstabulations.

The **Add-on modules** used in the preparation of examples in this book include:

Statistics Module: which contains the complete range of statistics including *t*-tests, nonparametric tests, *anova*, correlation, regression, factor and cluster analysis.

The following modules are also available, though were not used in the final preparation of this book:

Advanced Statistics: contains multivariate techniques such as discriminant, MANOVA, loglinear analysis, non-linear and logistic regression, probit, logit and survival (life-tables) analysis.

Trends: a time-series and forecasting tool featuring curve-fitting, exponential smoothing, methods for estimating autoregressive models, Box-Jenkins (ARIMA and X11ARIMA) and spectral analysis.

Categories: correspondence analysis (optimal-scaling) converts a complex multi-dimensional table into a simple two-dimensional map. Conjoint (trade-off) analysis measures consumer preferences about the attributes of a product or service.

Tables: displays data in tabular form with control of format and layout including nesting and concatenation. A wide range of statistics can also be produced. The module can use all types of data including multiple responses.

Data Entry II: allows the user to create and edit data files in SPSS, Lotus, dBase or ASCII format. Other options allow for the use of spreadsheet and customised form layout with validation rules, ranges and routing.

Mapping and graphics and other modules also exist, though are not discussed here and readers are advised to consult SPSS (address below) for further information. The minimum machine requirements to run the systems are:

1. An IBM PC/XT, PC/AT, PS/2 or compatible computer.
2. The machine system should be DOS version 2.0 or later.
3. 640K of RAM and a hard disc with 3.9Mb of free memory for installation of the Base System. The entire SPSS system with all available options will require 16.2Mb.
4. A maths coprocessor and a mouse are strongly advised for versions up to 4.0.

Further information can be obtained from either:

SPSS Inc
444 N. Michigan Avenue
Chicago, Illinois 60611
USA

or

SPSS(UK) Ltd
SPSS House
5 London Street
Chertsey
Surrey KT16 8AP, England

In common with MINITAB a version of SPSS now exists which can be run on the Apple Macintosh range of computers. The SPSS system is also available for Microsoft Windows versions 3.0 and higher. It follows that the machine requirements for the latter are more demanding and the appropriate literature should be consulted.

APPENDIX XIII

This appendix contains the names and addresses of the organisations and their publications sections cited as data sources in Chapter 3.

The Meterological Office, Educational Service
Johnson House, London Road
Bracknell, Berkshire RG12 2SY, UK

The Meteorological Office
Saughton House, Broomhouse Drive
Edinburgh EH11 3XQ, UK

Belfast Weather Centre
1 College Square East, Belfast BT1 6BQ, UK

The National Water Archive, Inst. of Hydrology
Maclean Building, Crowmarsh Gifford
Wallingford, Oxfordshire
OX10 8BB, UK

Chadwyck-Healey Ltd
Cambridge Place, Cambridge
CB2 1NR, UK

Her Majesty's Stationery Office
Publications Centre,
51, Nine Elms Lane
London SW8 5DR, UK

National Climatic Data Center
Federal Building, Ashville
North Carolina 28801-2696, USA

NOAA Educational Affairs Division
Room 105, Rockwell Building
11400 Rockville Pike, Rockville
Maryland 20852, USA

US Geological Survey
Superintendent of Documents
U.S. Government Printing Office
Washington D.C. 20402, USA

Earthinfo Inc, 5541 Central Avenue
Boulder, Colorado 80301-9755, USA

United States Bureau of the Census
Congressional information Service
4520 East-West Highway, Bethesda
Maryland 20814, USA

Index